AF692125

The Last Steel Warrior
The U.S. M14 Rifle

By Frank Iannamico

Disclaimer

The information in this publication is believed to be reliable. However, because of the possibility of human or mechanical error by our sources, Frank Iannamico, Moose Lake Publishing LLC, or others. Moose Lake Publishing LLC does not guarantee the accuracy, adequacy, or completeness of any information and is not responsible for any errors or omissions or the results obtained from use of such information. This book is not intended as a legal guide or publication.

All correspondence should be addressed to:
Moose Lake Publishing LLC
631 N. Stephanie St. #562
Henderson, NV 89014
Phone: 702-565-2926
Fax: 702-558-1728
E-Mail: http://www.smallarmsreview.com

Cover Photo
Mark Hamilton, a veteran of the US Army's famed "Old Guard" 3rd Infantry Regiment, portraying an Army sniper with an XM21 rifle during the Vietnam War. Although virtually identical to the version used in Vietnam, the Leatherwood scope pictured here is the improved ART II model. Copyright 2002 Robert Bruce Military Photo Features

Acknowledgments

The author has made a concerted effort to insure the accuracy and completeness of the material presented in this book. Although there were many contributors to this book, the author resumes full responsibility for the material presented herein. Hopefully this book will help generate new interest and additional information will surface as a result. If you have any information, documents or photographs that you would like to share or contribute for possible inclusion in future editions of this book please contact the publisher.

A very special "thank you" to all of those who contributed...

Robert Bruce for the many great photographs including the cover photo; Bill Dormody of Springfield Armory Inc. for his research efforts on the M1A™; Scott Duff, for his contributions and access to his vast library of research material; Al V. Houde of the U.S. Marine Corps Museum, Quantico, for his gracious assistance and patience during my visits; Mark Keefe and Philip Schreier of the National Rifle Association for their gracious hospitality and assistance with both material and allowing me to photograph the NRA's M14 s/n 0006. Walt Kuleck of the Fulton Armory for providing information and photographs; John McCabe of the Springfield Armory NHS for his continued assistance; John M. Miller for his wealth of M14 knowledge and access to his collection; Bill Ricca for inviting me to his home to photograph his unique collection of U.S. militaria, and sharing his many years of collecting experience, also for sharing his knowledge of TM manuals; Dan and Debbie Shea of Moose Lake Publishing LLC for once again publishing my work; Ron Smith and Bill from Smith Enterprises Inc. for their time and photographs; Don G. Thomas who I can always rely on to provide me with documents and information from his vast resources; Jodie Creen Wesemann of the Rock Island Arsenal Museum for her dedication and efforts; Jeff Zimba, once again for the great book cover designs.

Individuals and Organizations

Lou Biancospino, LRB Arms; Gene Barnett, Barnett Barrels; Steve Brousseau; Tony "Hawk" Diaz; Lee Emerson; David Fassinger; Drew Paskovitch; Jim Pecori; John Kline, Sage Int; Kent Lomont; Dale Rader, Springfield Armory Inc.; Robert Silvers; Dominic Spedicci; Dennis Todd; Prokopios Ziros. The U.S. Marine Corps Museum; The National Archives; U.S. Army Historical Center; U.S. Navy; Fulton Armory; Springfield Armory Inc; Smith Enterprises Inc.; The Springfield Armory Museum, National Historical Site; Ohio Ordnance; The Rock Island Arsenal Museum; National Rife Association Firearms Museum.

This book is dedicated to
Dennis William Hippo
Killed in Action 17 January 1969
U.S. Army 1st Air Cavalry
Silver Star, Bronze Star, Air Medal, Purple Heart
Posthumously Awarded

Updated 4th edition of ***The Machinegun Dealers Bible*** **by Dan Shea.**

Small Arms Review Magazine
A monthly magazine dedicated to class III firearms.

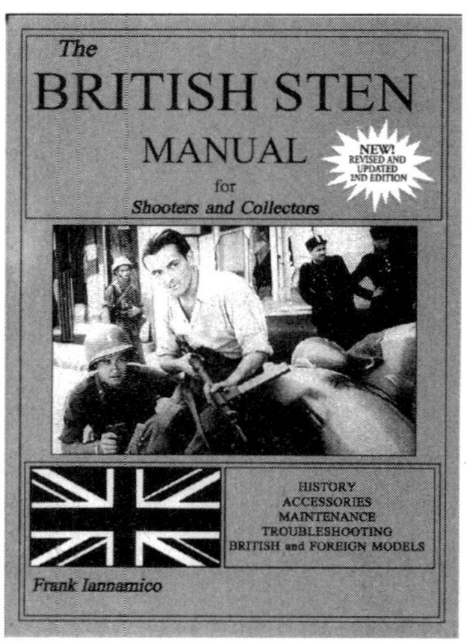

Other titles available from Moose Lake Publishing LLC (702-565-2926)
website:www.smallarmsreview.com

The Reising
Submachine Gun
Story
Includes automatic and semiautomatic
☆ Models 50-55-60-65 MC-58 ☆
History
Development
Manufacturing
Trials and Testing
Parts and Accessories
Collecting and Shooting
Police and Military Models
Prototypes and Experimental Models
Frank Iannamico

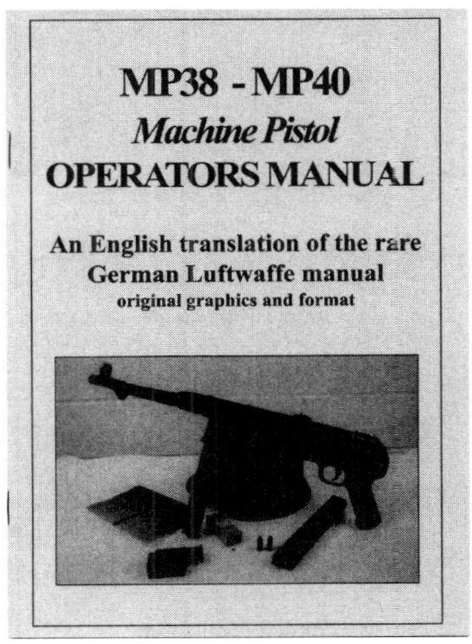
MP38 - MP40
Machine Pistol
OPERATORS MANUAL
An English translation of the rare
German Luftwaffe manual
original graphics and format

THE U.S. M3-3A1
SUBMACHINE GUN
THE
"Grease Gun"
HISTORY
PRODUCTION
RARE MODELS
TRIALS AND TESTING
PARTS AND ACCESSORIES
COLLECTING AND SHOOTING
Frank Iannamico

Contents

Chapter 1
Introduction

Army recruits on the range with the M1 Garand rifle. Troops often trained on the M1 during basic training and were later issued the M14 rifle. (NARA)

The great success of the M1 Garand rifle during World War II virtually assured that its successor would be of a similar design. Although the M1 rifle was revered by the Army there were several changes desired in the weapon to include; lighter weight; a full-automatic capability and a detachable large capacity magazine. A program was begun to produce such a variation of the M1 as early as 1944.

A study of U.S. military rifles often reveals that progress often gave way to convenience, tradition or cost. One example was the M1 Garand Rifle. The M1 was chambered for the World War I 30'06 cartridge mainly because of huge stock piles of the ammunition that were left over from the First World War.

Another example was the 7.62 NATO cartridge. The Germans introduced a midrange 7.92x33 mm *Kurz* cartridge for their revolutionary *Strumgewehr* assault rifle near the end of World War II. The new round and weapon proved able to fill the gap between the full-power battle rifle and the limited range of the submachine gun, providing a capable dual-purpose weapon. The weapon had sufficient range to engage most targets in combat, and controllable full-automatic capability for street fighting or ambush situations. Due to the less powerful 7.92x33 mm round the weapon could be lighter and more compact than the full power battle rifles then in use. One other advantage was that the infantryman, due to lighter weight of both the weapon and cartridges could carry a larger ammunition load. The Russians were quick to grasp the concept and introduced their own 7.62x39 cartridge and the SKS semiautomatic rifle in 1945. The British also began experimenting with their own midrange cartridge. The United States Ordnance Department certainly was in favor of a smaller cartridge, but insisted that it have the same power as its 30'06 predecessor, making the concept of a lightweight rifle with controllable full-automatic fire unfeasible.

The New Light Infantry Rifle

The M14 rifle was originally intended as a multi-purpose weapon, to replace the M1 rifle, BAR, M1 carbine and the submachine gun. Along with the M60 machine gun, the needs

The Russian 7.62x39 mm cartridge (right) and the German 7.92x33 mm round (left) are intermediate-power rifle cartridges. The mid-range power of these rounds allows controllable automatic fire from shoulder weapons.

of the infantryman would be met using one common cartridge.

When the T44E4 rifle (the M14 prototype) was pitted against its main competitor the Belgian FN FAL (designated as the T48 rifle), there were several reasons given for choosing

Introduction

A U.S. soldier holds an early M14 rifle, on the table in front of him are the four weapons the M14 was intended to replace. From top; M3A1 submachine gun; M1 carbine; M1 rifle and the BAR. (NARA)

The belt-fed M60 General Purpose Machine Gun, along with the M14 rifle, were intended to fulfill all of the infantryman's small arms requirements using the same 7.62 mm NATO cartridge. (NARA)

the T44E4 over the T48. One was that the T44E4 (M14), could be manufactured on existing M1 Garand machinery, and many of the internal parts were interchangeable between the two rifles. The U.S. chose the T44E4 designating it as 7.62 mm M14 Rifle. Virtually all of the other countries in the NATO alliance chose the Belgian designed FAL.

Project SALVO

During 1948 the U.S. Government established the Ordnance Research Office. One of the functions of the organization was to study battlefield casualties and how they were inflicted. An enormous number of case studies were performed. At the conclusion of the project it was determined that; most combat actions occur at close ranges; the enemy was generally engaged by mischance than by plan,

The U.S. T44E4 rifle (bottom) and its main competitor the Belgian T-48. There were several reasons given for choosing the T44E4 over the T48. One was that the T44E4, could be manufactured on existing M1 Garand machinery, and that many of the internal parts were interchangeable between the two rifles. (U.S. Marine Corps Museum, Quantico)

and when an action ensued the unit having the most firepower would usually emerge as the victor. Also discovered by the ORO study was that combat injuries from rifle projectiles were random, accurate aiming had little to do with inflicting casualties. One conclusion was that the number of casualties inflicted upon the enemy was directly related to the number of bullets that were fired at him. The increasing use of personal body armor by infantrymen was also studied.

The findings basically revealed that the infantryman needed a large ammunition load to inflict casualties on the enemy during firefights. A lighter weapon using smaller caliber ammunition was suggested. It was stated that a select-fire weapon firing four random shots full-automatic into an area twenty-inches in diameter would have a higher hit probability than firing a single shot from a semiautomatic weapon.

Another concept that was proposed by the SALVO project was a weapon that could fire a large number of projectiles at once. At this time Irwin Barr of AAI introduced a shotgun shell containing thirty-two steel darts called flechettes. Following the multi-projectile concept a number of similar experimental weap-

Allies of the United States used their FN FAL rifles during the early years of the Vietnam War. This Australian soldier is armed with the 7.62 mm Belgian designed weapon. (NARA)

ons were devised, including multi-barrel weapons and conventional-type rifle cartridges that contained two or more projectiles.

The Ordnance Department became disturbed at many of the suggestions being offered by the Ordnance Research Office that was basically a civilian agency. In an attempt to discredit ORO's findings the Ordnance Department requested a study be conducted by the Aberdeen Proving Center on the combat effectiveness of the full-power .30 caliber round versus a smaller 5.59mm cartridge. After the study concluded in 1955 it was determined, much to the chagrin of the Ordnance Department, that a 5.59 mm projectile would be as effective as the larger 7.62 mm bullet in most situations, and that the hit probability would be increased with a higher rate of controllable fire possible with a lighter projectile. Based

A small 5.56 mm caliber cartridge (right) was adopted for the AR-15 rifle. The small round replaced the 7.62 NATO cartridge (left).

on the initial findings of the small caliber concept, additional funding was requested to continue the research and development of a small caliber cartridge. The request for additional funding was denied and the project shelved. Just a short time later the small caliber concept would be resurrected in the form of the 5.56 mm cartridge.

The Special Purpose Individual Weapon. Acting on the effectiveness of multiple projectile weapons reports of the Ordnance Research Office, the U.S. Ordnance Corps had been developing a new weapon designated as SPIW. SPIW was an acronym used to designate the Special Purpose Individual Weapon. The SPIW was a lightweight multi-function,

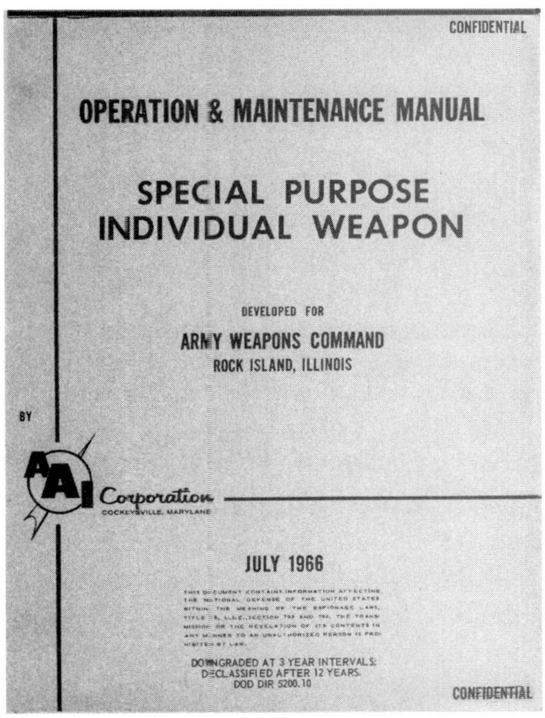

CONFIDENTIAL

OPERATION & MAINTENANCE MANUAL

SPECIAL PURPOSE INDIVIDUAL WEAPON

DEVELOPED FOR

ARMY WEAPONS COMMAND
ROCK ISLAND, ILLINOIS

BY

AAI Corporation
COCKEYSVILLE, MARYLAND

JULY 1966

DOWNGRADED AT 3 YEAR INTERVALS;
DECLASSIFIED AFTER 12 YEARS.
DOD DIR 5200.10

CONFIDENTIAL

A rare manual for one version of the SPIW weapon. (Courtesy of Donald G. Thomas)

The Special Purpose Individual Weapon or SPIW was a long-awaited development. The weapon was made in many prototype forms, but the concept just could not deliver all that it had promised. In the end the Army turned its attention back to conventional rifles. (Courtesy of Donald G. Thomas)

infantry weapon designed for point target engagement. In place of conventional rifle ammunition the SPIW fired flechettes (tiny steel arrow-like projectiles) as well as having the capability to launch 40 mm grenades. The basic concept of the SPIW was increased hit probability in combat situations. The weapon was conceived from exhaustive studies of combat scenarios. The studies were incorporated into a simulated combat computer analysis of the SPIW system, and it appeared that the weapon would render all conventional infantry small arms obsolete. While the project looked good on paper, the reality was the SPIW couldn't deliver what it promised.

Chapter 2

The M1 Garand Rifle

No complete story about the U.S. M14 rifle can be told without first mention of its famous ancestor the M1 rifle. The M1 is more commonly known as the Garand rifle, so named after its designer John C. Garand.

Even before the First World War the United States began to casually cast about for a semiautomatic service rifle, and established a set of specifications it desired in such a weapon. Unfortunately there was little money available at the time to fund the project. The U.S. entry into World War I temporarily sidelined the search for a semiautomatic rifle, while all of the United States' resources were shifted to producing the current Standard bolt-action M1903 rifle. The United States fought World War I armed with both M1903 and the M1917 "Enfield" rifles. The M1917 Enfield rifle was an adaptation of the British .303 caliber Pattern 14 bolt action rifle that was being manufactured in the U.S. under contract with the British. Due to a shortage of M1903 rifles the Enfield Pattern 14 rifle was reworked for the U.S. standard 30'06 caliber cartridge. At the end of World War I there were more M1917 rifles in the inventory than M1903 models, however the Army retained the "American" designed M1903 rifle as its Standard infantry rifle.

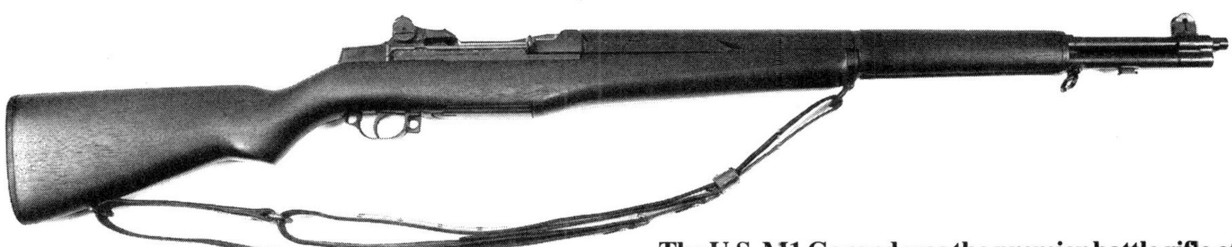

The U.S. M1 Garand was the premier battle rifle of World War II, its stellar reputation insured that its successor would be of a similar design.

After the war ended the Ordnance Department resumed its search for a semiautomatic

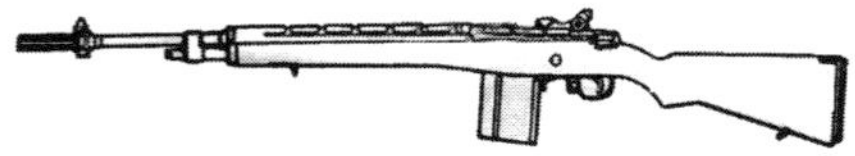

rifle, but once again post war funds were limited. There were quite a number of semiautomatic designs in the pre and post World War I period; however all were unreliable and prone to breakage.

John C. Garand was born in Canada but as a young man gravitated south to the United States' New England area. When World War I broke out young Garand moved to New York City where he was employed as a design engineer for a precision instrument firm. At the time there was a great controversy regarding the United States' lack of a suitable light machine gun. In an apparent act of desperation the United States Army began to issue the dreadfully inadequate French Chauchat light machine gun. Use of the French weapon probably caused more allied causalities than it was able inflict on the enemy, when it worked.

Intrigued by all the newspaper articles on the subject Garand set out to solve the problem himself. After receiving financial backing from Mr. John W. Kewish, a New York financial broker, Garand began work to design his own light machine gun design. Garand's creation was a primer-actuated weapon that resembled a service rifle, the biggest problem was getting the government to notice his new weapon. After much frustration and effort, the Ordnance

The M1 rifle served as the Standard U.S. infantry rifle during the Korean War. (NARA)

The Browning Automatic Rifle, the BAR, was quite an effective weapon and well-liked by the troops. The major drawback was that the BAR was too heavy. (NARA)

Department finally agreed to examine Garand's light machine gun. Unimpressed Ordnance officers submitted a report on the weapon stating that it offered little improvement over existing designs. After a conflict between Garand and Kewish on just who owned the weapon, Garand was eventually issued a patent in 1919 for his design. The Army eventually adopted the Browning Automatic Rifle, the BAR, to fulfill the U.S. Army's need for a light machine gun.

Garand's efforts to develop a suitable light machine gun were not totally in vain. An Ordnance officer who became familiar with Garand's work was impressed with his talents, and arranged to have him assigned to the Springfield Armory. Garand's primary assignment at the Armory was to develop a self-loaded infantry rifle. On 14 November 1919 young Garand began his career as an employee of the Springfield Armory. John Garand would soon discover that working for the government was filled with frustration and red-tape.

Most of Mr. Garand's early semiautomatic rifle designs were based on a primer-actuated action. This principle relied on the slight rearward movement of the cartridge's primer for function. After firing a round, the primer would move slightly out of the cartridge case and move the firing pin rearward at the same time. Garand used the rearward action of the firing pin to operate the breech mechanism. The primer-actuated design required specially designed cartridges to operate. The Ordnance Department was opposed to supplying special cartridges for rifles and other standard ones for machine guns. Two other factors ended Garand's primer-actuated concept. One was the switch to progressive gunpowder, that would not work with primer actuated weapons, the second factor was when the U.S. Ordnance Department adopted crimped primer pockets for all military cartridges.

In 1924 John D. Pederson, a respected firearms designer, was contracted by the Ordnance Department to develop a self-loading rifle. One of the first things that Pederson suggested was that a smaller less powerful cartridge was needed to develop a successful semiautomatic rifle, the 30'06 round being simply too powerful. After stating this Mr. Pederson offered his own 7 x 53 mm, .276 caliber cartridge for consideration. At a meeting of the Ordnance Committee held on 27 December 1927 it was decided to follow Mr. Pederson's advice and soon John Garand and Pederson were instructed to begin designing new rifles around the .276 caliber Pederson cartridge. Over the next few years both the Pederson and Garand rifle designs were pitted against each other in a seemingly endless succession of tests, also participating in the testing were the occasional rifle designs submitted by private companies and individuals.

The Infantry was not excited over the possible adoption of a new cartridge, particularly one with less power and range than the 30'06, that had proven itself more than capable of

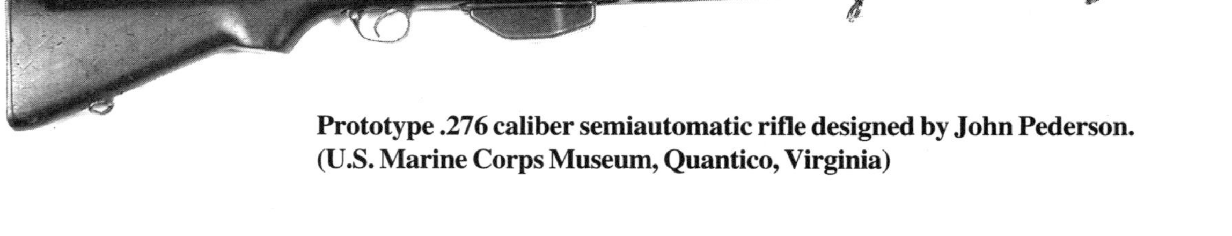

Prototype .276 caliber semiautomatic rifle designed by John Pederson. (U.S. Marine Corps Museum, Quantico, Virginia)

John Pederson's .276 caliber (7mm) rifle competed with John Garand's design for the U.S. semiautomatic infantry rifle. Garand's rifle had proven to be the better of the two and was adopted in 1936. (U.S. Marine Corps Museum, Quantico, VA)

John Garand's T3E3 rifle chambered for the Pederson .276 cartridge. After the .276 round was rejected, both Pederson and Garand resumed work on a .30 caliber semiautomatic rifle. (U.S. Marine Corps Museum, Quantico, Virginia)

doing the job during World War I. Another concern of the Infantry was that accuracy would suffer from the use of a lighter projectile, and that effective long range firing from a machine gun would not be possible. The .276 versus the 30'06 cartridge would become the subject of much disagreement, controversy and conflict. Tests performed on live animals failed to convince anyone opposed to the adoption of the .276 round. This would not be the last "caliber controversy" within the Army hierarchy.

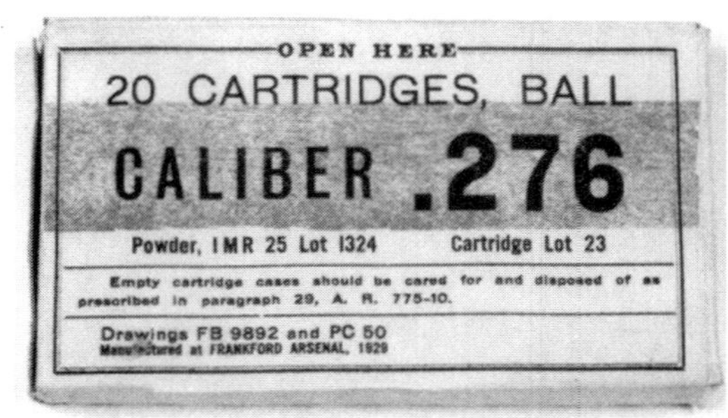

The .276 Pederson cartridge was shelved because of existing stockpiles of .30 caliber ammunition, and the U.S. Army's insistence on using a full-power .30 caliber round. (John M. Miller)

On 4 January 1932 a board found Garand's rifle to be the better of the two designs submitted, and recommended the adoption of the Garand rifle in the Pederson .276 caliber. The Ordnance Committee agreed with the recommendation. Problems occurred when Army Chief of Staff General Douglas MacArthur intervened and refused to approve the recommendations. General MacArthur complained that reduced postwar funds from Congress made adoption of a new caliber impossible. He used the huge existing stockpiles of .30 caliber ammunition and weapons in the U.S. inventory to further reinforce his position. Making his point, the development efforts were redirected back to development of a .30 caliber semiautomatic service rifle. During the caliber debacle Garand had noted the resistance to the .276 cartridge and had anticipated that the United States would retain the World War I 30'06 round. As a result of Garand's foresight he had been quietly working on an alternate rifle design using the Standard .30 caliber cartridge. Finally in 1936 Garand's rifle was adopted as the Caliber .30, U.S. M1 Rifle, it would prove to be a wise choice. The M1 Garand rifle would prove itself to be the outstanding infantry rifle of World War II. The ruggedness of the M1 rifle and its nearly flawless semiautomatic performance gave the U.S. soldier and Marine a decided advantage over his enemy.

Ordnance personnel firing the M1E3 rifle in a simulated rain storm, the "rain" ran through a series of pipes above the shooter's heads. One of the problems encountered with the M1 rifle was that in extreme wet conditions the action would seize. Adding a roller to the bolt of the M1E3 experimental rifle solved the problem. The roller was not implemented into the M1, but was later used on the M14 rifle. (APG)

From the time the M1 rifle was first adopted there were ongoing upgrades and improvements being implemented into the original design. Problems reported from the field as well as methods for easier manufacture were the primary reasons for the changes.

During World War II there were several basic small arms issued by the U.S.; the full power M1 rifle, the Browning Automatic Rifle, the M1 carbine firing the small .30 caliber round and the .45 caliber submachine gun. Each weapon had its own attributes and limitations, as well as (in most cases) using different ammunition. The M1 rifle was semiautomatic-only, too heavy and limited by its eight-round en-bloc clip. The BAR had the reliability, accuracy desired, but at over twenty-pounds it was far too heavy. The M1 carbine was lightweight and handy, but its small .30 (7.62x33) caliber cartridge was inadequate for many situations. Submachine Guns possessed the firepower, but they were heavy and their .45 caliber pistol caliber cartridges had a very limited range.

The E Series M1 Garand Rifles

After the M1 rifle was successfully proving itself on the battlefield during World War II, the Springfield Armory continued its efforts to improve its performance. New variations were created to include sniper rifles and short barreled carbine versions. A number of experimental rifles were built using a variety of different gas systems to operate the action. The E suffix added to the M1 designation indicated experimental status of the weapons. Many of the experimental features of the E series rifles later appeared on the M14 rifle

M1E1 Rifle

One of the early problems encountered with the M1 rifle was that it would jam under extremely wet conditions or during a heavy rain. The problem was traced to water washing away the lubricant from the area where the lug of the bolt rode in the operating rod groove. Without the presence of a lubricant in this area the metal to metal contact between the lug of bolt and the groove in the operating rod would cause the parts to overheat and gall. To address the problem a number of solutions were attempted including new cam angles, increasing the size of barrel's gas port, fluting the chamber and others. Eventually the cam angles between the bolt and operating rod were modified and successfully tested.

The M1 with the modified bolt and operating rod angles was designated as the M1E1 rifle. The problem was that the modified parts were not easily adaptable to existing M1 rifles already in service. Further studies found the quickest and easiest fix of the problem was to issue a different type of grease that resisted being washed off of the parts. The new lubricant, Lubriplate 130-A grease, was issued in small plastic containers along with instructions of where and how to apply it. The lubricant was packed in the buttstock along with the cleaning kit. Research for a more permanent solution continued at the Springfield Armory.

M1E2 Rifle

The M1E2 was the first sniper version of the M1 rifle using a prismatic telescopic sight. The 3-power scope was mounted offset to the left to allow normal top loading and ejection of the M1 rifle, but the eyepiece was centered. This design was evaluated against a normal straight tube type sight using a cheek pad to align the shooter with the eyepiece. Although neither combination was found suitable, certain characteristics of mounting a scope on the M1 would prove useful in later designs.

M1E3 Rifle

In March 1943 a report was received from the Springfield Armory announcing an experimental bolt that incorporated a roller in place of the lug. A new operating rod with a redesigned groove was used to accommodate the bolt roller. In November a test of the roller equipped bolt was performed. In one phase of the test the bolt had all traces of lubricant removed and the rifle endurance fired. The weapon designated as the M1E3, fired nearly 14,000 rounds without the bolt freezing. Although the roller bolt was never implemented into the standard M1, it was however used later in the design of the M14 rifle.

M1E4 Rifle

One area of the M1 rifle that was the focus of much attention was the gas system that operated the action. The gas system used on the M1 rifle was the gas impingement type. In this design high-pressure gas tapped from a port near the end of the barrel acts directly on the end of the operating rod. The operating rod

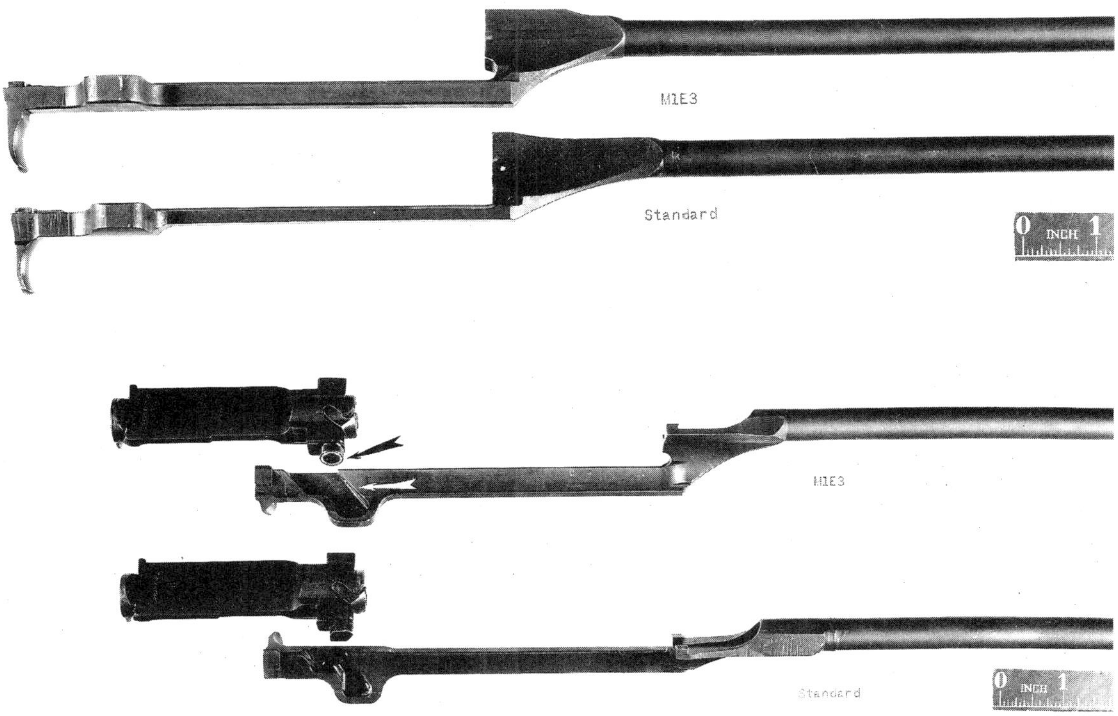

A13513 10-21-44 ABERDEEN PROVING GROUND ORDNANCE DEPT.
Project No. 4937(134-S10-214). TOP: Operating Rod, U.S. Rifle, Cal. .30, M1E3. Note increased size of operating rod arm. BOTTOM: Bolt and Operating Rod U.S. Rifle, Cal. .30, M1E3. Note roller on right

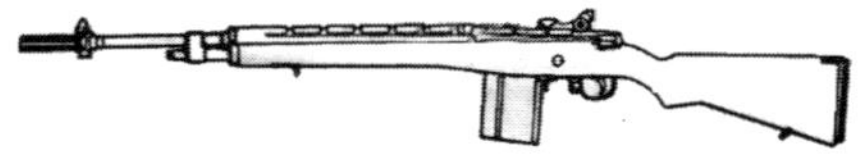

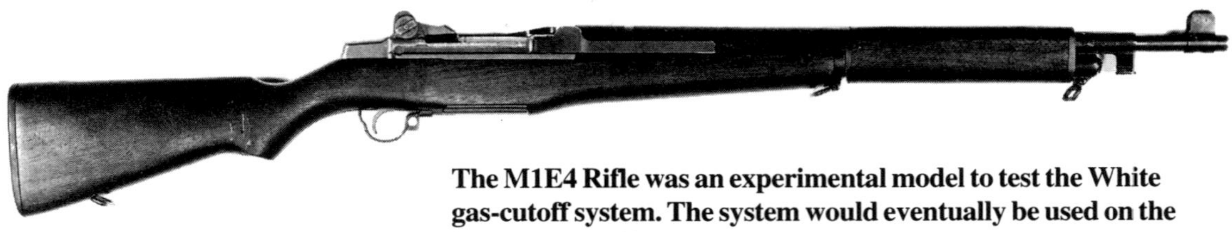

The M1E4 Rifle was an experimental model to test the White gas-cutoff system. The system would eventually be used on the M14 rifle. (NARA)

designed for the M1 was nearly twenty-four inches in length with a slight bend. The long operating rod proved difficult to manufacture correctly and was easily damaged.

One of the experimental rifles developed at the Armory was the T25. During his preliminary research, the designer of the T25, Earle Harvey, had studied a number of gas systems that he could possibly incorporate into his experimental rifle. One design that showed promise was the "gas cut off and expansion system". A patent on the system had been issued to Joseph C. White who had used it in a .276 caliber rifle he designed for submission to the Ordnance Department in 1929. The White system used a gas port in the barrel to direct gas into a closed chamber, located in the chamber is a hollow gas piston. High-pressure gas from the barrel port is directed into the chamber through a hole in the piston. As the expanding gas fills the chamber the piston is propelled rearward and at the same time cuts off the gas supply. A modified M1 rifle designated as the M1E4 was built using an adaptation of the White gas-cutoff system. Testing had proved that although the initial velocity of the operating rod was reduced there was no advantage to the incorporation of the white gas system. The basic White gas system design was eventually adopted for use in the M14 7.62 mm rifle.

M1E5 Rifle

The Ordnance Committee recommended that a short-barreled folding stock carbine based on the M1 rifle be developed. This concept was derived from the study of the FG42 select-fire German paratrooper rifle. An M1 rifle was modified into a carbine version designated the M1E5, with a shortened 18-inch barrel and an under-folding pantograph-type stock. The muzzle blast from the shortened barrel proved excessive and tests indicated the need to incorporate a pistol grip. A pistol grip was added to the design after the initial recommendation. Later the folding stock was replaced by a standard full stock. The short barreled

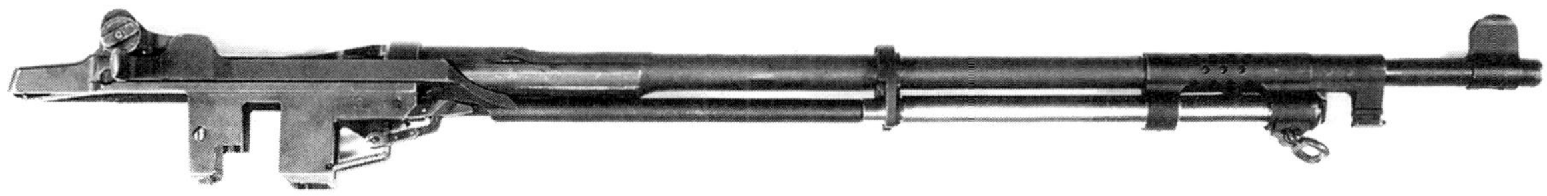

Details of the M1E4 White gas-cutoff system are shown in this photograph. The piston used in this experimental rifle was much longer than that adopted for use in the M14 rifle. (NARA)

M1 with the full stock was designated as the T26. An order was planned to procure 15,000 of the T26 rifles for use in the Pacific Theatre, but was cancelled when the war ended. The T26 rifle is often (incorrectly) referred to as the "tanker" Garand for issue to tank crews.

M1E6 Rifle

The M1E6 was another sniper prototype variation of the M1 that mounted a telescopic sight on the left side of the receiver. In addition there was an M1903A3 ramp style iron sight adapted to the receiver. The purpose was to allow the rear iron sight to be used without having to remove the scope. The problem was that the M1E6 required extensive modifications to the heat-treated receiver, making the arrangement undesirable.

M1E7 Rifle

The M1E7 was the prototype of the M1C sniper rifle known as the Griffin and Howell design. The receiver had a dovetailed pressure plate attached to it, that was secured by cam pressure. The undesirable feature of this design was that it required modifications to the receiver that had been hardened during its original manufacture. The M1E7 was Standardized as the M1C Sniper Rifle in June 1944.

M1E8 Rifle

The M1E8 was also a sniper's weapon that was the prototype of the what would become the M1D sniper rifle. To avoid any modifications to the standard M1 receiver a special steel block was added to the receiver end of the barrel. A shortened hand guard was fitted to accommodate the steel mounting block on the barrel. The M1E8 was adopted as the M1D rifle in September of 1944.

M1E9 Rifle

A later prototype rifle designated as the M1E9 was designed with a modified variation of the White gas system. The M1E9 used a short stroke gas piston that only contacted the op-

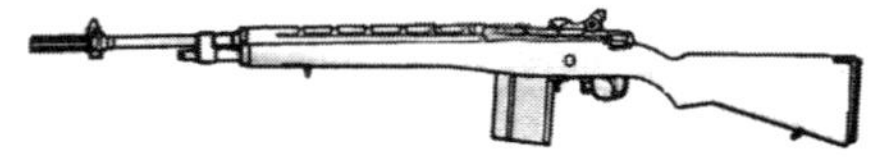

erating rod for a brief period generating a "push" that allowed the operating rod to continue on its own momentum. Although the design didn't warrant enough merit to justify a redesign of the M1 rifle, a modified variation of the gas system would be incorporated into the final M14 rifle design.

M1E10 Rifle

The M1E10 rifle was an experimental M1 rifle, which used the gas system from the Swedish Ljungman rifle to replace the impingement system of the Garand. The Ljungman system used a pipe to direct gas pressure to a contact point on the bolt instead of using a mechanical means such as the operating rod. No known examples were built. The M16 rifle used a similar arrangement.

M1E12 Rifle

The M1E12 was a M1 rifle adapted to use the 7.62 mm NATO cartridge. Because the 7.62 mm cartridge has an overall length that is .5 inch shorter than the older 30'06 round used in the M1 rifle, a new barrel chambered for the 7.62 mm round would normally be re-

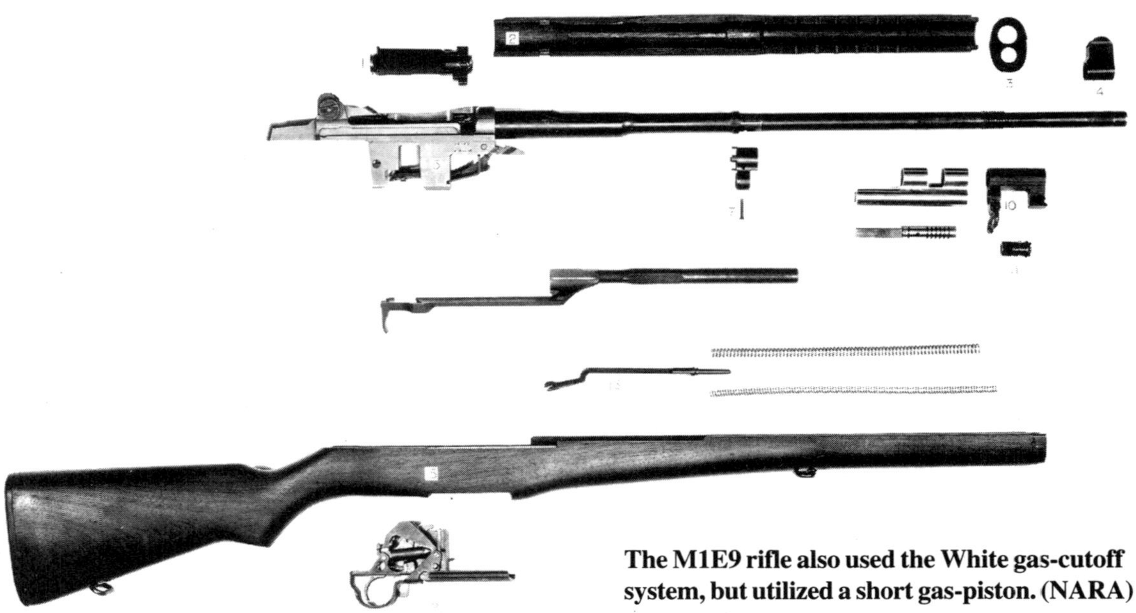

The M1E9 rifle also used the White gas-cutoff system, but utilized a short gas-piston. (NARA)

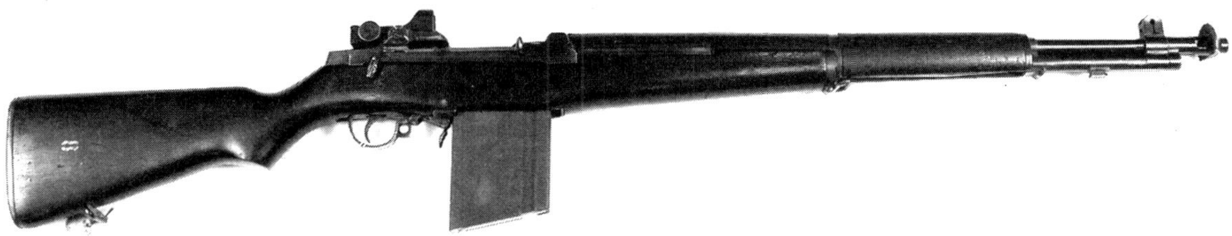

The T20 series of rifles was an attempt by the Springfield Armory to design a select-fire version of the M1. Aside from internal modifications, new features included a bipod, muzzle brake and a twenty-round box magazine. (Courtesy of the U.S. Marine Corps Museum, Quantico, Virginia)

quired. However in an attempt to avoid the expense of manufacturing and installing a new barrel an experimental bushing was designed to fit into the existing M1 barrel's chamber. With the bushing in place a 7.62 mm round could be used. Problems occurred when the bushing was accidentally ejected with a spent cartridge case. The operator of the weapon was usually unaware what had happened and the subsequent 7.62 mm round would be chambered into the 30'06 barrel. The 7.62 mm round could be jammed into the barrel and the succeeding round could strike the primer of the stuck round causing an out-of-battery discharge. This would present a very dangerous situation for both the shooter and the weapon. An M1 Garand was produced with a new manufacture barrel chambered for the 7.62 mm round. This M1 was designated as the 7.62 mm T35 rifle.

The Select-Fire M1 Garand

When the Infantry formally requested an automatic version of the M1 rifle in September of 1944, the Ordnance Department responded by informing the Infantry Board that they had already begun to develop a select-fire version of the M1 rifle back in May of 1944. The select-fire M1 under development at the Springfield Armory was designated as the T20 rifle.

A separate contract had been awarded to the Remington Arms Company to concurrently develop a select-fire version of the M1 designated as the T22. Some of the desired features listed in Remington's contract were: the weapon was to weigh no more than 9.0 pounds; have select-fire capability; be equipped with a folding bipod; folding buttstock; be able to launch grenades and have a magazine with a capacity of twenty rounds,

Receiver of T20E1 serial number 8. This rifle is fitted with an experimental rear sight. (U.S. Marine Corps Museum, Quantico, Virginia)

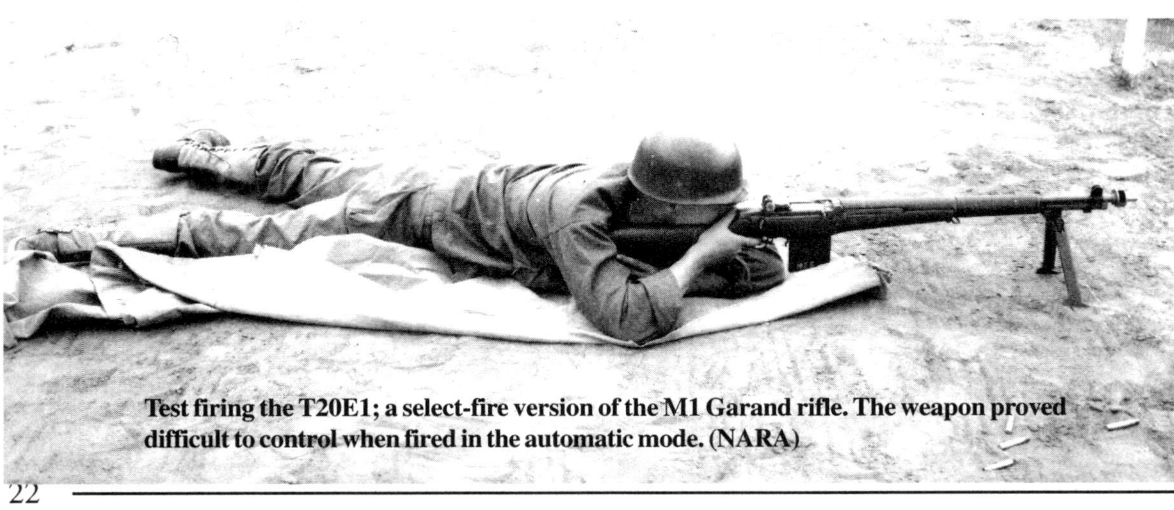

Test firing the T20E1; a select-fire version of the M1 Garand rifle. The weapon proved difficult to control when fired in the automatic mode. (NARA)

HEADQUARTERS ARMY GROUND FORCES

MEMO SLIP

File No. GNRQT-11/35442
Subject: Rifle, Caliber .30, T20E1

From	To	Date	Memoranda - Use full width if lengthy
(1)RQTS-11	RQTS 6 8 9 10 11	20 Jul 45	1. This headquarters has recently requested standardization of the Rifle, .30 Caliber, T20E1, with modifications.

2. Subject weapon has been service tested by the Infantry Board and determined to be satisfactory in operation and function.

3. General characteristics of the weapon compared to the standard M1 are:

		T20E1	M1
a.	Wt of weapon complete w/o ammo:	10.84 lbs	10.03 lbs
b.	Operation:	Gas	Gas
c.	Feed:	Box Mag (20 rds)	Clip (8 rds)
d.	Cyclic Rate:	686 RPM	

e. Appearance: Outwardly the same except for magazine and change lever.

f. Stripping and Assembly: In general, the same.

g. Accuracy: Very similar when fired semi-automatically; automatic fire compares very favorably with the BAR.

4. Further details of the results of test may be found in the attached report of test.

5. The subject weapon is now classified as limited procurement with a contract for 100,000. Deliveries are contemplated to commence on or before 1 February 1946.

6. The following represents a proposed basis of issue of the subject weapon for the infantry, mountain, and airborne.

For inter-office use only - please use reverse side

AGF FORM NO. 2

25- 1900 4-50M

After a few minor changes were implemented the T20E1 rifle described in this document was approved for limited procurement as the T20E2.

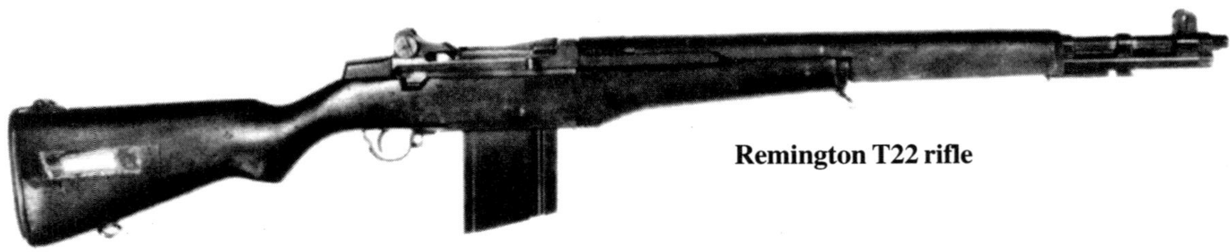

Remington T22 rifle

While the Springfield Armory was developing the T20 series of select-fire M1 rifles, the Remington Arms Company was concurrently engaged to develop a similar rifle designated as the Remington T22 series. (NARA)

the standard M1 rifle weighed 9.5 pounds unloaded. As it would soon be realized, such a weapon was simply not feasible without using a cartridge that had less power than the M1 rifle's .30 caliber (7.62x63mm) round.

Back at the Springfield Armory John Garand was busy attempting to transform his M1 rifle to a select-fire configuration. By October the Springfield Armory had built the first T20 rifle according to Garand's specifications. The Armory shipped the rifle to the Aberdeen Proving Center for testing in November. Shortly after Remington Arms' version of the select-fire M1, the T22, was also shipped to Aberdeen for tests, both of the rifles far exceeded the original 9.0 pound weight requirement. When the rifles were fired in the full-automatic mode they both operated from an open-bolt position. Magazine problems with both rifles were encountered. To ensure the consistent feeding of cartridges, an extremely strong magazine spring was required. It was feared such a strong spring would damage the cartridge follower. Another problem was the retention of the magazine's floorplate, which would be difficult with a stiffer follower spring. Rather than attempt to redesign the magazine, Garand addressed the problem by lengthening the receiver just .3125-inch. The alteration allowed for much better cartridge feeding, but required a newly designed receiver. The Aberdeen test did prove that a select-fire rifle using Garand's M1 action was workable.

Other problems plagued the T20 rifle, one was excessive heat generated by automatic fire. The wooden foregrip would get hot enough to char during extended firing. Special grooves were machined into the barrel to aid in heat dispersion. Excessive recoil generated by the powerful .30 caliber cartridge proved to be a major problem. To aid in control of muzzle

The T20E2 select-fire version of the M1 rifle was considered sufficiently developed in 1945 that 100,000 were ordered as a Limited Procurement Item, U.S. Rifle, M2. The war ended before production could begin and the order was cancelled. (U.S. Marine Corps Museum, Quantico, VA)

Muzzle brake of the T20E2 rifle was designed to help control muzzle rise. (U.S. Marine Corps Museum, Quantico, VA)

CONFIDENTIAL

C- 370
-1-

CONFIDENTIAL

ARMY SERVICE FORCES
Office of the Chief of Ordnance

EGCooper/mfh/3085
12 May 1945

FROM: Sub-Committee on Small Arms

TO: Ordnance Technical Committee

SUBJECT: RIFLE, U.S., CALIBER .30, T20E2 - Recommended for Limited Procurement of

1. REFERENCES:

a. O.C.M. Item Number 25245, dated 28 September 1944
b. O.C.M. Item Number 25411, dated 12 October 1944
c. APG Firing Record No. S-42679, dated 31 Oct. 1944
d. APG Firing Record No. S-42727, dated 4 Dec. 1944
e. APG Firing Record No. S-42728, dated 22-26 Dec. 1944
f. APG Firing Record No. S-42879, dated 29 March 1945
g. APG Firing Record No. S-42889, dated 9 April 1945
h. File OO 400/24404 dated 17 April 1945

2. DISCUSSION:

a. Based upon the recommendation of Headquarters, Army Ground Forces, by reference 1a. above, Ordnance Committee Action was taken recommending the development of a selective semi and full-automatic rifle in accordance with the following characteristics:

(1) The present Rifle, U.S., Cal. .30, M1, be modified to include selective semi and full-automatic fire.

(2) Such automatic fire by means of a bipod be equally as effective as the Browning Automatic Rifle as concerns dispersion, cone of fire, and ballistic characteristics.

(3) The mechanism to be simple; capable of field stripping with operative features which will reduce stoppages to a minimum.

(4) A suitable twenty round bottom feed magazine and a detachable bipod mount be developed for use with this weapon.

CONFIDENTIAL

rise John Garand designed a series of muzzle brakes. Though they proved effective in controlling muzzle climb, there were several problems with Garand's muzzle brakes. The first muzzle brake design required a redesigned gas cylinder lock, another problem was that with the device installed on the barrel a bayonet or a grenade launcher could be not be used on the rifle. Subsequent T20 variations had redesigned muzzle brakes that did permit the attachment of a bayonet or grenade launcher.

One of the major challenges facing the personnel attempting to develop a select-fire M1 was they were limited to only minor design changes. Any major redesigns would hamper the current production process of the M1, which was critical to the war effort. Another problem with the program was that both Remington and the Springfield Armory's primary mission was war production not research and development. Regardless, development of the T20 rifle continued and by the spring of 1945 a weapon designated as the T20E2 emerged.

The T20E2 could fire either semiautomatically or full automatic with a cyclic rate of 700 rounds per minute. The rifle used the elongated M1 receiver designed by John Garand. The bolt was modified for ease of feeding and extraction of cartridges. The T20E2 weapon was fitted with a bipod, muzzle device and a detachable telescopic sight. Full automatic fire was achieved by a connector assembly, which was actuated by the movement of the operating rod handle. In turn, a sear release or trip was actuated, which with the trigger held in a rearward position, disengaged the sear from the hammer lugs immediately after the bolt was locked or in-battery position. With the connector assembly disengaged, the rifle could

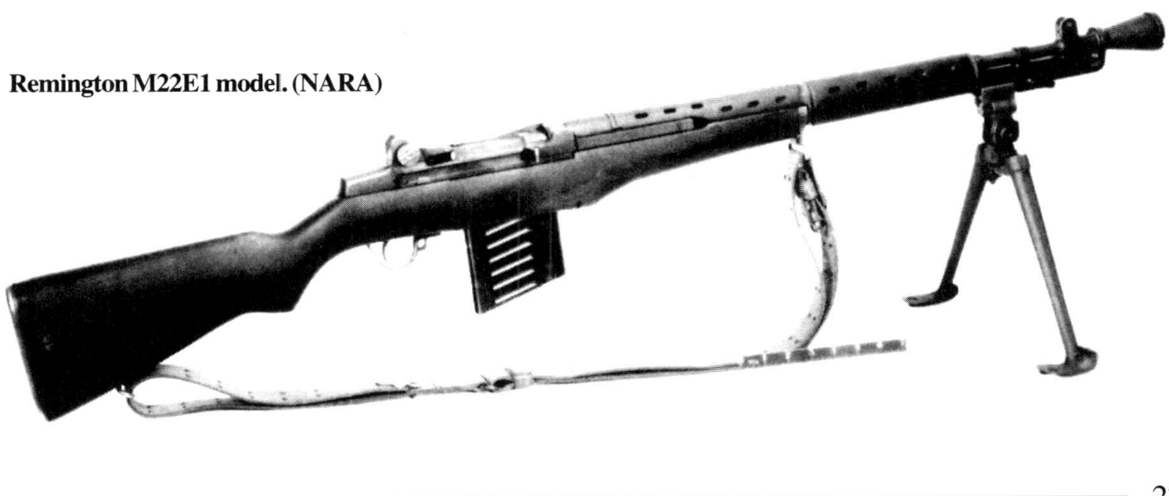

Remington M22E1 model. (NARA)

only be fired semiautomatically and functioned in a manner similar to that of the M1 rifle. The weapon greatly exceeded the initial 9.0 pound weight requirement with a loaded weight of over 12.5 pounds. The M1 based T20E2 was tested and considered sufficiently developed to place into production. In May of 1945 the T20E2 was designated as a Limited Procurement Type. On 17 May 1945 the Ordnance Technical Committee recommended the procurement of 100,000 T20E2 rifles, designated as the Rifle, Caliber .30 M2. The war ended before any of the rifles could be produced and delivered. However, the end of the war did not end the development of the T20 series, scant post-war funds merely slowed it down, and with the war now over time was no longer an issue.

T20E3 Rifle

There were heavy-barrel variants of the T20E2 intended as weapons to fill the role of the squad automatic rifle. The heavy barreled rifles were designated as the T20E3. One version utilized an inline stock design and pistol grip.

T22 Rifle

The development of the T22 rifle was initiated during early 1944 by the Remington Arms Company, at the same period as the T20 was under development at the Springfield Armory. In the T22 design the full automatic function was accomplished in the open bolt position, while semiautomatic operation was from a closed bolt position. The cyclic rate of the T22 was 500 rounds per minute. The open bolt feature did not effectively prevent the cook-off of cartridges. The T22 rifle featured a recoil check, flash-hider and bipod. The T22 used a 20-round box-type magazine. Minor modifications were required to improve functioning and usability of the magazine release and retention of the bolt in the open position. The T22 project was terminated in March 1948. The T22 weighted 9.6 pounds without accessories and had an overall length of 43.6-inches.

T22E2 Rifle

The T22E2 was another evolution of the T22 and T22E1 Remington designed rifles. The T22E2 incorporated a slight change in the design of the trigger group to simplify manufacture and an improved magazine catch. The major advantage of the T22E2 rifle was its adaptability to remanufactured M1 rifles as a non-critical peacetime operation. This was attained by the use of the standard length M1 receiver, rather than the increased length receiver used on the T20E2 rifle. The T22E2 project to modify the M1 rifle was terminated in March of 1948.

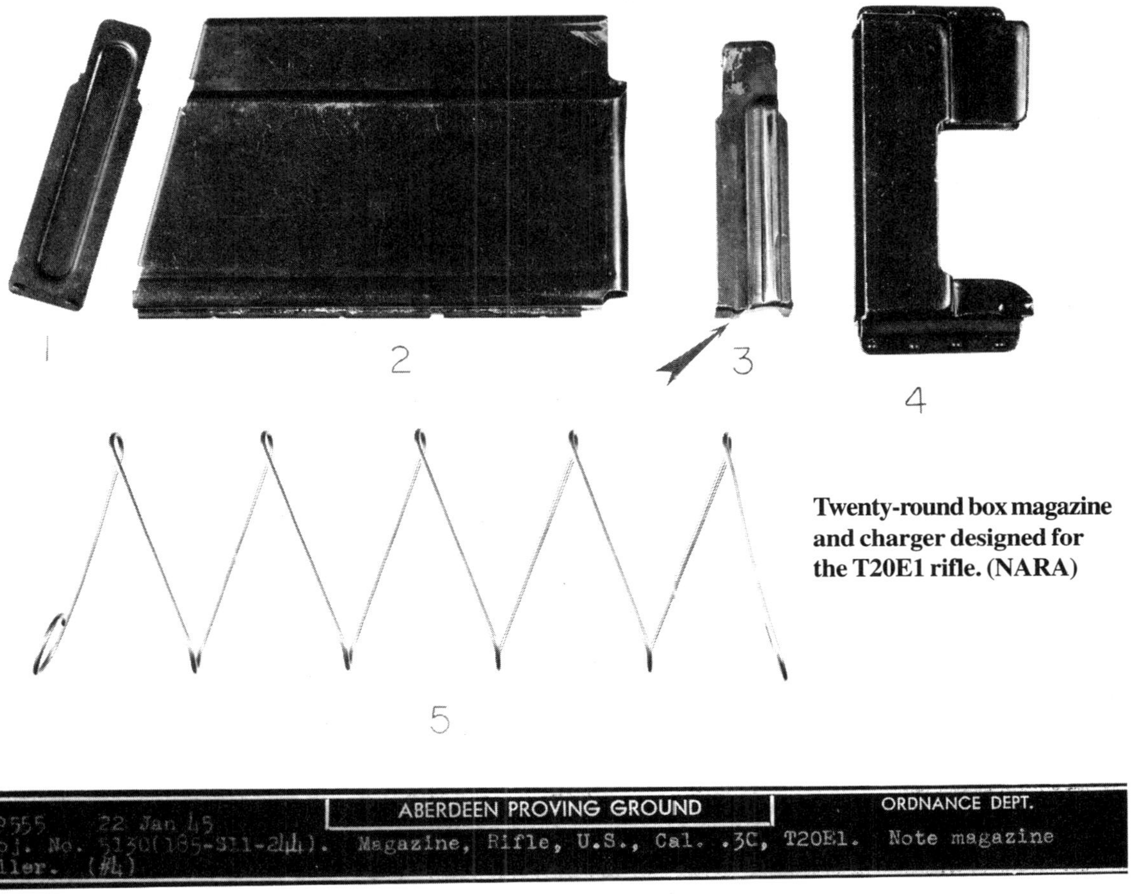

Twenty-round box magazine and charger designed for the T20E1 rifle. (NARA)

The T23 Rifle

The T23 rifle was a modification of the M1 rifle to provide full and semiautomatic fire. The automatic fire feature in this rifle was provided by an independent hammer release. The T23 was advantageous from the standpoint of design, durability and the minimization of functional stress. Because of the mechanism's timing this model fired from an open bolt only 20-percent of the time. The T23 had a cyclic rate of 500 rounds per minute. Initially the T23 was fed from the 8-round M1 Garand clip. Testing of this rifle indicated to the designers

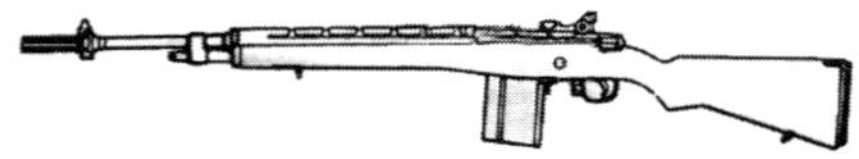

the desirability of firing from the closed bolt position. The T23 was later modified to utilize the 20-round box magazine from the BAR. Experience gained with the T23 indicated that a new proprietary magazine design was needed, and that a device to increase the weapons stability during automatic fire was required. This project was initiated in October of 1944 and terminated in March of 1945.

The T24 Rifle

The T24 rifle was another in a series of modifications to the M1 rifle to provide reliable full and semiautomatic fire. The T24 project was initiated simultaneously with that of the T23 rifle. The T24 fired from a closed bolt position during both semiautomatic and full automatic operation. The T24 provided the minimum time and cost considerations for implementing an automatic fire feature into the existing M1 rifle. The T24 had a unique high-comb stock. The T24 fed from the same 8-round clip as the M1 rifle. The automatic cyclic rate was 500 rounds per minute.

Post War Developments

The program for development of infantry rifles after World War II revealed three definite trends. The first reflected a decision to provide the infantryman with a reduced weight weapon, but as accurate and effective as previous infantry rifles. The second trend was the development of an acceptable rifle with selective semiautomatic and automatic fire. The last trend was simplification of logistical and training problems by developing two almost identical rifles (M14 and M15) to replace four radically different weapons, the M1 rifle, the carbine submachine gun and the Browning Automatic Rifle.

There was no requirement for post-war U.S. development of a mid-range assault rifle similar to the *Sturmgewehr* fielded by the German Army during the closing months of World War II. While the Russians took a profound interest in the 7.92 x33 mm cartridge, the United States displayed little interest in the concept. The infamous AK47 assault rifle introduced to the world in 1949 was the weapon that descended as a direct result of the Russian efforts.

There were numerous prototype rifle designs developed, manufactured and tested. In the end the Ordnance Corps and the Springfield Armory felt that they had achieved their goal of developing a suitable lightweight rifle. The result of their efforts that began in 1944, was the 7.62 mm M14 rifle. The program to develop the M14 rifle as well as the rifle itself would soon come under severe criticism.

The T25 Rifle

The T25 was the first in the new generation of lightweight rifles to fire the improved type of T65E2 ammunition (7.62 mm) The T25 project was first initiated in September 1945 just after World War II had ended. The weapon was designed for both semiautomatic and full automatic operation. Full automatic was performed in the open-bolt position. The T25 had an in-line style stock and elevated sights, which because of the design of the stock, were necessary to align the shooter's eye with the sights. A box magazine with a 20-round capacity was used. Full automatic cyclic rate was 750 rounds per minute. The front sight and bayonet lug were designed integral with the flash suppressor. The gas cutoff system and front end design were eventually incorporated into the T44 rifle. The T25 was light in weight at 7.5 pounds, the overall length was 43.25 inches.

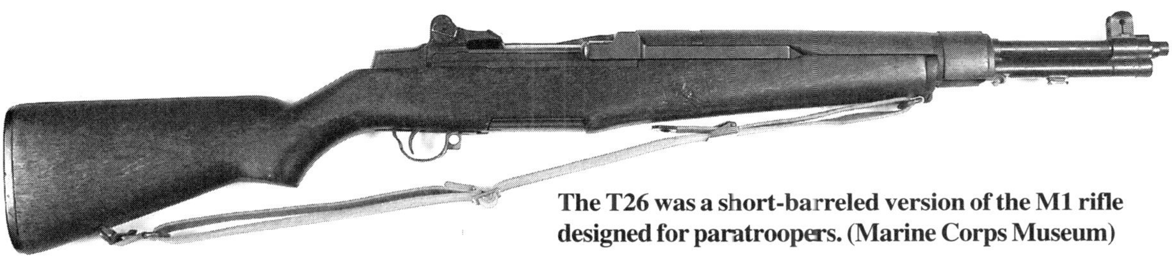

The T26 was a short-barreled version of the M1 rifle designed for paratroopers. (Marine Corps Museum)

The T25 project was terminated during November 1951. Testing of the T25 indicated that it had above average performance, effectiveness and accuracy. The action was extremely strong, and both straight line and conventional style stocks were used with the T25. The main problem of the T25 was its weight, it was simply too light to be controllable. By 1947 the T25 had received numerous mechanical refinements that included quick field-stripping and assembly, and easy removal of the 20-round magazine. An evolution of the T25 was dubbed as the T47. The T47 weighted 8.5 pounds and was fitted with a conventional style stock. Accessories included a bayonet, flash hider, grenade launcher and bipod.

Subsequent to the development of the T25 design the Springfield Armory undertook the development of a lightweight select-fire weapon, which could utilize some of the parts and tooling used for the M1 rifle. The resulting rifle was designated as the T36. The T36 weighed 8.5 pounds with an overall length of 41.25 inches, the weapon fed from a 20-round detachable box magazine. Further modifications in the bolt and trigger mechanism resulted

The T27 rifle was a Remington Arms-modified M1 Garand designed to fire the T-65 (7.62 mm) round. (U.S. Marine Corps Museum, Quantico)

in yet another rifle, the T37. The T37 was fitted with a conventional drop type stock and was slightly lighter than the T36 at 8.0 pounds. The weapon's cyclic rate was 750 rounds per minute. By 1951 the T37 had evolved further into the T44 rifle.

T26 Rifle

The T26 was a semiautomatic-only M1 Garand rifle shortened for airborne troops. The T26 was designed during World War II. The T26 evolved from the M1E5 rifle described earlier. In July of 1945, U.S. forces in the Pacific Theater requested 15,000 of the shortened M1 rifles, but the war ended and the order was terminated.

T27 Rifle

The Remington Arms T27 rifle project was initiated in April of 1946 after World War II had ended. The T27 was a modified M1 rifle with both semiautomatic or full automatic capability that fired the T65 cartridge, later adopted (in a modified form) as the 7.62 mm NATO round. The T27 featured an auxiliary device that could be installed in the field. Once the device was in place the rifle could only be operated in the automatic mode of fire. A later redesign allowed the weapon to be fired ei-

Muzzle and bipod of the 7.62 mm T27 rifle.

ther semiautomatic or full automatic with the device in place. The T27 used the 8-round clip from the M1 rifle. The project was terminated in March 1948.

T28 Rifle

The T28 light rifle development program was initiated during October 1946. The rifle was a 6.88 pound lightweight, select-fire 7.62 mm weapon designed to replace the M1 rifle, M2 carbine, M3A1 submachine gun and the Browning Automatic Rifle (BAR). The full automatic cyclic rate was 600 rounds per minute. The T28 rifle with its inline stock was designed to explore the feasibility of low-cost fabrication techniques. Sheet metal stampings and simplified forgings were used extensively in the design of the weapon. (The Germans began manufacturing weapons from sheet metal stampings during World War II) The mechanism of the T28 had insufficient structural rigidity for satisfactory function and durability. The action operated on the gas impingement system. The breech mechanism was an adaptation of an experimental Mauser design, as was the trigger design. The T28 rifle proved that an infantry rifle requiring screw drivers or other tools for field stripping was unacceptable. Development of the T28 rifle was suspended in late 1950.

T31 Rifle

The program for 7.62 mm T31 rifle was begun during March of 1947. This weapon was an 8.75 pound lightweight, semiautomatic and full automatic rifle with an inline stock. The full automatic cyclic rate was 600 rounds per minute. The primary intention of the T31 was to replace the M1 Garand rifle, M2 carbine, M3A1 submachine gun and the Browning Automatic Rifle (BAR). The T31 was a bullpup design regarded as a novel approach to infantry rifle and had unusually low stripping forces, requiring a redesign of the box magazine used in earlier T series weapons. A magazine of similar design was eventually adopted for use with the M14 rifle. One of the main objectives behind the T31 design was to reduce recoil and eliminate muzzle flash and blast, however the weapon was unable to achieve these goals and the program was terminated in the late 1950.

T33 Rifle

The program for development of the T33 rifle began in March of 1949. The T33 was a lightweight, semiautomatic and full automatic rifle that weighed 7.875 pounds with an overall length of 40.75 inches. Automatic cyclic rate was 600 rounds per minute. There were several variations of the T33 rifle, one using a unique short-stroke gas piston design to operate the action. A private inventor, Howard

Test firing the T47 rifle under Arctic weather conditions in Alaska.

R. Clark, originally submitted the design with guidance from the Office of the Chief of Ordnance. The T33 was chambered for the T65E3, 7.62 mm ammunition. Work was suspended late in 1950 primarily because the weapon lacked durability.

T34 Rifle

The weapon designated as the T34 was a Browning Automatic Rifle (BAR) modified to fire 7.62 mm ammunition. The T34 weighed 19.40 pounds with an overall length of 47.8 inches. The box magazine was redesigned slightly for the shorter ammunition. Development was suspended in late 1950.

The T47 rifle was one of the few U.S. designs that didn't use the turn-bolt action of the M1 Garand. The T47 used a tilting breech-lock system. (NARA)

T35 Rifle

The T35 development program had begun in June 1944. This weapon was a basically an M1 rifle designed to fire the new 7.62 mm ammunition. The T35 was a semiautomatic-only weapon. Firing tests indicated that the T35 firing the new 7.62 ammunition was more accurate than the M1 rifle firing the older .30 caliber round. Development was suspended during 1950.

T36 Rifle

The T36 was produced during 1949 by modifying a T20E2 rifle to fire 7.62 mm (T65) ammunition. The T36 fired either semiautomatic of full automatic at a cyclic rate of 600 rounds per minute. The weapon operated from the closed bolt position in both modes of fire. A modified T25, 20-round magazine was designed for the rifle. Other modifications included a one-piece handguard and a redesigned buttplate. Work on the T36 was suspended during the latter part of 1950.

T37 Rifle

The T37 was another design based on the T20E2 and the T36. The weapon was designated as the T37 in November of 1949. The 8.25 pound weapon fired 7.62 mm ammunition in either semiautomatic or full automatic at a cyclic rate of 750 rounds per minute. The rifle had a lightweight 22-inch barrel with gas port located approximately four inches from

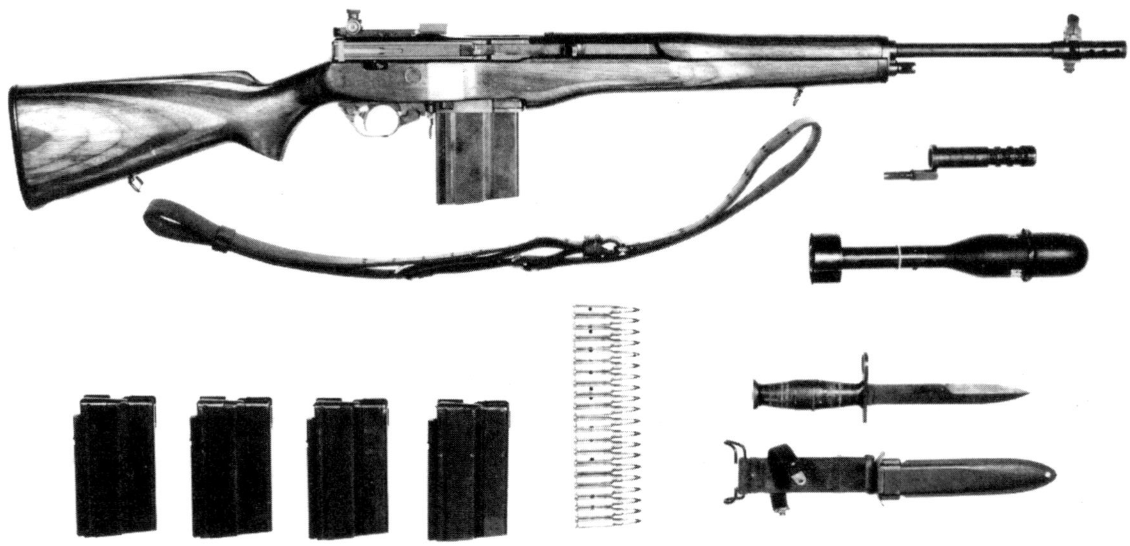

The U.S. T47 7.62 mm rifle and its accessories. (NARA)

the muzzle, and a lightweight wooden stock. Overall length was 41.25 inches. The design used the T20E2 receiver, but with filler blocks located both fore and aft of the magazine. Further modifications included a lightweight stabilizer/flashhider and a bolt buffer to prevent overtravel of the bolt in the long receiver. Recommendations were made for further development of a lightweight rifle that could be manufactured with existing M1 rifle tooling. The T37 was further developed into the first of the T44 model series.

T47 Rifle

During October of 1951, as successor to the T25 rifle was designated. The new rifle was the 7.62 mm T47. The T47 rifle was equipped with a lightweight barrel, and unlike the T25, the T47 fired in both semiautomatic and full automatic from the closed bolt position. The T47 had a conventional style stock and sights. The T47 rifle utilized the gas expansion and cut-off system with 1.5-inches of piston travel. The rifle was designed to be fitted with a bayonet, flashhider, recoil check, grenade launcher and bipod. The T47 was capable of semiau-

Test firing the T47 rifle. This weapon was a strong contender for the new U.S. infantry rifle, but the T44 rifle with the Garand type action was considered superior. The T47 development program was terminated in April of 1953. (NARA)

tomatic and automatic fire at a cyclic rate of 850 rounds per minute. The rifle fed from a 20-round box magazine. There was a heavy barrel version designated as the T47E1 rifle. The bolt of the T47 was locked and unlocked by the tilting action of the breech lock. This was the primary feature that distinguished the T47 from the T44. The T47 rifle was a strong contender for the new U.S. infantry rifle, but the T44 design was considered superior and the T47 development program was terminated in April of 1953.

The T48 Rifle

The Belgian designed FN rifle was designated as the T48 during October 1951. The weapon was designed by the Fabrique National d'Armes de Guerre of Herstal, Belgium. The rifle was modified to fire the U.S. designed T65 7.62 mm cartridge and was ready for testing in 1952. The T48 was a lightweight, gas-operated, air-cooled rifle that could fire both semiautomatic and automatic. The automatic cyclic rate of the T48 was 660 rounds per minute. The weapon fed from a 20-round box magazine. The Belgian rifle was tested against the U.S. designed T47 and T44 rifles as a possible successor to the U.S. M1 rifle. The outstanding feature of the T48 was its ease of field stripping attributed to its hinged receiver, like that of a break-open shotgun. The gas cylinder was located on top of the barrel. A heavy barreled version, called the T48E1 was also made for the trials. In April 1953 testing of the T47 rifle was terminated, leaving the U.S. T44E4 to compete with the T48 for the next U.S. infantry rifle.

Test rifles from the top; The Belgium designed T48, the U.S. T44E4 (1953 model) designed by the Springfield Armory and the M1 Garand rifle. The M1 was used as a benchmark for the testing. Both the T48 and the T44E4 outperformed the M1. (NARA)

Chapter 3

The Experimental T44 Series of Rifles

The experimental T37 rifle was derived from the T20E2 select-fire M1 Garand rifle, but used the shortened experimental T65 7.62 mm cartridge cases. Early models used the T20E2 receivers with front and rear filler blocks fitted to improve feeding. Later versions used a shortened receiver especially designed for the T65 round. The M1 rifle's gas-impingement system was initially used on the T37, but the White gas cutoff system adapted from the T25 rifle provided longer operating rod life and was incorporated into the T37. Work on the T37 rifle was terminated in the fall of 1950.

A U.S. soldier test fires a T44E4 rifle, the prototype design of the U.S. M14. (NARA)

Early T44 rifle with T20 style "long" receiver. Note screw inserted in the back of the receiver used to hold a filler block in place. (U.S. Marine Corps Museum, Quantico, Virginia)

The T44 series of rifles evolved from the T37 prototype weapons, and was the final development of the basic rotating bolt M1 rifle design. In June of 1954 an order was received at the Springfield Armory for production engineering studies, and manufacture of 500 caliber .30 T44 lightweight rifles.

The T44 rifles were built in both a light and heavy barreled configuration. A flash-hider device was incorporated with the standard M1 front sight. One version also used the rear aperture sight of the M1 while others used an open style rear sight. The T44 borrowed the front-end features of the T25, the breech system and magazine catch system of the T20E2 and the magazine design of the T31. The action was the basic M1 rifle design. The gas cut-off system (White System) was used on the T44. A few T44s were fitted with a gas pressure relief valve that would bleed-off any excess gas pressure not required to operate the mechanism. The main purpose of the relief valve was for variations in cartridge pressures, especially those not manufactured in the United States. Many foreign NATO nations did not

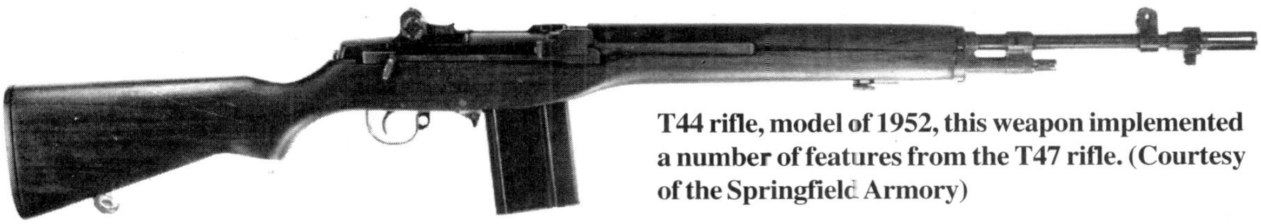

T44 rifle, model of 1952, this weapon implemented a number of features from the T47 rifle. (Courtesy of the Springfield Armory)

have the same strict quality control standards as the U.S. However, the gas cut-off system proved to be insensitive to pressure variations and the relief valve was removed from the design.

The T44E1 Rifle

prototype rifle introduced in 1951, was similar to the basic T44 rifle, but was designed as a squad automatic weapon possibly to replace the heavier BAR. The T44E1 was fitted with a heavier stock, a heavy barrel, two-position bipod and a hinged buttplate. An early version had an unusual semi-permanent 10-round magazine. The magazine could be reloaded by stripper clips through a guide on the receiver. Although the system was not adopted the receiver stripper clip guide was retained. In addition to a select-fire feature, the T44E1 also incorporated a rate reducer that allowed the operator to choose from two different cyclic rates. A fast rate of 735 rounds per minute or a slower rate of 550 rounds per minute. The T44E1 weighted 11.5 pound without accessories.

Receiver heel markings on a T44E4. After a few minor modifications the weapon was adopted as the 7.62 mm, Rifle, M14. (Courtesy of the U.S. Marine Corps Museum)

Seven short-receiver T44E4 rifles fabricated by the Mathewson Tool Company were prepared for cold weather testing. The rifles performed favorably throughout the extremely cold Arctic testing conducted at Big Delta, Alaska. (U.S. Army)

T44E2 Rifle

The Mathewson Tool Company completed and delivered the first T44 rifle with a reduced length receiver and bolt. Other improvements were made to the magazine latch, the bolt catch, and to the magazine. Another contract was consummated for a second rifle using the gas impingement system from the M1 rifle. The gas impingement rifle was designated as the T44E2. Initial testing of the T44E2 rifle, although being limited, resulted in a satisfactory performance.

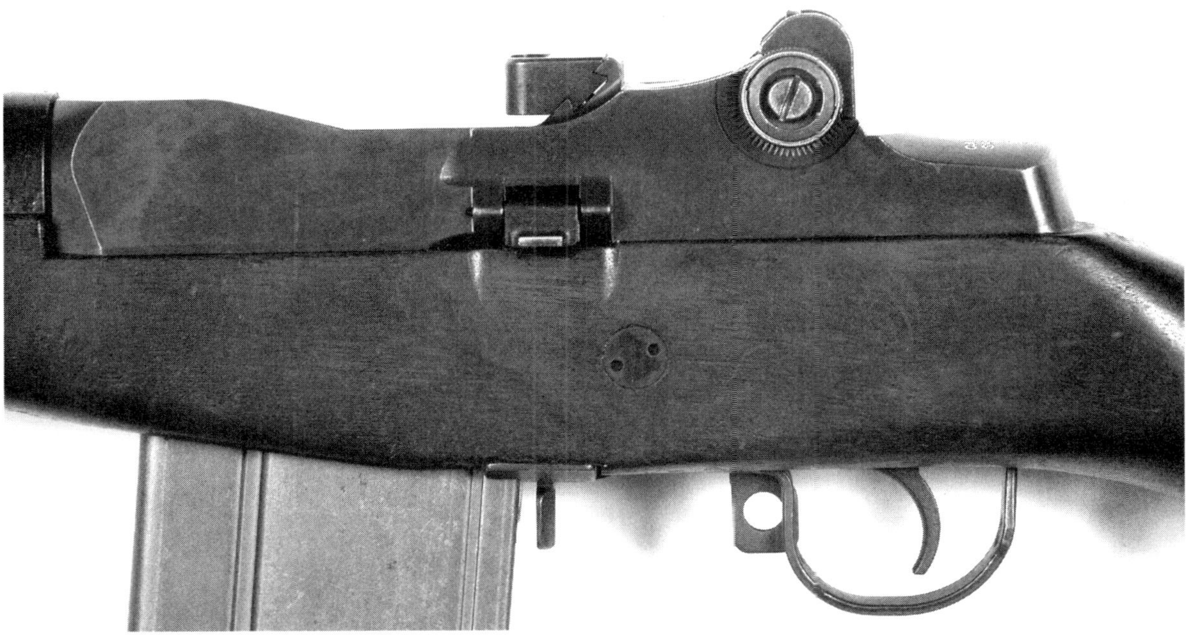

Short-receiver T44E4 rifle. Note the absence of the threaded hole and grooves for mounting a telescope. The magazine in the weapon is an aluminum prototype. (U.S. Marine Corps Museum)

The T44E5 rifle was a heavy barrel version of the T44E4 series using the gas-cutoff system. Note the early "flip up" style buttplate. A bipod was also used on the T44E5, but is absent from this weapon. (U.S. Marine Corps Museum, Quantico, Virginia)

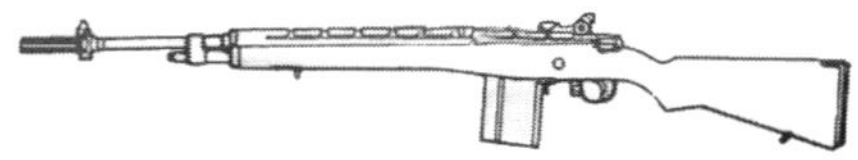

An early long-receiver T44 rifle. There were a number of T44 rifle variations, each one was slightly different than the previous one. Most were the same weapons that were continually being upgraded and modified by the Springfield Armory. (U.S. Marine Corps Museum, Quantico)

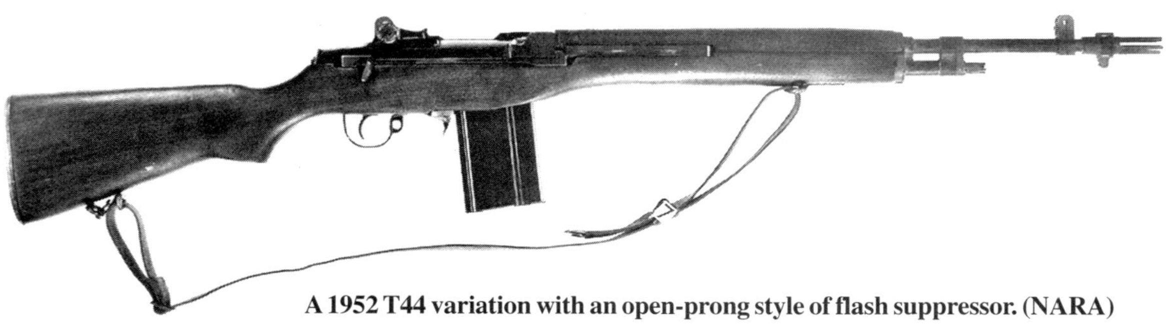

A 1952 T44 variation with an open-prong style of flash suppressor. (NARA)

T44E3 Rifle

A heavy barrel squad automatic variation of the T44E2 rifle, again using the gas impingement system of the M1 rifle.

T44E4 Rifle

A prototype rifle featured a lightened magazine charger, a gas cylinder shut-off valve and a redesigned gas piston and cylinder. A special stock and handguard incorporating a protective cover for the operating rod handle were fabricated for the Arctic testing at Big Delta. The rifles performed favorably throughout the extreme cold Arctic testing conducted at Big Delta, Alaska. The rifles were then forwarded to Fort Benning, Georgia to determine the effects of a temperate climate on the rifles that had been modified for the cold weather tests. Testing continued at Fort Benning on the center of impact stability with a change in barrel length. The barrels were being shortened in .125-inch increments and then tested.

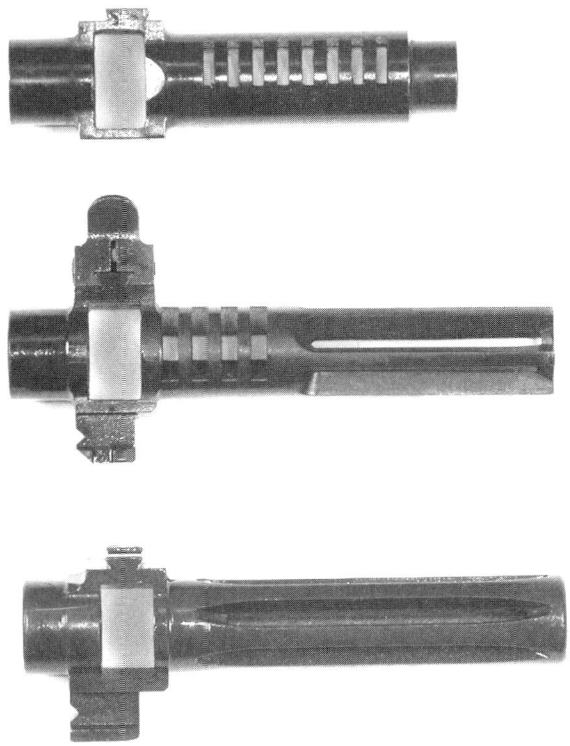

There were a number of flash-eliminators designed by the Springfield Armory. From the top; T44 1951 model, T44 1952 model and the final T44E4 type. (Courtesy of Bill Ricca)

Seven additional short-receiver T44E4 rifles fabricated by the Mathewson Tool Company were tested and prepared for Arctic testing.

Work at the Armory began preparing Ordnance drawings for the in-house manufacture of 500 caliber .30 T44E4 rifles. Spare parts requirements were coordinated with the Rock Island Arsenal, for the quantities of parts that would be required to maintain or repair the rifles during the testing period. Engineering studies and revisions were made to improve manufacture, reduce cost, improve parts interchangeability and functioning of the rifles. A study of the gas cylinder and valve assembly resulted in the elimination of eight components replacing them with four simpler parts.

The T44E5 rifles were constructed by modifying existing T44E4 rifles. A heavy barrel, bipod and flip-up buttplate were added. The T44E5 was eventually adopted as the M15 rifle as a replacement for the BAR. The M15 was cancelled just a few months after it was adopted, no production models were ever built. (U.S. Marine Corps Museum, Quantico, Virginia)

Progress was delayed by problems resulting in numerous changes in dimensions, tolerances and designs. A shortage of qualified personnel assigned to the T44E4 project added to the delay.

The testing for the Springfield Armory's T44E4 rifle and the Belgian-designed, U.S. manufactured, T48 rifle was scheduled to begin 1 September 1954. The stakes were high, it was expected that one of the rifles would be chosen as the new U.S. Infantry weapon, replacing the M1 rifle. The first production T44E4 rifle was test fired at the Armory on 18 April 1955. The first eighty rifles were function fired for acceptance using FA 6-T104E1 ammunition. Early prototype magazines had to be used because functioning problems had delayed the delivery of production magazines. Performance and accuracy was considered to be exceptional. Soon it was learned that all future tests would be conducted using M59 (T104E2) ammunition. Production magazines were received but functioning problems with the magazines continued. The change in the ammunition contributed to further functioning and accuracy problems. A new magazine spring (C7267078) resulting from the engineering study seemed to solve the magazine functioning problem. During July of 1955 the manu-

A T44E4 receiver remarked to indicate the conversion to the heavy-barrel T44E5 (M15) configuration. (Photo courtesy of the Springfield Armory)

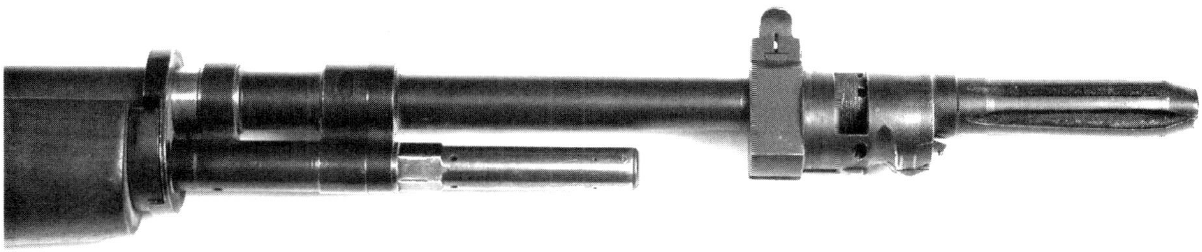

Barrel of the T44E5 rifle, the large round surface just behind the flash suppressor was for the mounting of a bipod. (U.S. Marine Corps Museum)

facture of 510 T44E4 rifles and accompanying spare parts were completed at the Armory, but early preliminary testing revealed a few malfunctioning problems with the rifles, delaying shipment. The entire lot of T44E4 rifles were held at the Armory until July awaiting revised drawings covering the modifications needed to resolve the problems. A total of 101 engineering change orders were received that affected 420 dimensions. In some cases original parts were deemed obsolete after the rifles were assembled, requiring disassembly to replace the affected parts. Deliveries were delayed further by a last minute modification to the bolt. After all of the problems were seemingly solved, the rifles were inspected and packaged for shipment. The last T44E4 rifle was packaged on 13 October 1955. All of the T44E4 test reports and data were prepared and cataloged; all of the drawings were updated to reflect the design of the rifles that were packaged. The completed rifles and spare parts were stored at the Armory awaiting disposition orders from the Chief of Ordnance's Office. Of the 510 T44E4 rifles, 8 were destined for endurance testing. A number of the T44E4 rifles were shipped to CONARC Board Number 3 for the temperate phase of the testing.

Early style buttplate with a hinged shoulder rest. Although not initially used on the first production M14 rifles, a similar type of hinged flip-up buttplate was added soon after production began and was retrofitted to earlier rifles. (U.S. Marine Corps Museum)

During preliminary testing of the T44E4 problems were discovered with the ejection pattern of the rifles. This was addressed by modifying the stripper clip charger guide and a heavier ejector spring. The extractor from the M1 rifle was used, but this would later prove problematic when firing blank ammunition. Other problems began to surface as well, the clips for retaining the roller on the bolt were failing and the rollers began to fall off of the bolts. This was found to be caused by interference between the operating rod and the roller. New rollers with modified grooves and heat treated roller retaining clips were fitted. Operating rods with deeper cam cuts were also tested. The results from the redesign initially proved favorable.

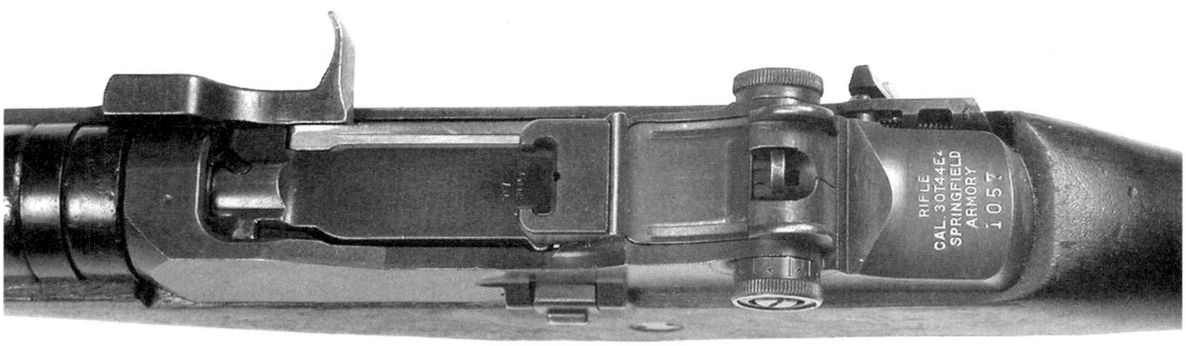

Springfield Armory T44E4 rifle serial number 1057. Note the early style full-automatic connector rod. (U.S. Marine Corps Museum, Quantico, Virginia)

Barrel of the T44E4 rifle. (U.S. Marine Corps Museum)

The 1953 version of the T44 rifle. Note the elongated gas cylinder that contained a pressure relief valve to compensate for variations in 7.62 mm cartridges and for grenade launching. This was the first of the T44 rifles to use a metal stock liner to keep the wood from breaking while launching grenades. (NARA)

A program to modify the gas plug was initiated for easier removal and to lessen the chance of cross-threading the plug upon reinstallation into the gas cylinder.

Two different quick-detachable bipod models featuring adjustable legs were completed and shipped to the Arctic Board for evaluation.

A further development effort resulted in the rear sight elevating knob being recalibrated from yards to meters for T104E2 (7.62 mm) ammunition.

T44E5 Rifle

During October 1954 a new heavy barrel rifle based on the best features of the earlier heavy barrel weapons the T44E1 and T44E3 was introduced. The new rifle designated as the T44E5 weighted 13.75 pounds without accessories and had an overall length of 44.14 inches. The latest heavy barreled rifle was developed to eliminate the use of modified M1 components used on the T44E1 rifle. The T44E5 used a new short receiver designed for the 7.62 mm cartridge. The weapon included all of the basic improvements that were imple-

Displayed on the table is an early Belgian manufactured .280 British caliber lightweight rifle and an early variation of the U.S. designed T44 rifle. (NARA)

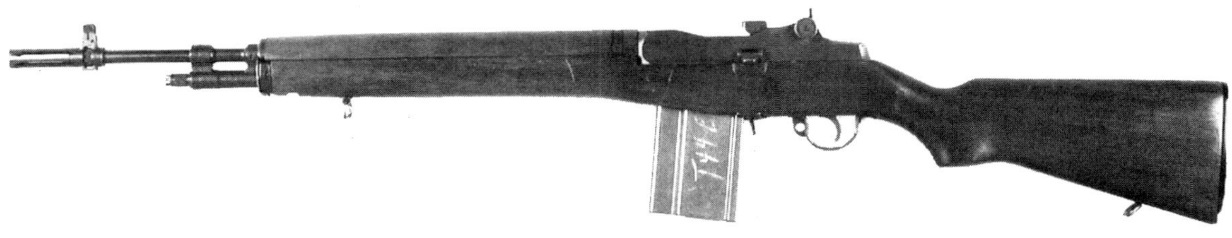

The experimental T44E6 rifle was designed in an attempt to reduce the weight of the T44E4 rifle. The T44E6 used a shortened 18-inch barrel and had no bayonet lug or capacity for the launching of grenades. Even the side ears of the rear sight were eliminated to save a few ounces of weight. An aluminum alloy 20-round magazine was also used. The result of the efforts reduced the rifle's weight by less than a pound. The side effects included an increase in felt recoil, as well as excess muzzle flash and noise. (Photo courtesy of the Springfield Armory Museum)

mented on the T44E4. A new fully adjustable bipod that could be folded back along the barrel was fitted. The heavy barrel T44E5 was tested both with and without the rate reducer, the test showed no significant difference in accuracy. The T44E5 was type classed Standard as the M15 rifle intended to replace the BAR in June of 1957, only a few prototype models were ever built before the weapon was declared obsolete in December 1957. Successful testing of the T44E4, adopted as the M14, fitted with a bipod, slotted handguard and hinged buttplate proved very successful and was instrumental in the demise of the M15 rifle.

T44E6 Rifle

One problem that was encountered with most prototype rifles chambered for the 7.62 mm T65 cartridge was their weight. It was proving difficult to produce a true "light rifle" using the T65 round. An experimental rifle designated as the T44E6 was built in an attempt to shave excess weight from the T44 rifle. The overall length of the T44E6 rifle was approximately 2.5-inches shorter than the T44E4. The T44E6 used a shortened 18-inch barrel and had no bayonet lug or capacity for launching of grenades. Even the side ears of the rear sight were eliminated to save a few ounces of weight. An aluminum alloy 20-round magazine was also used. The result of the efforts reduced the rifle's weight by about a pound, the side effects included an increase in felt recoil, as well as excess muzzle flash and noise. Due to the disadvantages outweighing the advantages, the project was terminated.

The study of the U.S. rifle evolution from the M1 to the M14 rifle is a complicated subject involving numerous prototype rifles and designs. The T44 is a typical example where there are numerous variations within the same T44 designation. (NARA)

Chapter 4
The T48 Rifle

After many years of research, development and testing by the U.S. Ordnance Corps for a new lightweight U.S. Infantry rifle, in the end it came down to a choice between just two weapons. The U.S. T44E4 that was developed at the Springfield Armory and the Belgian designed T48, better known as the Fusil Automatique Leger or FAL rifle. Both of the rifles were chambered for the U.S. T65E3 cartridge standardized as the 7.62 mm NATO round in 1954. Former contenders for the light rifle honors, the U.S. developed T25 and the British E.M.2 rifles were both dropped from further consideration in 1952.

The Belgium FAL rifle as originally manufactured in Europe was produced using the

A U.S. soldier fires a Belgian designed / U.S. made 7.62 mm T48 rifle. (NARA)

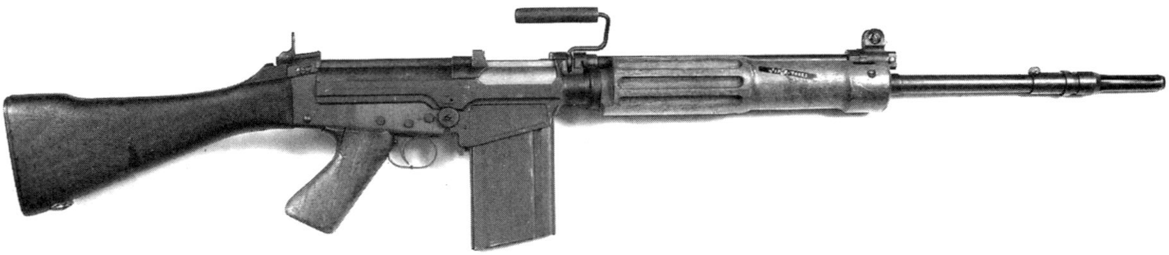

During 1953, the United States ordered 3,000 standard and 200 heavy barrel metric FAL rifles from Fabrique Nationale, Belgium for troop testing. The Belgian made rifles were designated as the T48. While the United States began work on making an inch-pattern U.S. manufactured FAL rifle, the Belgian weapons were issued to U.S. troops for field testing. (Courtesy of the U.S. Marine Corps Museum, Quantico)

metric system of measurement. Canada, Great Britain and the United States all used the Imperial or English system of inches. Any rifle produced in these countries would need to adhere to their standard of measurement.

One of the obstacles encountered with the Belgian rifle was that all of the factory drawings were done using the metric system. The Canadian Government was the first to begin the task of converting the drawings from metric to inches. Canada was also the first NATO country to adopt the FAL rifle in July of 1953. Later that year the United States ordered 3,000 standard and 200 heavy barrel metric FAL rifles from Fabrique Nationale, Belgium for troop testing. The Belgian made rifles were designated as the T48. While the United States began work on making an inch-pattern U.S. manufactured FAL rifle, the Belgian weapons were issued to U.S. troops for field testing.

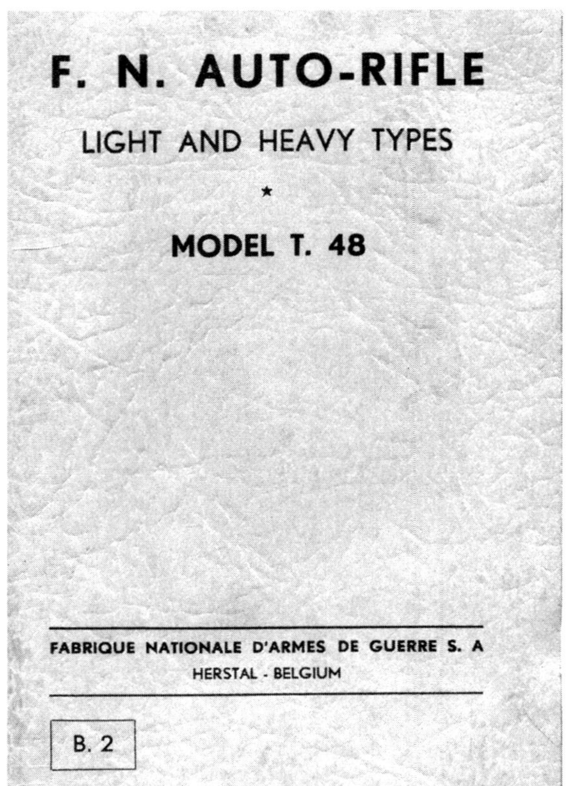
F. N. AUTO-RIFLE

LIGHT AND HEAVY TYPES

★

MODEL T. 48

FABRIQUE NATIONALE D'ARMES DE GUERRE S. A

HERSTAL - BELGIUM

B. 2

With some reluctance from the European countries, the U.S. developed 7.62x51 mm T65E3 cartridge was officially adopted as the Standard small arms round of the NATO alliance in 1954. In January 1954 Great Britain purchased a number of the FAL rifles chambered for the new 7.62 NATO cartridge for testing. Before long many other countries began to seriously consider the FAL rifle. Being that most of the NATO countries had reluctantly adopted the U.S. cartridge, they expected the U.S. to also adopt the FAL rifle so that not only would NATO have a common cartridge, but a common service rifle as well. However, there was no NATO requirement established for a standard infantry rifle, each country was free to adopt whatever rifle it wished.

The U.S. Ordnance Department charged the Springfield Armory with the job of converting the FAL rifle from the metric to the English system of inches. After a brief study it was determined that the converting the Belgian drawings to inches would not be an easy task. Other concerns were U.S. Ordnance specifications for materials, heat-treating procedures, manufacturing tolerances and other manufacturing operations. The Springfield Armory was busy with the T44E4 rifles and numerous other projects, decided to turn the task over to a commercial entity. Bids were taken for the project and the contract was eventually awarded to the High Standard Company. As part of the agreement High Standard was to manufacture twelve functional FAL T48 rifles from their final drawings. Canada had adopted the inch-pattern FAL then assisted High Standard and the Springfield Armory with their drawing conversions from metric into inches. The rifles were to be delivered by September of 1954. The U.S. FAL- the T48- used the Canadian magazine design that included a reinforced magazine lip. This modification

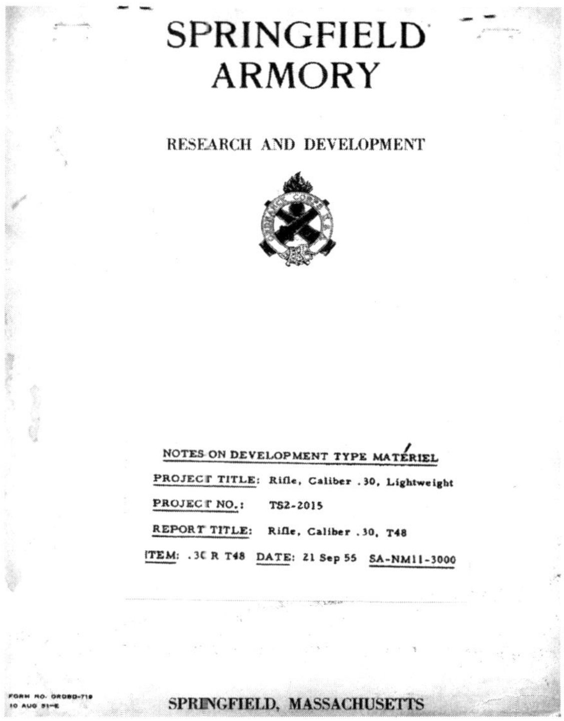
SPRINGFIELD ARMORY

RESEARCH AND DEVELOPMENT

NOTES ON DEVELOPMENT TYPE MATÉRIEL

PROJECT TITLE: Rifle, Caliber .30, Lightweight

PROJECT NO.: TS2-2015

REPORT TITLE: Rifle, Caliber .30, T48

ITEM: .30 R T48 DATE: 21 Sep 55 SA-NM11-3000

SPRINGFIELD, MASSACHUSETTS

T48 Rifle Research and Development manual from the Springfield Armory. One of the major problems encountered by the Armory was the conversion of the drawings from a metric to an inch pattern. (Courtesy of Bill Ricca)

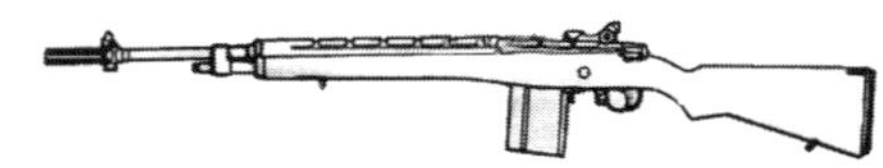

made to all inch-pattern FAL rifle magazines, made the magazine non-interchangeable with metric FAL rifles. Early testing found the Belgium FAL rifle proving superior to the U.S. T44E4. The first U.S. made T48 FAL rifle was fired at the Springfield Armory on 9 May 1955. Despite the fact that there was no formal contract between Fabrique Nationale and the United States, the company granted the U.S. at no cost the complete rights to produce the FN T48 rifle domestically.

The Mathewson Tool Company delivered the first U.S. designed short receiver T44E4 that summer. The Ordnance Corps authorized the procurement of 500 each of the T48 and T44E4 rifles for testing. The rifles were to undergo Ordnance testing in 1955. To circumvent a conflict of interest problem it was decided that a commercial company be awarded the contract for the T48 rifle rather than the Springfield Armory, who had developed the T44E4.

A meeting was held at the Office of the Chief of Ordnance on 10 May 1954 in order to brief

Receiver detail of a Belgian manufactured FN FAL T48 rifle. A number of Belgian made rifles were purchased for preliminary testing of the weapons by the United States. (Courtesy of the U.S. Marine Corps Museum, Quantico)

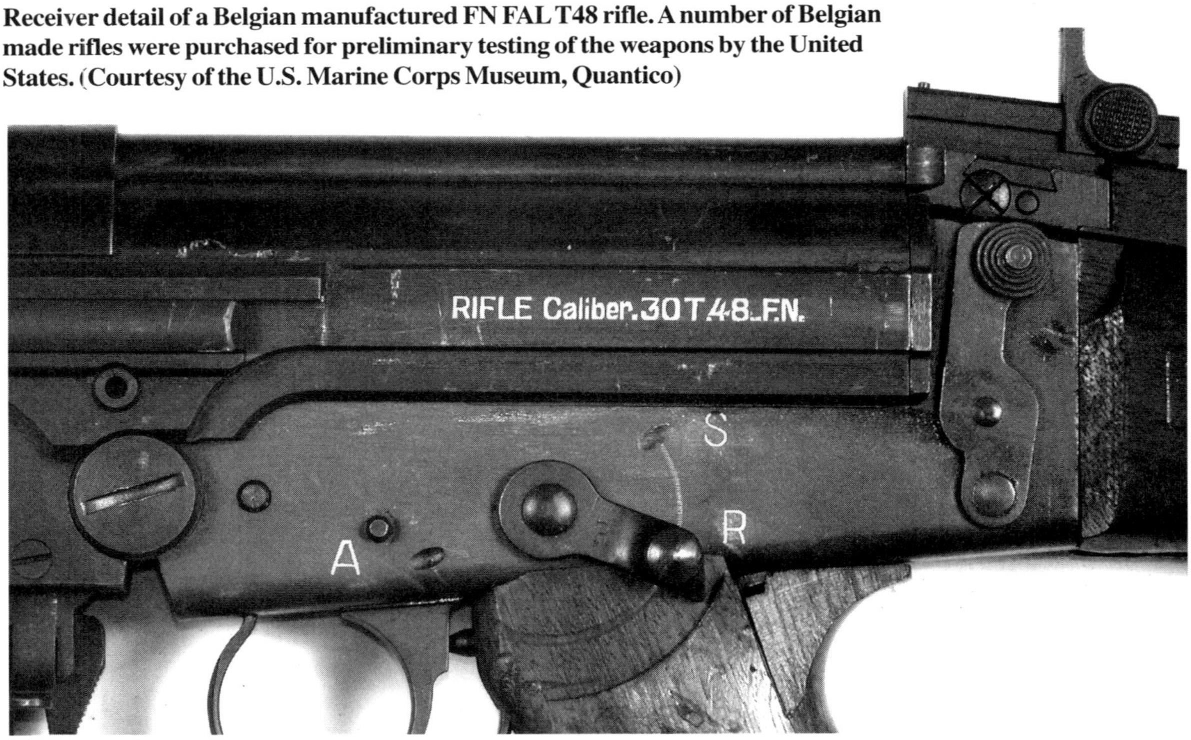

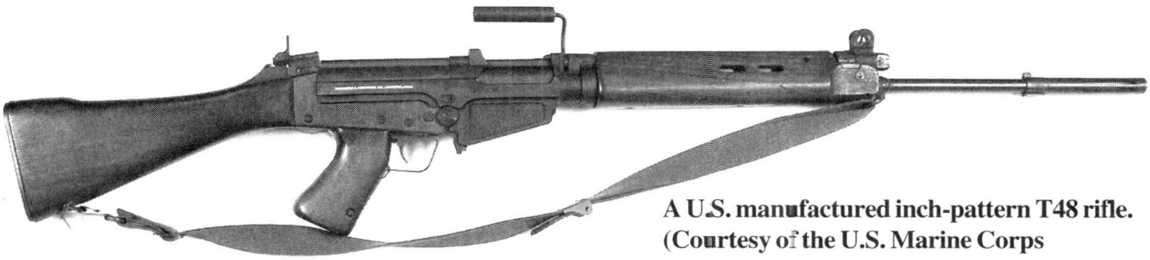

A U.S. manufactured inch-pattern T48 rifle. (Courtesy of the U.S. Marine Corps Museum, Quantico)

representatives of the Springfield Armory on the policy for the T48 rifle program. On 26 May 1954 Armory officials attended a conference held at the head office of Canadian Arsenals Limited. The purpose of the Canadian meeting was to participate in what was called the American-British-Canadian Program (A-B-C) to convert the metric FN FAL drawings to American Standards. The group distributed translated drawings and manufac-

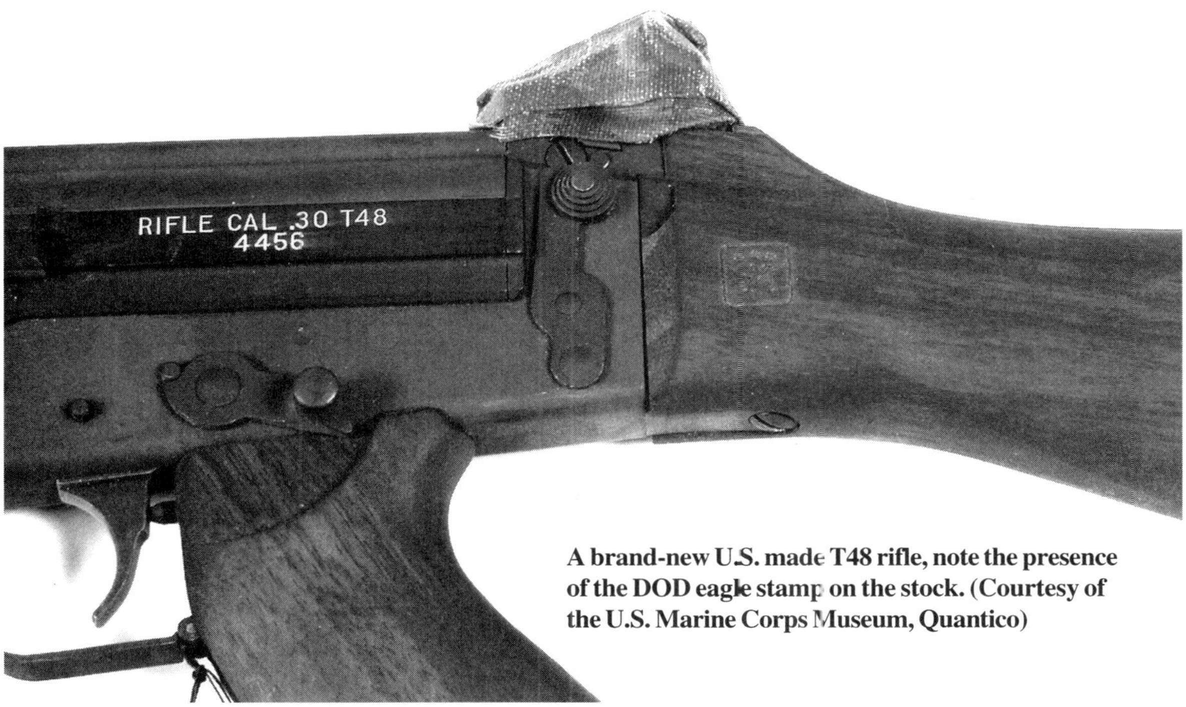

A brand-new U.S. made T48 rifle, note the presence of the DOD eagle stamp on the stock. (Courtesy of the U.S. Marine Corps Museum, Quantico)

turing data to prospective contractors who were invited to submit bids for production studies. It was established that engineering information would be exchanged between the United States, Great Britain and Canada with Canadian Arsenals LTD acting as the clearing house. Spare parts requirements for a normal one-year maintenance program were compiled for the 500 T48 rifles to be manufactured in the United States. At the same time seven new accessories for the T48 were engineered and Ordnance drawings prepared.

The Harrington and Richardson Arms Company and International Harvester both submitted bids to manufacture the T48 rifle. Harrington and Richardson was chosen to produce the T48 rifles while the Springfield Armory would manufacture the T44E4 design, as well as the heavy barreled T44E4, the T44E5 rifles for the trials.

The Hand Arms and Equipment Unit of the Armory was ordered to prepare Ordnance drawings, provide manufacturing information and technical data to the Boston Ordnance District, for use in administering the contract number DA-19-020-ORD-3436 for the caliber .30 T48 rifle to H&R.

One of 500 7.62 mm T48 rifles manufactured by the U.S. firm of Harrington & Richardson for testing against the Springfield Armory designed T44 series of infantry rifles. (Courtesy of the U.S. Marine Corps Museum, Quantico)

An updated 1970 era production 7.62 mm FN FAL rifle with plastic furniture (top). Shown below the Belgian FN rifle is a production U.S. M14 rifle with a fiberglass stock and handguard. The FN rifle was adopted by over 90 countries.

A representative from the Springfield Armory attended a four-day conference held from 16 May 1955 through 20 May 1955 at the Royal Small Arms Factory in Enfield, Middlesex England. The conference initiated by the British government was the first in a series to promote cooperation and eventual tri-partite NATO standardization of the Belgian FN rifle.

Both the T48 and T44E4 rifles were initially tested under Arctic conditions during the winter of 1954-1955. Both of the rifles were also concurrently undergoing testing at Fort Benning, Georgia. Results from both locations reported that the T44E4 rifle was proving superior. Fort Benning reported that the T48 rifles were experiencing functioning problems under dusty conditions. Similar problems were encountered with the FAL rifles during desert testing by Great Britain. To address the problem a small delegation of Americans who represented the U.S. contractors, visited the Royal Arms Factory in England and the Fabrique Nationale factory in Liege, Belgium. The problem was traced to the FAL's close tolerances between the bolt carrier and the receiver. The Belgium engineers suggested the incorporation of "sand cuts", or longitudinal grooves machined into the bolt carriers, this eventually solved the problem.

After the first 100 U.S. made T48 Harrington and Richardson FAL rifles were ready in September 1955, 10 were randomly selected for

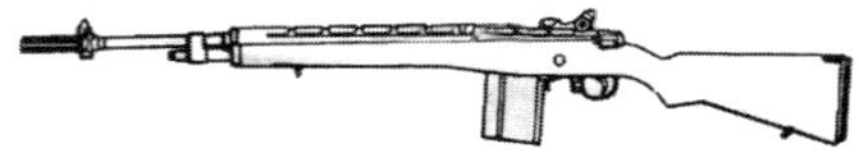

testing at the Aberdeen Proving Center. The Harrington and Richardson T48 rifles were initially experiencing a few problems with small parts failing and stoppages. After these problems were addressed and solved testing resumed. Subsequent trials found that the T44E4 had less malfunctions than the T48 rifles even after the modification of adding sand cuts to the bolt carriers of the T48. The Harrington and Richardson T48 rifles were unable to qualify in the accuracy test. This problem was traced back to excessive machining tolerances on the barrels, partially as a result of a drawing error.

A U.S. made T48 rifle fires a burst of 7.62 mm ammunition during testing at the Aberdeen Proving Center. (NARA)

After most of the early problems had been solved a final head to head full-scale test be-

A number of U.S. made T48 rifles in storage. Many of the weapons shown here appear to be unfired. (Courtesy of the U.S. Marine Corps Museum, Quantico)

tween the U.S. designed T44E4 and the Belgium designed, U.S. built T48 rifles was scheduled for the spring of 1956. However due to time restraints an abbreviated test was administered at both Fort Benning and at the U.S. Marine Base, Quantico, Virginia. As a control weapon for the program a U.S. M1 Garand rifle was tested and the results compared to the T48 and T44E4 weapons. There was a great deal of skepticism on behalf of those involved in the testing about the United States Army adopting a foreign designed infantry weapon for its forces. Many felt that the U.S. was simply going through the motions of testing the FAL rifle to appease Canada, Great Britain and other NATO members, who were pressuring the U.S. to adopt the Belgian FAL rifle. On the other hand there

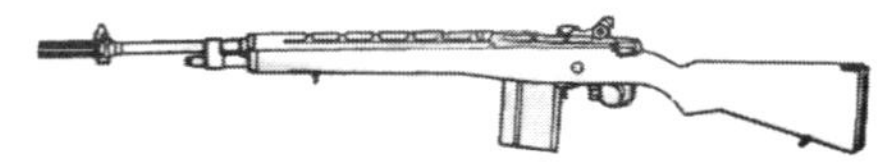

was great political pressure from the U.S. and the Springfield Armory to adopt the U.S. designed T44E4. Back in the 1940 to 1960 era there was a lot of pride in American produced products, after all the U.S. had been the "Arsenal of Democracy" during World War II when weapons designed and made in the U.S.A. "saved the day" for the Allies.

After the trials were completed, Fort Benning reported that both rifles were suitable for army use, failing to decisively choose one design over the other. The Marine testing at Quantico slightly favored the Springfield Armory's T44E4. It seemed that no one was prepared to select one rifle over the other, as it was becoming a very sensitive political issue. The only absolute was that both rifles out performed the U.S. M1 rifle that was tested along side of the T48 and T44E4 as merely a control weapon. Others were outspoken about the adoption of either rifle stating that neither one represented any real improvements over the old M1 rifle, emphasizing that the search

Many of those involved with the T48 program felt that the U.S. was simply going through the motions of testing the Belgian rifle to appease Canada, Great Britain and other NATO members, who were pressuring the U.S. to adopt the Belgian FAL rifle. On the other hand there was great political pressure from the U.S. and the Springfield Armory to adopt the U.S. designed T44E4.

for a true modern, lightweight infantry rifle had simply not been realized.

Nearly a year after the final trials, the U.S. Government finally concluded that the T44E4 was; a full pound lighter than the T48; contained fewer parts and had a single-port gas system that didn't require adjustments. The similarities between the T44E4 and the M1 Garand rifle would ease both training and manufacture requirements. Despite pressure from the other NATO countries for the U.S. to adopt the Belgian designed T48 rifle, on 1 May 1957 Secretary of the Army Wilbur H. Bruckner announced that the T44E4, and T44E5 were being adopted as the Standard U.S. service rifles as the M14 and M15 respectively. An Ordnance Committee meeting held on 14 November 1957, finalized the adoption of the M14 and M15 rifles as Standard. At the same meeting the M1 rifle, the Browning Automatic Rifle, the M1 carbine, and the M3A1 submachine gun were classified as Limited Standard items.

The ability to launch a rifle grenade was one of the requirements the U.S. Ordnance Corps had for any infantry rifle to be considered for adoption. A soldier demonstrates the launching of a grenade from a T48 rifle. (NARA)

The Belgian FAL rifle in slightly varying configurations eventually was adopted as the Standard service rifle by over ninety countries. The FAL would become the most prolific rifle among the NATO alliance. On the other hand the U.S. M14 rifle would have one of the briefest tenures of any previous U.S. service rifles. Although officially adopted in 1957 no deliveries of the rifle were made until 1959.

Within a few years after the last M14 rifle was made, the weapon began being phased out of U.S. service, replaced by the small caliber 5.56mm M16A1. Many of the NATO alliance countries were somewhat bewildered by the United States' decision to adopt the diminutive 5.56 mm cartridge. Great Britain was especially shocked in light of the U.S. Army handily rejecting their .280 cartridge back in 1954, on the grounds that it did not have sufficient power for a U.S. infantry weapon.

Chapter 5

U.S. Rifle 7.62 mm M14

"I don't want no teenage queen, I just want my M14"...
from the Hollywood film *"Full Metal Jacket"*

On 12 February 1957, after a develop ment period that spanned ten years and an expenditure of over 100 million taxpayer dollars, the M14 rifle was presented to the press and public during a ceremony held at the Aberdeen Proving Center. Almost immediately criticisms were voiced over the "modern", "lightweight" infantry weapon, which looked remarkably like the M1 Garand rifle and weighed only slightly less. More criticism would be forthcoming from numerous production problems that would be encountered with the new rifle.

The M14 rifle was formally adopted on 1 May 1957 to replace the M1 Garand rifle as the "Standard A" infantry weapon of the United States armed forces. The M14 that was very similar to the M1 rifle, evolved from a long series of experimental and prototype weapons. The developments that lead to the adoption of the M14 actually began in 1945 with the T20 series of M1 rifles. The T20 rifles were M1 Garands modified for both semiautomatic and full automatic operation. The M14 was chambered for a new cartridge known as the 7.62 mm NATO. The 7.62 round was .5 inch shorter in its overall

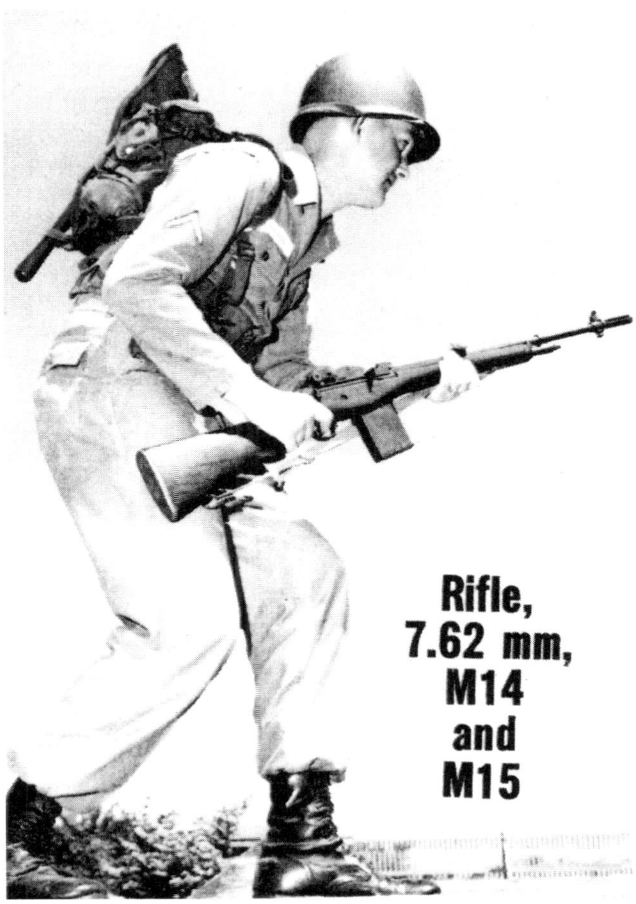

This Ordnance Corps photograph illustrated a U.S. soldier with an early M14 rifle with a Garand-style buttplate. (U.S. Army)

U.S. Rifle 7.62 mm M14

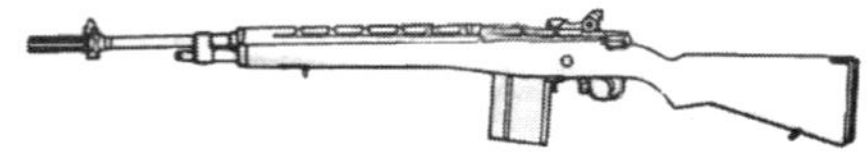

The M1 Garand rifle (shown right) is compared to the newly adopted M14 rifle held by the soldier on the left. The 7.62 mm M14 rifle and the M60 machine gun were intended to replace all previous infantry small arms. (U.S. Army)

length over the previous .30 caliber M2 (30'06) round used in the M1 rifle, but had nearly the same ballistics.

With its new "short" 7.62 mm cartridge the M14 was expected to displace the Browning Automatic Rifle, the M1 rifle, the M1-M2 carbine and the M3A1 submachine gun. Originally there were to be special variations of the M14 produced to replace these weapons. The special heavy-barrel M15 was to supercede the BAR. A compact version of the M14 rifles with a folding stock was intended to replace the carbine and submachine gun. Due to limited budgets, the special purpose M14 rifles never evolved past the prototype stage.

The M14 rifle along with the newly adopted M60 7.62 mm general purpose machine gun, were expected to fulfill all of the needs of the

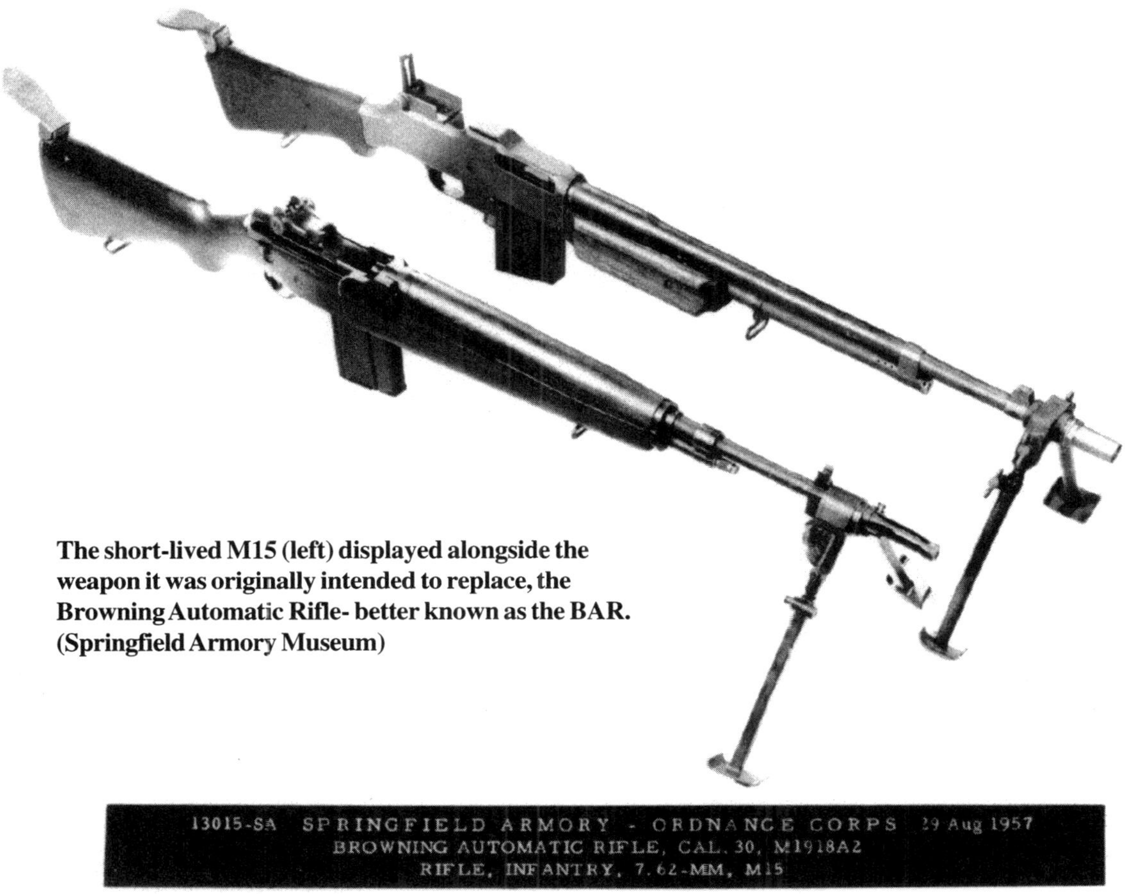

The short-lived M15 (left) displayed alongside the weapon it was originally intended to replace, the Browning Automatic Rifle- better known as the BAR. (Springfield Armory Museum)

infantryman while using one common cartridge. The M14 as originally configured weighed 11.25 pounds with a loaded twenty-round magazine. The overall length with the early M1-style buttplate was 44.14 inches.

The Official Description

The Rifle 7.62 mm M14 is an air-cooled, lightweight, gas-operated, magazine-fed shoulder weapon, designed primarily for semiautomatic fire. The weapon is chambered for the NATO 7.62 mm cartridge. The rifle is similar to the M1 rifle in appearance and operation. The M14 "action" (i.e. receiver, bolt and trigger group) is basically the same as that of the M1 rifle. The differences in nearly all instances are required to accommodate the shorter cartridge and the 20-round box magazine. The receiver,

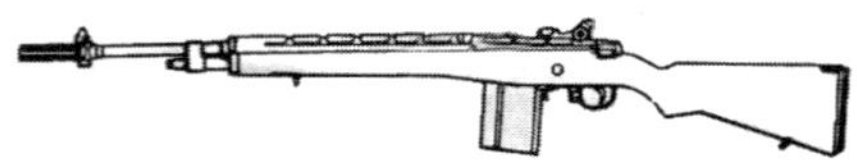

bolt and firing pin are all shorter than those in the M1. The bolt has a stud fitted with a roller that contacts the track in the operating rod. The trigger group, with the exception of the sear is basically the same as the M1, but a new housing and trigger guard are used. The trigger housing floor-plate is cut back to allow for the insertion of the box magazine, while sufficient surface area has been maintained to provide proper clamping of the stock. The magazine latch is located in a recess in the forward face of the trigger housing. The bolt lock is mounted on the left side of the receiver. The rear sight group, mounted on the receiver, is composed of M1 parts except that the elevation knob is calibrated in meters rather than yards. The sear release, selector shaft, and connector assembly, mounted on the right side of the receiver and positioned by the selector lock, are inoperative during semiautomatic fire. The operating rod is a welded assembly, consisting of a handle and capped short tube. The tube houses the operating rod spring and guide. A sliding lock pin secures the spring guide in place. The operating rod tube slides in the operating rod guide which is pinned to the barrel. The end of the operating rod tube abuts the gas cylinder piston, when the rod is in the forward position. The gas cylinder, unlike the M1 system, contains a hollow gas piston and is mounted on the barrel with the gas port located approximately eight-inches from the muzzle. The gas cylinder positions the front band against a shoulder on the barrel and is locked in place by the looped gas cylinder lock and the gas cylinder plug. The front and rear bands secure the top handguard. The flash suppressor is secured in place on the barrel by a flash suppressor nut, which is locked by a socket set screw. The flash suppressor is the five-prong type with a closed front end. The lower prong is wider than the others, to provide a moderate amount of compensation for muzzle climb.

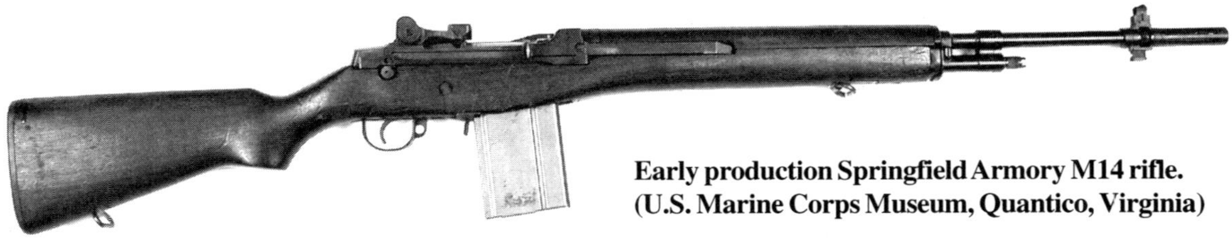

Early production Springfield Armory M14 rifle. (U.S. Marine Corps Museum, Quantico, Virginia)

The front sight is mounted on the flash suppressor. A lug mounted on the underside of the flash suppressor provides a means to attach a bayonet or a grenade launcher. The barrel and receiver assembly is "bedded" in a conventional drop stock. The rear sling swivel is secured by one of the hinged buttplate screws. The bracket for the front sling swivel is riveted to the forearm of the stock. The receiver well opening and the recoil abutments in the stock are reinforced by the metal stock liner, (the liner is not used on the synthetic stock) which is secured to the stock by two special fasteners. The trigger group locks the stock to the barrel. The M14 rifle weighs 8.33 pounds empty and 11.25 pounds with a loaded magazine, sling and cleaning kit. Overall length is 44.3-inches when the flip-up buttplate is fitted. The barrel length measures 22 inches in length, rifling is 4-groove, with a right-hand twist. Automatic cyclic rate is 750 rounds per minute. Muzzle velocity is 2800 feet per second. Maximum range is 3,725 meters, maximum effective range 460 meters. When equipped with the M2 bipod (1.75 pounds) the effective range is 700 meters.

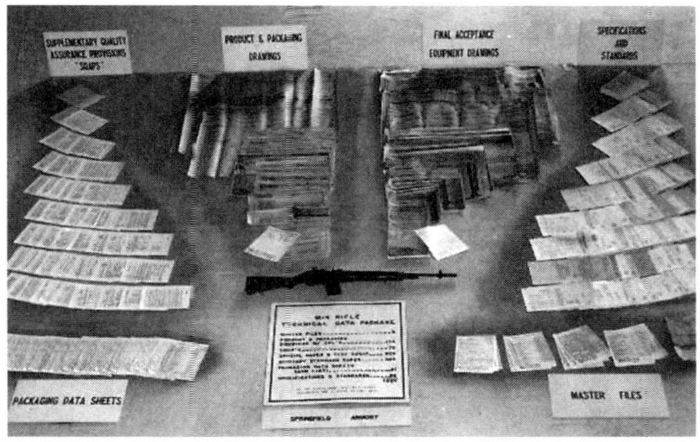

M14 Rifle is displayed with the massive technical data package required for its manufacture. The Springfield Armory developed the data package during their pilot line production program. The data package was then provided to the commercial firms who successfully bid on the contracts to manufacture the M14 rifle. (Springfield Armory Museum)

As originally Standardized the M14 rifle consisted of approximately 110 component parts, 29 of which were common to the M1 rifle.

The Springfield Armory

The Springfield Armory was not intended to be a large producer of the M14 as it had been with the M1 Garand rifle, manufacturing millions of them during World War II and Korea. Instead the Armory was directed to work out a pilot production line and provide technical support to commercial contractors that would place bids on the job of manufacturing M14 rifles for the government. This decision caused a lot of concern for the civilian employees of the Armory. There was much trepidation about this decision among both Armory employees and the surrounding community as the Armory was a large employer and economically vital to the area. A civic organization of concerned citizens was named The Springfield Rifle for Springfield Committee. The goal of the committee was to secure work for the Armory, and with the assistance of Congressman Edward P. Boland, petitioned to have the nomenclature of newly adopted M14 rifle changed to the "Springfield '57 Rifle". Their efforts proved futile.

Springfield Armory M14 rifle and 3,080 7.62 mm spent cases fired during an endurance test of the weapon. (Courtesy of the Springfield Armory Museum)

There were a number of delays in getting the newly adopted M14 rifle into production after it was officially adopted. Procurement funds could not be requested from Congress until the entire weapon's Standardizing documents were written. After the rifle was adopted in May 1957 it was announced to the Senate Appropriations Subcommittee that the adoption had come too late to procure funding for initial production of the rifle in the 1958 Fiscal Year budget.

Although the new M14 rifle was adopted as Standard in 1957, no production was initiated until April of 1958. After funding was granted there was pressure to quickly get the M14 into production after the initial delays. The Armory had to prepare up to date drawings that would include the latest design changes and specifications. Another daunting task was development of new tooling and inspection equipment. The engineers at the Springfield Armory cautioned against rushing into production until adequate preparations were made.

The funds for a preproduction engineering study were granted in March of 1958. The Armory was directed by the Ordnance Weapon Command to complete the task by

September. The September deadline was soon moved up to 15 August 1958. In order to help expedite the process the Armory was instructed to only include critical design changes. Changes that were initially omitted were a synthetic stock and handguard and design changes to improve the magazines. Problems would later be encountered because of the magazine design causing the Armory to alter them so that they would function during initial testing of the first M14 rifles off the production line.

Brand new M14 rifles travel down a conveyer belt to be packaged and shipped. (Springfield Armory Museum)

The United States Ordnance Corps issued an order to establish a pilot production line at the Springfield Armory for the manufacture of 15,669 M14 rifles. The pilot line was used to establish the best manufacturing procedures and to solve technical problems that could not be foreseen until production actually began. The Armory's experience gained during this phase would eventually be used to assist commercial contractors as they came on line. The initial order of rifles was to be completed by April, 1960. At this time it was stated that the Springfield Armory would organize a pilot production line, prepare Technical Data Packages and produce M14 rifles, but a large majority of the M14 rifles would be contracted out to commercial companies through a bidding process.

By December of 1958 the Springfield Armory had set up a pilot production line and began to manufacture some key parts such as the receiver, trigger housing, bolt and operating-rod. An initial delivery of fifty complete M14 rifles was scheduled for July of 1959. The estimated cost of the M14 was reduced to $126.59 during the period due to the overhead rate in the shops being revised. The Armory began to prepare a production technical data package for the commercial manufacturers that were being contracted to produce the rifle. The pilot lines for various parts were established in buildings 102, 103 and 104. The

Just one of the vast manufacturing facilities of the Springfield Armory used to manufacture the M14 rifle during 1962. (Springfield Armory Museum)

receiver production line was set up in building 104 to take advantage of the overhead conveyor system in the building. M14 barrel production was setup in building 201C. Direct labor hours for production were established at 10.4 hours per rifle. A number of problems were encountered with mass producing some of the more complex parts. All M14 parts produced at the Armory were marked with the letters SA to identify the manufacturer. Barrels manufactured at the Springfield Armory were also SA stamped along with the month and year of manufacture. Many parts were also marked with their Ordnance drawing numbers.

Further delays would be encountered. Due in part to lack of suitable funding, no large amounts of steel had been stockpiled for future production needs. A nationwide steel strike in 1959 made steel a critical and unavailable material. As a result the Springfield Armory had only completed a total of 4,245 M14 rifles by the April 1960 deadline. The final twenty-five rifles were eventually shipped in October of 1960. By this time the production rate of the M14 was up to 3,000 rifles per month. The first production M14 was presented to U.S. Army Master Sergeant George Ferguson in October of 1959 by Secretary of the Army Wilbur C. Brucker, during a special ceremony that was held at the Aberdeen Proving Center, Maryland.

In April of 1959 a few last minute improve-

The receiver production line located in building 104 at the Springfield Armory. At its peak the Armory was manufacturing 6,500 M14 rifles a month. (SANHS)

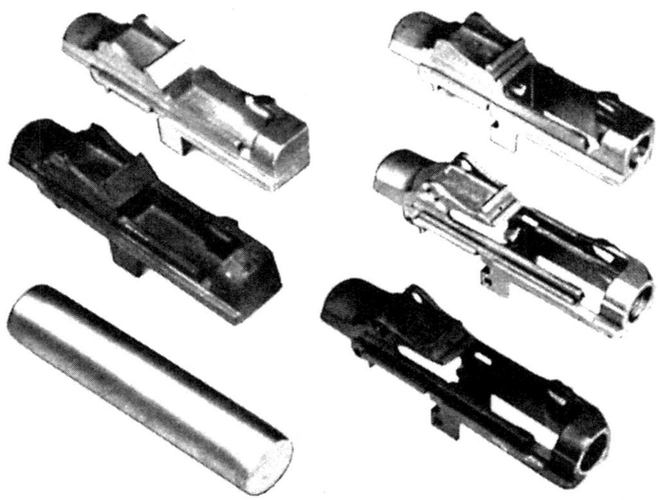

The complex receiver of the M14 rifle proved to be the most difficult part to manufacture. (SANHS)

ments were implemented into the design of the M14 rifle. The changes included; a longer connector lock-pin to facilitate disassembly while wearing Arctic hand gear; a modified gas piston to facilitate alignment of the piston during reassembly; providing the gas cylinder and flash suppressor with off-set splines to prevent incorrect assembly; modifications to the barrel and providing a pilot diameter to prevent cross-threading of the gas cylinder lock.

During the early months of 1960, a decision was made to fit all future M14 rifles with the newly developed ventilated plastic handguard, flip up style buttplate and M2 Bipod. All previously manufactured M14 weapons were to be retrofitted and updated with the latest design changes at the Raritan Arsenal in New Jersey. The features were previously used on the heavy-barrel M15 rifle. The "new" M14 was designated in the field manuals as the M14 (M) for modified.

After World War II had ended the United States Army began to convert over to the metric system. This was done in part to comply with NATO requirements. The Army converted all of its maps, technical manuals and associated equipment over to the metric system. During 1957 AR 700-75 dated 14 May 1957 indicated that "to comply with U.S. Army Regulations the rear sight of the rifle was changed to metric calibration and satisfactorily tested at Aberdeen". The rear M14 sight was then marked with an M to indicate metric, as not to confuse the sight with that of the M1 rifle, which was calibrated in yards. Likewise, M14 ammunition was designated in metric terms as the 7.62 mm cartridge. In October of 1958 it was reported by the Springfield Armory that the rear sight conversion was completed.

The Armory was instructed to increase the rate of M14 rifle production to 5,000 per month by March, 1961. An increase in manpower and overtime were authorized to attain the Armory's new goal. At the same time a study was begun to see what manpower, tools and funds would be required to step-up production to 10,000 M14 rifles a month. As the study was underway, a shortage of funds terminated the planned production increase.

Planning for the Fiscal year 1961 included new funds for an increase in M14 production. After the funds were received in June, production was to be stepped up for production of 6,500 rifles with spares per month. The Armory was given until September of 1961 to get itself up to speed. The order for M14 rifles was increased by 45,000 additional weapons. The area for manufacturing the M14 had to be rearranged in order to handle the increased production. The receiver line had to run 24 hours a day to keep up with the revised sched-

An Army general test fires a new M14 rifle at the Fort Benning range. (NARA)

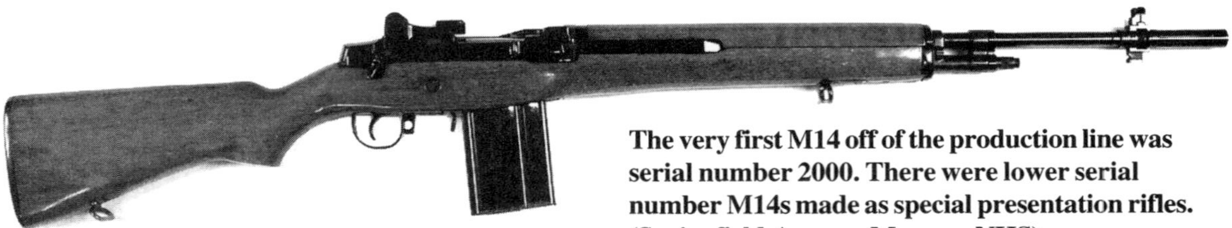

The very first M14 off of the production line was serial number 2000. There were lower serial number M14s made as special presentation rifles. (Springfield Armory Museum NHS)

ule. Further problems were encountered with the Armory's accelerated production requirements. As an interim measure to solve the problem, the Winchester and the Harrington and Richardson firms supplied the necessary parts to the Armory. During this period the walnut stocks for the M14 were being replaced by stocks made of birch wood. Walnut became an alternate material for M14 stocks.

During the period from September through December 1961 the production rate of M14 rifles at the Springfield Armory was 6,500 weapons per month. An amended schedule was received from Ordnance Weapon Command Headquarters during December to re-

Typical receiver markings on a standard production Springfield Armory M14 rifle. (Marine Corps Museum, Quantico, Virginia)

duce production to 5,000 rifles during January 1962 and 4,000 rifles per month starting in February. The reduction was expected to remain in place until September 1963. During June 1962 it was announced that a follow-on order for an additional 20,000 M14 rifles from the Armory would not be forthcoming.

The Quality Assurance Section of the Springfield Armory devised internal forms to document any feedback for problems encountered during pilot line production. The Section also checked all acceptance inspection equipment and gages on a regular basis.

Of special concern were complex and expensive items like the receiver. To reduce the scrap-rate, in-process station inspections were established along the line. The Quality Assurance Section services were extended to components, and subassemblies being purchased from commercial vendors. The Springfield Armory Administration and Training Branch was assigned the task of preparing Quality Assurance training manuals and inspection training courses for Ordnance District and Arsenal inspectors for the M14 rifle. Other duties tasked to the Armory were to conduct research and development for ways simplify parts and reduce the cost of manufacturing the M14. Any cost savings implemented by changes in the design were expected to be passed back to the government by the commercial manufacturers.

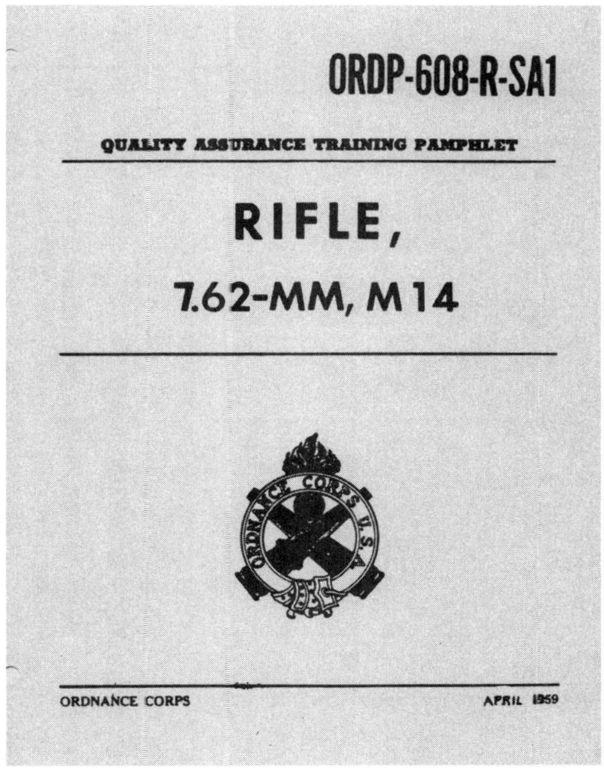
ORDP-608-R-SA1

QUALITY ASSURANCE TRAINING PAMPHLET

RIFLE,

7.62-MM, M14

ORDNANCE CORPS APRIL 1959

Cover of a 1959 dated Quality Assurance Manual. (Courtesy of John M. Miller)

One component of the M14 rifle that proved especially troublesome to manufacture was the flash suppressor. The flash suppressor was originally designed as a two-piece casting by the engineers at the Watervliet Arsenal. A number of problems were encountered with the casting of the part, and the suppressor was redesigned to be made from a one-piece forging. The new forging process resulted in a better quality part and reduced manufacturing costs. The cast suppressor had cost $10.01

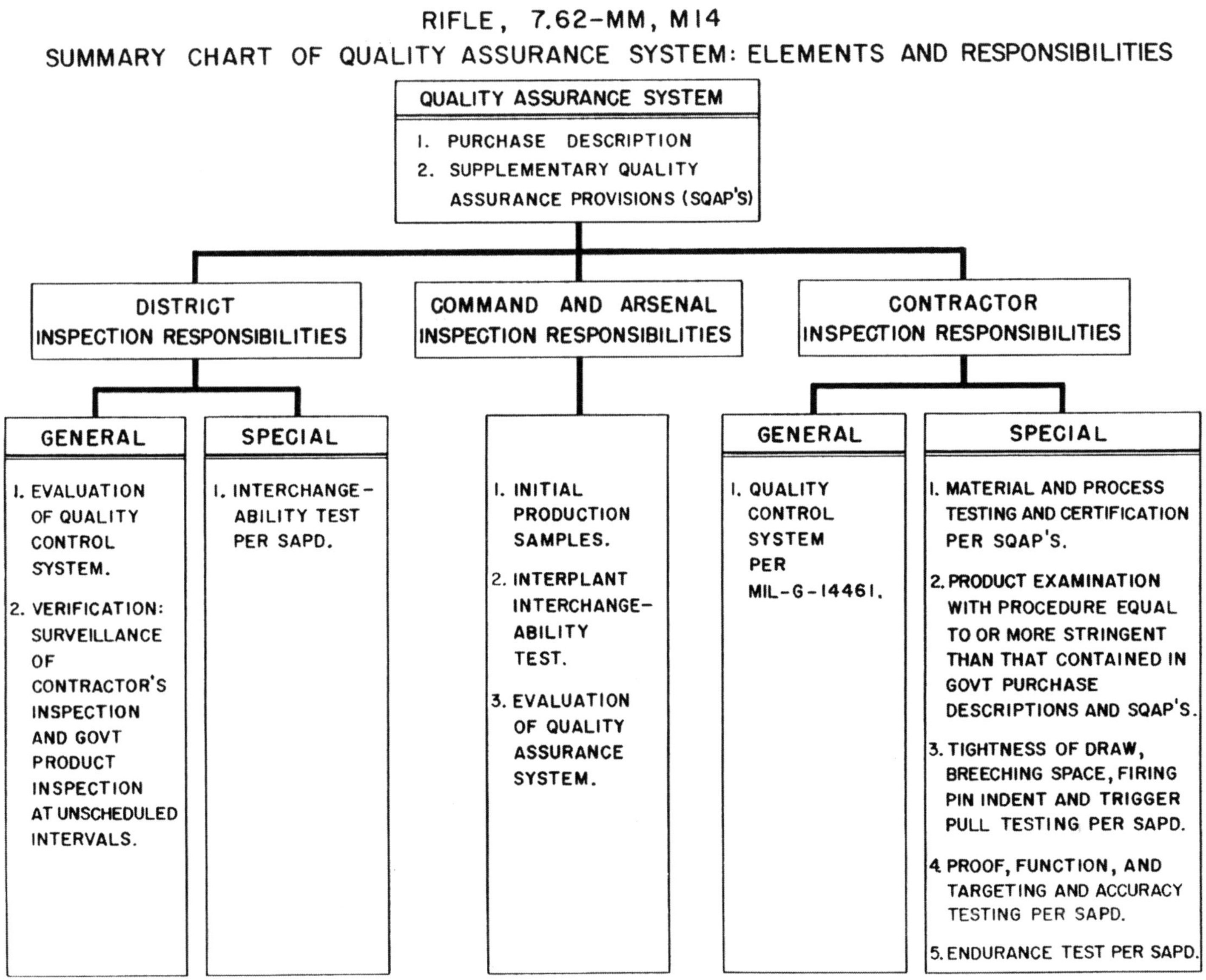
RIFLE, 7.62-MM, M14
SUMMARY CHART OF QUALITY ASSURANCE SYSTEM: ELEMENTS AND RESPONSIBILITIES
QUALITY ASSURANCE SYSTEM
1. PURCHASE DESCRIPTION
2. SUPPLEMENTARY QUALITY ASSURANCE PROVISIONS (SQAP'S)
DISTRICT INSPECTION RESPONSIBILITIES
GENERAL
1. EVALUATION OF QUALITY CONTROL SYSTEM.
2. VERIFICATION: SURVEILLANCE OF CONTRACTOR'S INSPECTION AND GOVT PRODUCT INSPECTION AT UNSCHEDULED INTERVALS.
SPECIAL
1. INTERCHANGE-ABILITY TEST PER SAPD.
COMMAND AND ARSENAL INSPECTION RESPONSIBILITIES
1. INITIAL PRODUCTION SAMPLES.
2. INTERPLANT INTERCHANGE-ABILITY TEST.
3. EVALUATION OF QUALITY ASSURANCE SYSTEM.
CONTRACTOR INSPECTION RESPONSIBILITIES
GENERAL
1. QUALITY CONTROL SYSTEM PER MIL-G-14461.
SPECIAL
1. MATERIAL AND PROCESS TESTING AND CERTIFICATION PER SQAP'S.
2. PRODUCT EXAMINATION WITH PROCEDURE EQUAL TO OR MORE STRINGENT THAN THAT CONTAINED IN GOVT PURCHASE DESCRIPTIONS AND SQAP'S.
3. TIGHTNESS OF DRAW, BREECHING SPACE, FIRING PIN INDENT AND TRIGGER PULL TESTING PER SAPD.
4. PROOF, FUNCTION, AND TARGETING AND ACCURACY TESTING PER SAPD.
5. ENDURANCE TEST PER SAPD.

to produce while the more reliable forged unit was estimated to cost $4.27 per piece. The redesigned flash suppressors were integrated into the production line during September 1960

The front sight was also redesigned to use drawn stock instead of the original investment casting. The total savings projected were 25-cents per unit.

Additional problems were encountered at the Armory with machining the complex receiver of the M14. Another problem encountered with the receiver was with distortion after heat treating. An Armory team headed by consultant Mr. E.W. Wood eventually solved the distortion problem. Many other problems encountered with production of the M14 components were solved by the relaxation of the strict machining tolerances.

During 1961, much difficulty was being encountered by the Springfield Armory and the commercial contractors with meeting the requirements specified for core hardness, free ferrite, case hardness and case depth on M14 receivers and bolts using the 8620H steel specified. The 8620 steel was found to have a wide harden-ability band, which varied from one heat mill lot to the next. Another problem was the difficulty in obtaining uniform results from the standardized heat-treating process.

In attempts to overcome the production prob-

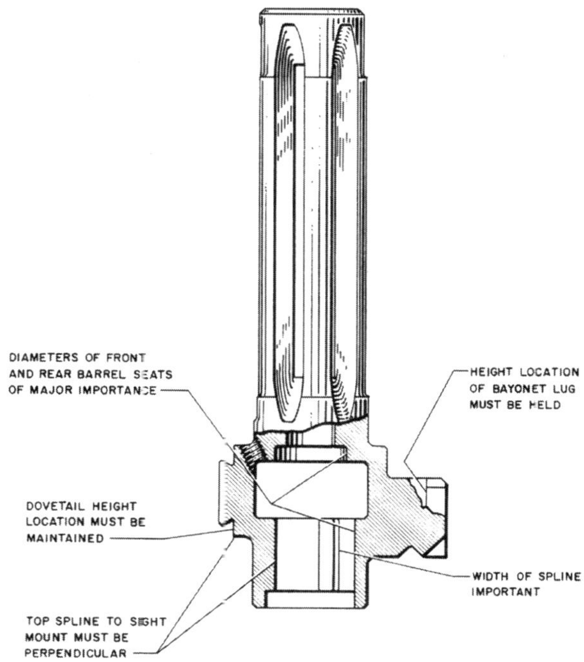

The M14 flash suppressor was a difficult part to manufacture. Early variations were cast, but were prone to failure. Later the part was redesigned as a forging. (John M. Miller)

lems with 8620H steel, the use of other steel alloys was investigated. 3130 steel was found to have good, consistent results as far as physical properties, but was found to be difficult to machine.

The Springfield Armory made arrangements with the Republic Steel Corporation to make up a steel designated as 47B10H, having the same alloying elements, but in slightly different proportions, with a narrow harden-ability band. Springfield Armory manufactured twenty-four M14 receivers and forty-eight bolts from the 47B10H steel. Heat-treating was accomplished by two methods. Half of the components were salt carburized, the other half by gas carburizing. The core hardness was well within specification and free ferrite was well below the 10-percent maximum allowed. Case hardness and depth were also within the allowable specifications. One problem encountered was that linear shrinkage was 60-percent greater than with 8620H steel. In general the new steel performed very well and was found to offer distinct advantages over the specified 8620H steel.

Further testing of the experimental M14 receivers and bolts produced from 47B10H steel was delayed until more results of the original testing were received and studied. By this time period the last order for the M14 rifle production at the Springfield Armory was to be completed in a few months. No further action was planned by the SA Engineering Division regarding the use of 47B10H steel.

During September 1961, Springfield Armory engineers introduced a design change in the M14 trigger guard. The earlier design consisted of several parts and a brazed subassembly. The part was redesigned to be stamped from a single-piece of steel stock. The new design would eliminate the pin, the stop and the assembly and brazing of the pin to the trigger guard. There was also a broaching operation eliminated as well as the drilling of the holes for the pin into the wings. The new trigger guards were implemented on the production line in May of 1962. There were problems with the alignment of the holes in the wings and the heat-treatment process. Tooling changes were made to solve the alignment problem.

Another part that was simplified during the period was the recoil spring guide. The guide previously had long milled slots in the side. The milling was replaced by a procedure that simply punched a series of short holes into the shaft of the guide. The procedure was implemented to reduce the cost of manufacturing the part.

A problem with the forming of the point of the firing pin was addressed by the procurement of a new Cincinnati Chucking Grinder. The use of the new equipment also eliminated the problem of compensating for heat-treating shrinkage of the firing pin. After the new procedure for producing the firing pins was being used there were reports of punctured cartridge primers being received. It was determined that the problem was being caused by excessive deformation of the pin's tip.

Engineering changes were made on both the configuration of the striker point and the hardness of the firing pin. The change in the hardness was to prevent deformation of the tip of

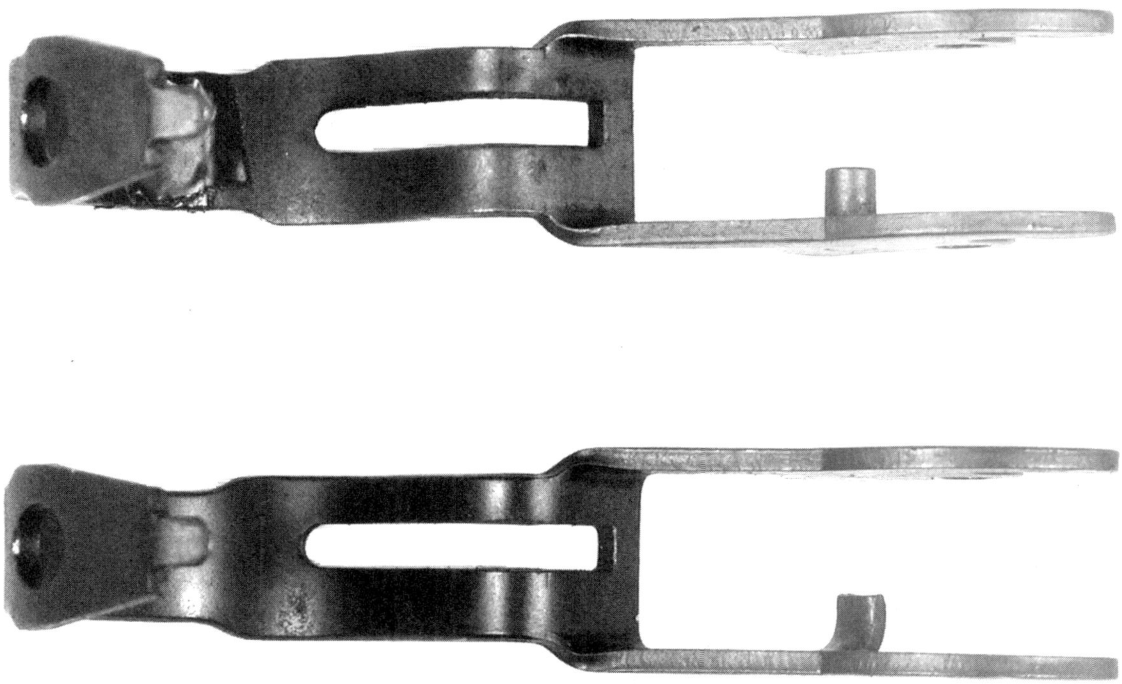

During September 1961, Springfield Armory engineers introduced a design change in the M14 trigger guard. The earlier design (top), consisted of several parts and a brazed subassembly. The part was redesigned to be stamped from a single-piece of steel stock. The new design (bottom) would eliminate the pin, the stop and the assembly and brazing of the pin to the trigger guard. (John M. Miller)

the pin. The contour of the pin tip was redesigned with an increase in the radii that joins the spherical contour to the cylindrical body of the tip. On 15 November 1962 a recommendation was made to change the blending radius from .010R + .015 to .025R + .010. Testing indicated that this change would prevent the firing pin from puncturing the cartridge primers.

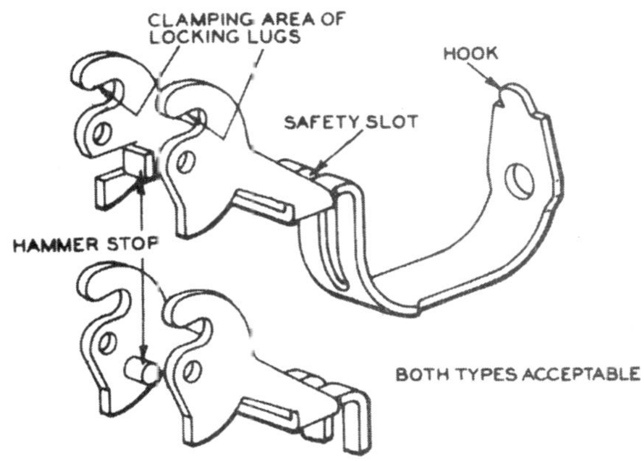

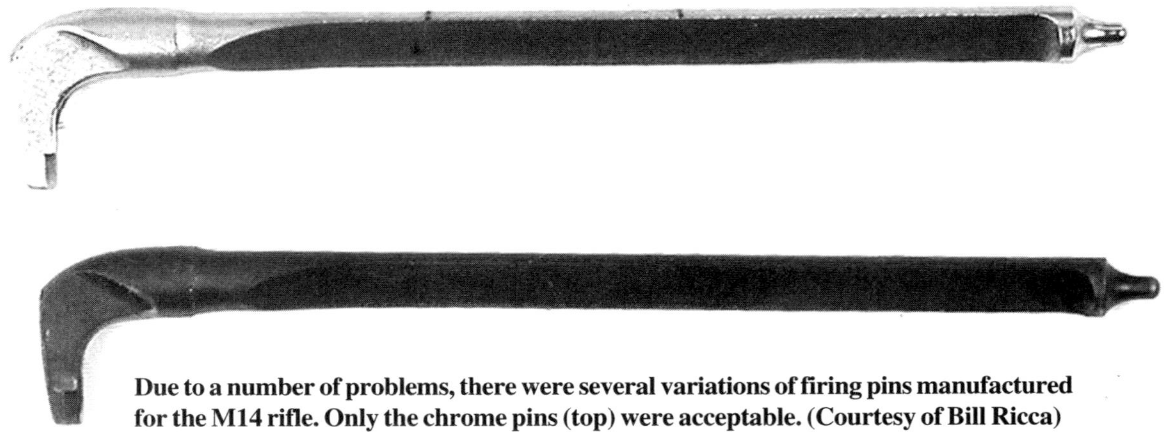

Due to a number of problems, there were several variations of firing pins manufactured for the M14 rifle. Only the chrome pins (top) were acceptable. (Courtesy of Bill Ricca)

Reports from the field had indicated that during certain bayonet maneuvers, the action of the rifle would be pulled away from the stock assembly. After a study of the defect was completed, an engineering change was made to the gas cylinder, which incorporated a supporting boss on the underside of the unit to retain the front band in position. Since the gas cylinder was a forged component a change had to be made to the forging die. This change was successful in securing the barrel to the stock.

While performing routine training exercises with the M14 it was discovered that the bolt would occasionally come apart when firing the rifle with M82 blank ammunition. The problem was quickly traced to the extractor. The case extractor is the primary part that holds all of the bolt components in place; once the extractor is removed the entire bolt will come apart.

The bolt problem usually occurred during a burst of fire in which an object would be observed flying upward out of the receiver, after which the weapon would cease to operate and jam. Although it was known that the cause of the problem was the extractor coming out of bolt, it was unknown just why it was happening. Further investigation using a high-speed camera revealed that the extractor was being cammed up and out of the bolt by blank rounds being fed from the right side of the magazine. The M82 blank round has a silhouette similar to that of a 7.62 mm standard ball cartridge; however it lacks the ogive of the 7.62 mm projectile.

The M14 magazine is a double-stack, double-feed design. When a cartridge is stripped from the magazine it is shifted and centered in order to feed into the chamber of the barrel.

When training blanks were used, the blunt end of the blank round would contact the chamber wall and instead of sliding into the chamber, would momentarily drag causing the base of the round to abruptly move left. This action would pry the extractor right out of the bolt and along with it the firing pin, ejector and spring. To solve the problem the extractor was redesigned with a small 30-degree flat machined at the lower front right-hand edge of the lip. Since the problem only affected the rifle's operation during training exercises the weapons were not retrofitted with the new extractors. The Ordnance drawing for the part was revised so that newly manufactured rifles and future replacement extractors would be of the new design.

Manufacturing problems with the operating rod were discovered to be caused by a combination of insufficient product design and discrepancies in final and in-process inspection tools. The Industrial Engineering Department at the Armory also designed a simplified and less costly automatic-fire connector assembly. The new design was incorporated into production rifles.

As previously stated, one of the Armory's missions was to assist and support the commercial manufacturers of the M14 rifle with production or technical problems. One such occurrence was a problem in machining barrels. The engineers at the Armory performed an exhaustive comparison test between chrome-molybdenum-vanadium barrels and barrels made from 4150 resulfurized alloy steel, which was much easier to machine. As a result of the testing the use of 4150 steel for M14 barrels was authorized by the Ordnance Weapons Command. This resulted in a substantial cost savings.

Through certain periods parts manufactured at the Springfield Armory were supplied to

An early M14 gas cylinder (top) is shown here with a later production part. Reports from the field had indicated that during certain bayonet maneuvers, the action of the rifle would be pulled away from the stock assembly. After a study of the defect was completed, an engineering change was made to the gas cylinder (bottom), which incorporated a supporting boss on the underside of the unit to retain the front band in position. (Courtesy of Bill Ricca)

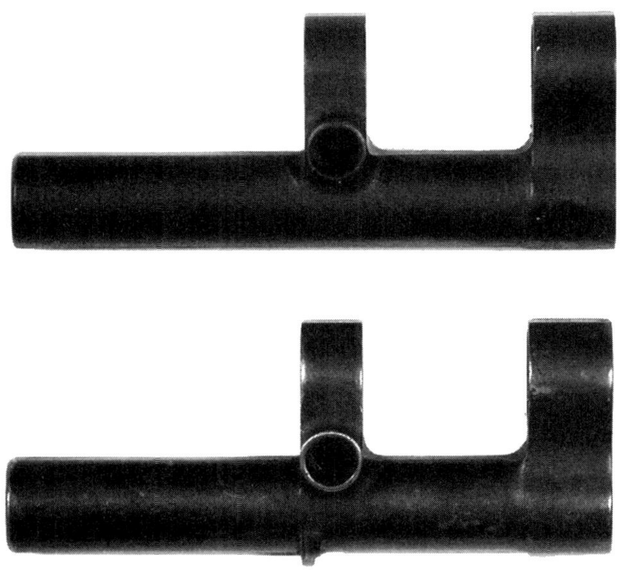

Test firing the M14 rifles at the factory range.(TRW)

both commercial contractors Harrington & Richardson and Winchester. Parts shipped included stocks, bolts and bolt rollers, gas cylinders, operating rods, trigger housings, firing pins, sear releases and flash suppressors. Although the parts opened up some bottlenecks at the contractor's plants, it began to hinder rifle production at the Armory.

During 1962 problems were encountered during the electropolishing cycle of the barrel process, wherein pockets were being produced on the inner surface of the barrel. This condition was at least partially solved after the chromium plating was applied to the barrels. Further investigation of the problem continued. Other barrel problems were addressed after the Marine Corps complained that accuracy of their new M14 rifles was not acceptable. During the subsequent study of the accuracy problem a few experimental changes were implemented in the stock, gas cylinder and front band, but only a slight increase in accuracy was achieved. During the study it was revealed that there was a significant rifle-ammunition relationship in regards to affecting accuracy.

Another less serious problem that was brought to the attention of Armory officials was the misalignment of the holes in the wooden buttstocks and the buttplates. This prevented the insertion or removal of the cleaning kits and tools. The problem was resolved by supplying reamers to armorers in the field. The Ordnance drawings were revised so that the problem would not reoccur in the future.

The Rock Island Arsenal engaged the Electronics Group of General Mills Inc. for an independent engineering study on the M14 receiver. Specifically in the areas of simplification, economy and function of the receiver. Representatives from the company visited the Armory on 18 August 1961 to be briefed on the functional areas of the receiver and the reason for the close tolerances required in some areas. After the study a report dated 22 June 1962 was received that basically revealed that no major changes in the basic design or dimensions would provide any significant improvement over the current design.

The M14 program had progressed to the point that the mission of the Armory's pilot production line had been accomplished. Through the coordinated efforts in all areas of the Springfield Armory, the numerous engineering and production difficulties that had been encountered were resolved. The Armory had proven that the M14 rifle could be economically manufactured. The current orders for M14 rifle production at the Armory would last until September 1963, during this period process improvements for manufacturing the M14 continued. Among the improvements was the tooling for the splining operation on the flash suppressor. Other methods of faster production and methods of extending the tool life were implemented.

There was an additional order received for 1,000 Special Match "M" M14 rifles, which were regular production rifles with standard production barrels, no glass bedding, the selector shaft lock welded in place and a letter M engraved on the receiver. The special "M" M14 rifles were later returned to the Armory for additional welding on the full-automatic selector shaft assembly, to meet the requirements established by the Tax and Alcohol Unit of the Internal Revenue Service. No major problems were encountered, allowing 1,009 rifles to be completed during June 1963.

There was a concern by Armory personnel that no further orders for M14 would be forthcoming, and their concerns were well founded. The last production M14 was made at the Springfield Armory during October 1963, the last 10,000 rifles were fitted with the new synthetic stocks. From July 1959 to October 1963 the Armory delivered 167,173 M14 rifles. After the last Springfield Armory M14 was shipped, the process of preparing the M14 tooling for long-term storage began.

Although M14 production at the Springfield Armory was terminated, the facility did implement a program for the rebuilding of unserviceable M14 rifles. Because of the unex-

An important reminder to H&R employees is written on the wall. (SANHS)

pected low mortality rate of the M14's receiver the Armory was able to deliver 1400 more rifles than the original order specified. The manufacture of major component parts for the M14 also continued at the Armory. During the Fiscal Year 1967 a substantial order for operating rods, bolts and bolt rollers was completed. An order to produce 52,700 standard M14 barrels was also received during the fiscal year 1967.

On 19 November 1964 Secretary of Defense Robert S. McNamara announced that the Springfield Armory was one of ninety-five installations located throughout the United States that would be phased out. The tentative date scheduled for the closing of the Armory was April 1968.

There was a strong reaction on a local level against the closing of the Armory, it culminated in a visit to the Springfield Armory by Mr. McNamara together with the governor of Massachusetts John A. Volpe. Also present during McNamara's visit were congressional leaders from the state, including Senators Leverett Saltonstall, Edward M. Kennedy and U.S. Congressmen Edward P. Bolen and Silvo O. Conte. Citizens turning out in large numbers included the city fathers, business, civic and social leaders. After the visit McNamara announced that the management consultant firm of Booz, Allen and Hamilton had been selected to review the facts presented to him by those opposing the closing of the Springfield Armory. Mr. McNamara announced the he would review the consultant's report and together with his own findings review the phase-out plan of the Armory.

The closing of the Springfield Armory was planned as a gradual phase-down, which consolidated areas as operations decreased. The general manufacturing area was the last be closed in October of 1967. The normal Springfield Armory mission functions of research, engineering and development were transferred to the Rock Island Arsenal. Many employees were afforded the opportunity to transfer to the Rock Island facility, although few of the eligible personnel accepted the offer. The transfer of responsibilities and equipment was completed during April of 1968. The Springfield Armory museum was retained and was operated by the city of Springfield until the facility was designated as a National Historical Site in 1978, at which time it was taken over by the U.S. Park Service.

Initial M14 Testing Fort Benning, Georgia

After the first fifty M14 rifles were completed at the Springfield Armory in August of 1959, ten rifles were randomly chosen and shipped to Fort Benning, Georgia for a thorough evaluation and testing by the U.S. Army Infantry Board. M14 rifle production was temporarily

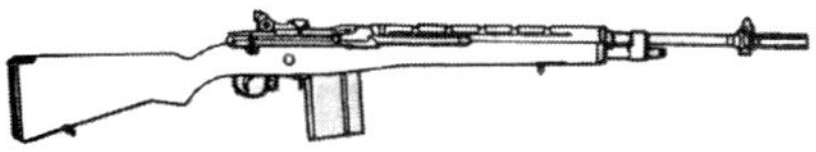

After the first fifty M14 rifles were completed at the Springfield Armory in August of 1959, ten rifles were randomly chosen and shipped to Fort Benning, Georgia for a thorough evaluation and testing by the U.S. Army Infantry Board. M14 rifle production was temporarily suspended until the results of the Fort Benning trials were documented. The initial trials proved successful and production resumed. (U.S. Army)

suspended until the results of the Fort Benning trials were documented.

Present to observe and provide technical assistance were; Mr. R.S. Henry Mechanical Engineer, Ordnance, Mr. A.A. Cole Design Engineer and Mr. G.V. Savioli, Mechanical Engineer Technician. A few problems were expected and subsequently encountered during the test phase. During an endurance firing test of the M14 a test weapon fired 3,080 rounds at a rate of one shot per second. The test was terminated after the chambered rounds began to cook-off. At no time during the test did the barrel fail to stabilize the projectile's flight. During the test the front end of the stock and the wooden handguard burst into flames. After the Infantry Board testing was completed in September 1959 production at the Springfield Armory resumed.

The very first M14 rifles manufactured were

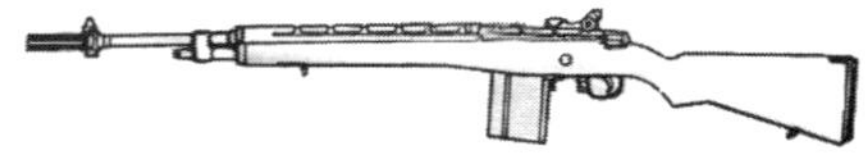

Stoic looking U.S. Army recruits march with their M14 rifles. (U.S. Army)

equipped with M1 rifle-style flat buttplates and an upper hand guard made of walnut. After the heavy-barrel M15 variation was cancelled a few features from the M15 were to be incorporated into the standard M14 rifle. One feature was the hinged flip-up buttplate. The incorporation of the new style buttplate required a revision in the design of the stock. Special tools and gages were designed to make the cut in the rear of the stock to accommodate the hinged buttplate. The earlier walnut handguard proved unable to withstand the heat generated by the barrel during sustained automatic fire. To address the problem a slotted fiberglass upper handguard was fitted. One last feature was the addition of the folding bipod that attached to the gas cylinder located under the barrel. This new variation was designated as the M14 (M) for "modified." The Springfield Armory did not have the capacity for the manufacture of the bipod, so the Ordnance Weapons Command had them manufactured commercially for assembly at the depot level.

Modifications were made to the M2 Bipod to provide positive locking of the legs in the open position by the incorporation of cylindrical plungers in place of the tapered plungers originally used. Problems were encountered at the contractor's facility with applying the finish on the bipod parts. Initial sample bipods were received from the Farmer Tool and Supply Company of Denver, Colorado on 7 March 1961. Testing of the bipod was conducted immediately after the bipods were received, and they were found to be acceptable. Confirmatory testing of the sample M2 bipods were conducted from 14 March 1961 through 17 March 1961 by the U.S. Army Infantry Board at Fort Benning. The Infantry Board found the bipods to be satisfactory. A Springfield Armory representative who observed the testing and was acting in behalf of the Ordnance Weapons Command recommended to the Chief of Research and Development in Washington, D.C. on 2 May 1961, that the Bipod, Rifle: M2 be considered suitable for Army use in temperate climates. Two experimental adapters were fabricated at the Armory to facilitate the use of the M2 Bipod on the M60 machine gun.

During May of 1960 the Ordnance Weapons Command sent a memo to the Springfield Armory requesting that they begin selecting two-hundred M14 rifles for a special program to develop a Match Grade version of the M14 rifle for the Army.

Also during the 1960 period crash procurement and production programs were established by the Armory's Industrial Engineering Department to obtain plastic M14 handguards, 20-round box magazines made from aluminum and a newly designed buttstock lubricant case. Other contracts were established for the commercial procurement of additional bipods, hinged buttplates and combination tools to keep up with increasing production.

The Commercial Contractors

During the fall of 1958 the Department of the Army began to cast about to locate a commercial manufacturers for the new M14 rifle. Bids were to be solicited from a number of firms. The successful bidder would be awarded an initial contract to manufacture 70,000 M14 rifles. The Army was moving cautiously on initial production as there were plenty of capable M1 Garand rifles in service, and the country was at peace. Original estimates projected that five-million M14 weapons would be needed to fulfill the needs of the military.

Although the government set the specifications and materials to be used in the manufacture of the M14 rifle, the actual manufacturing process and procedures themselves were left up to the successful bidder. Each of the commercial manufacturers that were engaged to produce the M14 used different manufacturing methods.

While the Army was seeking a single commercial supplier for the M14, the Small Business Administration with White House influence, concluded that there should be two commercial contractors placed on line. One contract was to be awarded to the low bidder, while a second contract given to a small business firm in a surplus labor area. Instead of one contract for 70,000 rifles there would be two contracts for 35,000 rifles each. In reality the contracts were small compared to the initial capital investment required to set up production lines. The only incentive for the successful bidder was the anticipation of larger contracts being awarded in the future. Some of the companies that were invited to submit bids were former contractors and subcontractors of the M1 rifle production during World War II and Korea.

After reviewing all of the bids and capabilities of the firms bidding, contracts were awarded. Winchester was the low bidder and awarded the first contract on 17 February 1959. The bid submitted by Winchester was for $68.75 per rifle. Winchester had produced the M1 rifle during the Second World War. The second commercial contract was awarded to the Harrington and Richardson Inc. H&R was given the contract because it was a small company in a surplus labor area, Worcester, Massachusetts. Harrington and Richardson Arms

Receiver markings of a Winchester manufactured M14 rifle. Winchester Arms was a division of the Olin Matheison Corporation. (Courtesy of the U.S. Marine Corps Museum Quantico, Virginia)

had successfully manufactured approximately 500,000 of the M1 rifles during the Korean War. They had also produced the Reising submachine gun for the Marine Corps during World War II, and the Belgian designed FN T48 rifles that had competed with the M14 prototype the T44E4 rifle in 1955.

Upon acceptance of their bid H&R had to agree to produce the M14 for Winchester's bid of $68.75. In reality the bid was far too low, but both companies were eager for work and the possibility of future contracts, both also had substantial experience in the manufacture of firearms. Each firm had its own ideas on how to reduce labor and overhead costs, but also counted on a revisal of the initial bid price.

After the commercial contractors of the M14 rifle had established their production lines the Springfield Armory inspectors would randomly choose six rifles from each plant and six rifles produced at the Armory. They would then conduct interchangeability tests of each of the rifle's parts to insure 100% compatibility between all concerns. At the same time a quality inspection of the weapons was conducted.

The Springfield Armory Training Section prepared and implemented an 80-hour training acceptance inspection-training course for government and contractor personnel. Other specific courses were implemented on the stock and the chrome plating procedures for the M14 barrels.

After the rifles were assembled they were required to be proof fired using a special high-pressure cartridge. A letter P was stamped on the stock to indicate that the test had been administered and passed successfully. The rifle would then be subjected to a functioning test

Receiver of an H&R M14 rifle.
(U.S. Marine Corps Museum)

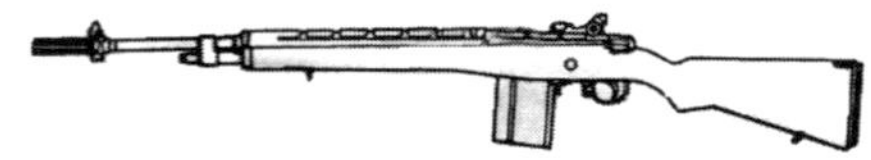

of semiautomatic, burst automatic and sustained automatic fire. The automatic cyclic rate was measured to be certain that it fell within the narrow limits specified by the Ordnance Corps. An accuracy test required that center of impact of each rifle must be within certain limitations at a range of 100 meters. A 5-shot group at 100 meters must fall within 5.6 inches using standard service ammunition. Any rifle that failed to meet the requirements was tagged and returned to the shop for corrective action. After the necessary repairs were made the rifle would be re-inspected and tested.

Winchester M14 receiver markings on an early production rifle. (U.S. Marine Corps Museum)

Winchester Division of the Olin-Matheison Corporation

The first commercial entity chosen to manufacture the new M14 rifle was Winchester, a well-known, and long time manufacturer of high-quality firearms. The company had previous experience manufacturing the similar M1 rifle during World War II. Winchester had placed a bid to produce the M14 for $68.75 per rifle. Winchester planned to automate the manufacturing process for the more complicated parts like the receiver. After an initial delay, the company was awarded its first contract on 17 February 1959. The first delivery of completed rifles was expected by April of 1960. The first contract was to be completed by March of 1961.

Winchester encountered a number of problems during their initial production of the M14.

The company had originally planned to produce M14 rifles during one shift and sporting arms on a second shift. During April of 1960 Winchester was given a second M14 contract for 81,500 rifles. Winchester agreed to a December 1961 delivery date.

Although Winchester had hoped to expedite production by the implementation of automated machines, manufacture of the complex receiver had proven more difficult then originally planned. Winchester had ordered two

Pallets of M14 receivers at the Winchester plant await machining. Winchester had hoped to expedite production by using machines that were especially designed to manufacture the M14 receiver. The machines automatically performed thirty-two critical operations. Unexpected delays were encountered in the delivery of the new machines and getting them set up for production. (SANHS)

expensive machines that were especially designed to manufacture the M14 receiver, by performing thirty-two critical machining operations. Unexpected delays were encountered in the delivery of the new machines and getting them set up for production. Winchester also encountered problems in barrel manufacture. Due to diligence on the part of the Winchester engineers and personnel the problems were eventually overcome and production was back on schedule. One of the most efficient new machines procured by Winchester was the sixteen-station stock inletting machine. The one piece of equipment was able to perform all the operations normally done by sixteen separate machines. The machine played a critical role in placing the company back on schedule.

The Winchester Division of the Olin-Matheison Corporation manufactured all of the major M14 components and many of the less-complicated smaller pieces in-house. Only the easily procurable parts were subcontracted out. Parts produced at the Winchester plant were marked with a numerical code 66118 while some parts were marked O-M indicating Olin-Matheison manufacture, Winchester's parent company. Barrels manufactured by Winchester were marked with the stylized Winchester brand name, the date of manufacture and the steel heat lot. The company was awarded a third contract on 28 September 1962 for 150,001 additional M14 rifles at a revised cost of approximately $104.50 per piece.

By 1961 most of the production difficulties had been resolved and the Winchester Company began meeting its monthly production goals on schedule. Full-time technical assistance from the Springfield Armory Team was phased out during March of 1962.

Harrington and Richardson Inc.

Harrington & Richardson Inc. was a well established firearms manufacturer and had previously manufactured weapons under contract for the U.S. Government. The company had planned to manufacture the rifle at low cost by utilizing much of the machinery and equipment they had remaining from their M1 rifle contract. The company also planned to use a large number of subcontractors to manufacture and supply parts. Parts produced at Harrington and Richardson Arms Company were generally marked HRA for manufacturer identification.

The first government contract was awarded to the Harrington and Richardson Arms Company during April of 1959. The first deliveries of completed M14 rifles from H&R were expected to begin during June of 1960. Harrington and Richardson Inc. had just started delivering their first completed rifles when they were awarded a second contract for an additional 75,000 M14 rifles at a cost of $97.70 each. The additional contract raised their production requirement to 10,000 rifles per month. The company then installed a second shift to increase production. While H&R did manage to substantially increase production, the rejection rate by Ordnance inspection personnel increased disproportionately. A dispute ensued with the government blaming inadequate quality control at the plant.

During extended range firing of several new M14 rifles at Fort Benning in December of 1960, three of the weapons experienced catastrophic failures that included one H&R receiver, serial number 73293, and two bolt failures. One of the bolts had a right lug shear off, the other bolt had a cracked lug. Upon initial inspection and investigation of the failed weapons it was quickly discovered that all of them were of H&R manufacture. The services of several government ordnance facilities were engaged to study and determine the cause of one receiver and two bolt failures. Personnel and engineers were immediately dispatched to the manufacturing facility in Massachusetts.

After looking into the company's procedures and manufacturing methods it was determined that a low grade steel had somehow been used to manufacture an unknown number of M14 receivers. The suspect receiver was made from AISI 1330 steel instead of 8620H steel. The bolts were discovered to have been improperly heat-treated, resulting in undesirable core hardness and microstructure. A process to establish nondestructive testing of the bolts was established. In February a revised Supplementary Quality Provisions (SQAP's) were introduced for the M14 bolt and receiver. The revised SQAP required an examination of the components for core defects, hardness and percentage of free ferrite. Also included were calibration standards for the magnetic comparator inspection.

A confidential summary report was written regarding the H&R receiver and bolt failures. There were also two other reports written on the methods used for testing receivers and bolts. The two reports were generated by the Springfield Armory. One was Report SA TR 19-1502, entitled; Properties and Methods of Non-Destructive Testing of Receivers for the 7.62mm, M14 Rifle. The second report compiled by the Armory was Report SA TR 19-1507, Properties and Methods of Non-Destructive Testing of Bolts for the 7.62mm, M14 Rifle.

The bolt heat-treating problem proved to be

U.S. Rifle 7.62 mm M14

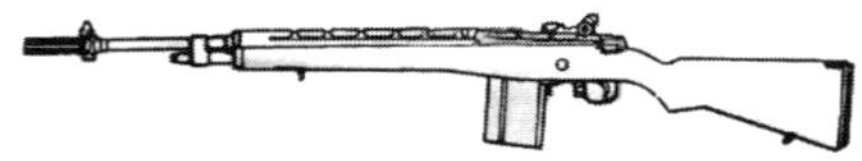

HARRINGTON & RICHARDSON, INC. has been a manufacturer of quality firearms since its founding in 1871. During most of these years its efforts have been concentrated on the design and production of sporting arms but more recent years have seen a major shift to the manufacture of military weapons.

Today, this Worcester, Mass., firm is known for its record production of the rifle, 7.62mm, M-14 for the Armed Forces of the United States. In three years since it started tooling in 1959 to produce this precision weapon, H&R has delivered over 350,000 of these rifles. This number is far in excess of the total delivered from all other sources.

The company presently has over 200,000 sq. ft. of floor space in its three plants devoted to the manufacture of firearms.

A newly constructed Wheeling, West Virginia facility will contain 60,000 sq. ft. of floor space at the start but has been designed so that it can be expanded to 120,000 sq. ft. or more, if needed.

In addition to its headquarters in Worcester, H&R's other plants are in nearby Rochdale and in Gardner, Mass. It has in operation another facility with 28,000 sq. ft. of floor space in Marlboro, New Hampshire.

More than 1,500 skilled workers are employed in the manufacture of military arms at H&R's present plants. In addition, the company has 140 suppliers of parts in 20 states whose total employment is in excess of 4,000 persons. Thus over 5,500 people are utilized by the company in the manufacture of firearms.

Harrington & Richardson, Inc., has the knowledge, the skills, the equipment and the facilities for large scale weapons production. It has recently produced as many as 30,000 M-14 rifles per month and is currently delivering rifles to the military at a rate of over 25,000 per month. Parts are produced to extremely close tolerances and undergo exhaustive tests before approval by the military authorities.

An excerpt about the M14 rifle from the 1963 Harrington & Richardson Research and Development manual.

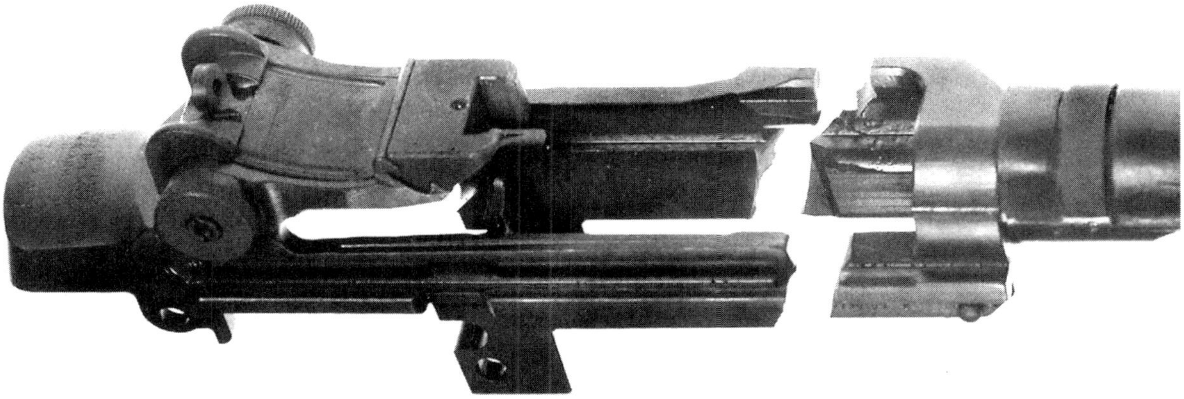

During extended range firing of several new M14 rifles at Fort Benning in December of 1960, three of the weapons experienced catastrophic failures that included one H&R receiver, serial number 73293. The receiver failure was traced to the use of the wrong steel by the factory. (Springfield Armory)

especially troublesome. Even after the process was revised, random testing was required by all three concerns currently manufacturing the bolts to insure that adequate controls were in place.

The receiver material mistake was blamed on poor quality control at the H&R factory. The problem was eventually traced to a mix up in their steel supplies, and the wrong steel was used for the receivers. The bolts had been improperly heat-treated. Technical teams were organized at the Springfield Armory and dispatched to the all of the contractor's plants to provide full time in-house assistance. The presence of the teams would remain until the contracts ended or the Ordnance Weapons Command decided to recall them back to the Armory. During June of 1961 more than 1,200 man-hours of inspection time was spent assisting the commercial contractors in resolving production problems with the M14 rifle.

Following the Ordnance Corps investigation new stringent polices were put in place regarding quality control. The Ordnance Weapons Command put out an order in January 1961 providing new inspection procedures. Commercial production by both H&R and Winchester were temporally halted until the new guidelines were in place. The entire lot of 33,808 suspect M14 rifles that had been manufactured by Harrington and Richardson Inc.were shipped to Raritan Arsenal to undergo an inspection for improper steel. During the process 1,784 M14 rifles were found to have defective receivers and other critical components. Other problems were discovered with inadequate heat-treating of bolts. Former

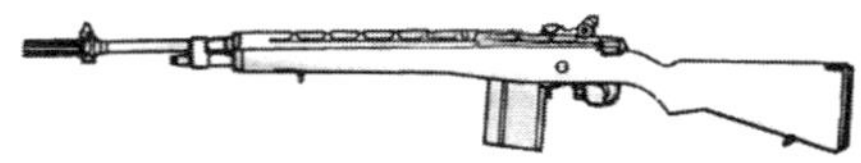

The 7.62 mm M14 rifle saw extensive combat use during the early years of the Vietnam War. The M14 was eventually phased out and replaced by the 5.56 mm M16A1 rifle. (NARA)

specifications of many parts were revised and new controls were implemented into the manufacturing procedure to circumvent any future occurrences of such serious problems. Although this proved to be a substantial setback for H&R they rebounded quickly and were able to complete their second contract on schedule.

There were a total of 33,808 questionable M14 bolts still impounded at the Raritan Arsenal waiting to undergo testing to determine if proper heat-treating procedures had be applied to them. In the interim, the Watervliet Arsenal in New York was given a temporary order to manufacture M14 replacement bolts while the investigation of the questionable H&R bolts continued.

A test method combining Rockwell C readings, magnetic permeability readings and oscilloscope waveforms were developed to test the serviceability of the bolts. Of the total 26,848 were found acceptable and reassembled to weapons. The 6,960 remaining bolts were scrapped.

By August 1961 Harrington and Richardson was producing a quality weapon that easily passed all of the inspections implemented by the Springfield Armory team assigned to the plant. The decision was made to terminate the Springfield Armory contractor assistance at the H&R facility. Only periodic inspections were to be scheduled in the future. The company began their third contract in September 1961 at an increased production rate. At the time of the third contract H&R had approximately 1,000 employees devoted to M14 production.

Because of the receiver and bolt failure problems experienced with Harrington and Richardson M14 rifles during 1960, the U.S. Army Audit Agency conducted a review from 23 May 1966 until 9 June 1966 entitled "Audit of Costs Incurred-Receiver, M14 Rifle-Inspection and Segregation Program". The report was to develop cost data to be used in the overall claim against Harrington and Richardson Inc. under contract DA-19-020-ORD-4921 ASBCA 9839.

Despite early problems the Harrington and Richardson Arms Company would become the largest producer of U.S. M14 rifles with a final total of 537,512 manufactured between 1959 and 1963.

In an effort to improve and expedite M14 rifle production the U.S. Secretary of Defense Robert S. McNamara dispatched a team of Army personnel to investigate the facilities manufacturing M14s. Findings of the team led to the selection of a project manager.

Brigadier General Elmer J. Gibson of the Ordnance Weapons Command was assigned as project manager in charge of the M14 rifle program. He also served as Chairman of the M14 Rifle Integration Committee. General Gibson had the responsibility of assisting with any problems that may be encountered, and he had the authority to cut through the red tape when necessary.

Packaging and Shipping of the M14 Rifle

An intensive study was undertaken at the Springfield Armory in order to establish packaging and bulk shipping procedures of the M14 rifles to prevent damage to the weapons during storage and delivery. Instructions for packing and controlling humidity were developed and distributed. The study found that a savings of approximately $1.36 cents per rifle could be realized if the packaging was changed from A to B level and using C/H Conex containers for shipping.

The shipping procedure established at the Armory was to preserve two M14 rifles in a single sheet of Volatile Corrosion Inhibitor (VCI) paper and packaged in a standard carton containing cardboard end blocks, and a separator to secure the weapons. The ends of the cartons were then sealed with three pieces of nylon tape. A single Conex container could accommodate 225 cartons (450 rifles). After the cartons were placed into the container it was brought up to the proper humidity level and supplied with sufficient dunnage and desiccant, then closed and sealed.

Transit authority was established with the New

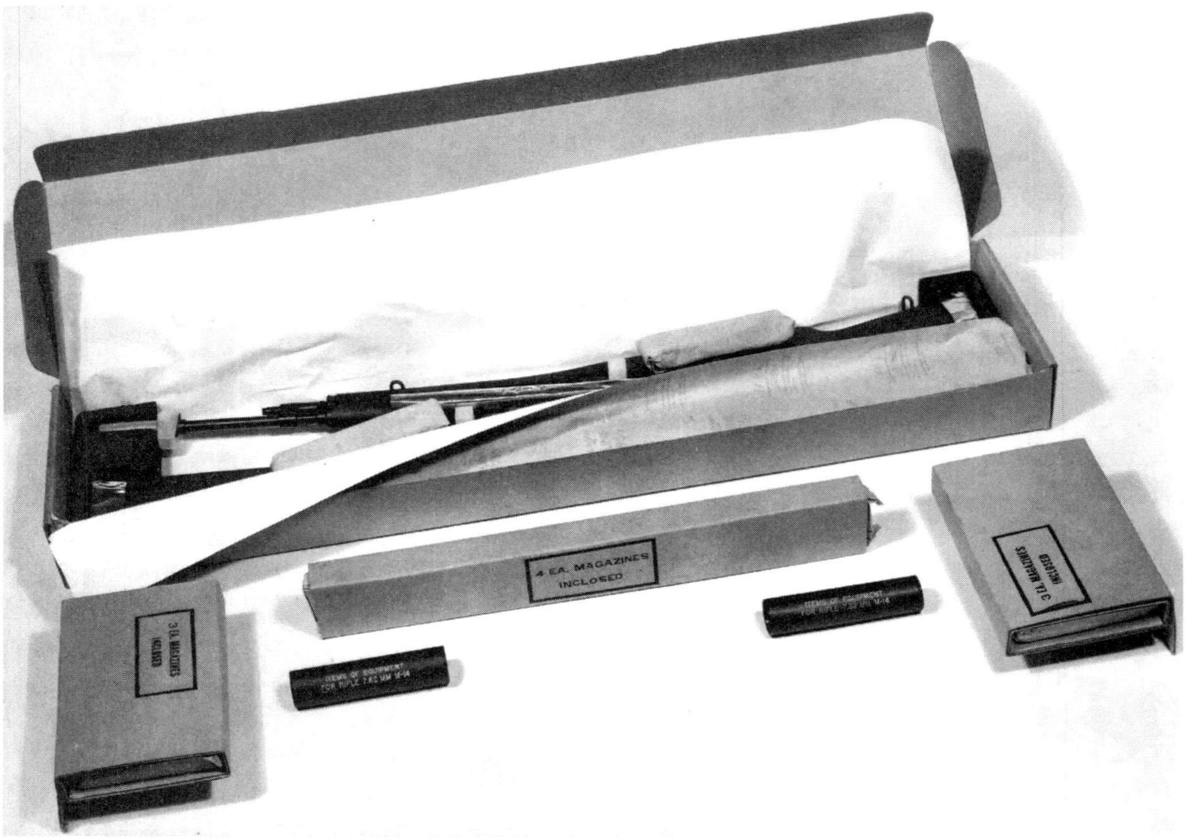

19-058-933/ORD-59 SPRINGFIELD ARMORY 4 Nov 1959
LEVEL B AND C PACKAGE FOR RIFLE, 7.62-MM, M14 WITH EQUIPMENT
A sheet of VCI treated paper replaces the VCI lined barrier bags for domestic shipment storage.

York Central Railroad for receipt and forwarding of the Conex containers of M14 rifles. The transit authority allowed a special rate privilege to the containers being shipped to the Armory and a re-forwarding of the containers to the H&R plant in Worcester and the Winchester factory in New Haven.

By 1961 a third contractor was being sought for producing M14 rifles. However the contract would be based upon competitive negotiations instead of the fixed-price bid procedure used earlier. Fixed-price bids led to low bids that could not be met by the contractor and usually resulted in a price adjustment. Under the newly implemented system a bonus of $2.50 per rifle was authorized for each weapon produced ahead of schedule, although there was a ceiling of $250,000.00 imposed. Another stipulation in the new contracts was that the contractor was required to reveal which components were made in-house and which were being subcontracted.

Thompson Ramo Wooldridge, Inc.

The search for a third commercial contractor to produce the M14 rifle by the Headquarters, Ordnance Weapons Command was supported by the Springfield Armory, who sponsored seminars along with tours of the M14 production line at the Armory. There were approximately fifty-five prospective bidders for the contract. The two-day seminars were designed to provide the bidders with comprehensive details of manufacturing, assembling, acceptance and proof firing as well as explanations of the drawings and specifications. A Technical Data Package was prepared by the Armory for a meeting of the third source bidders on 12 May 1961. A total of forty-two companies submitted bids for the contract.

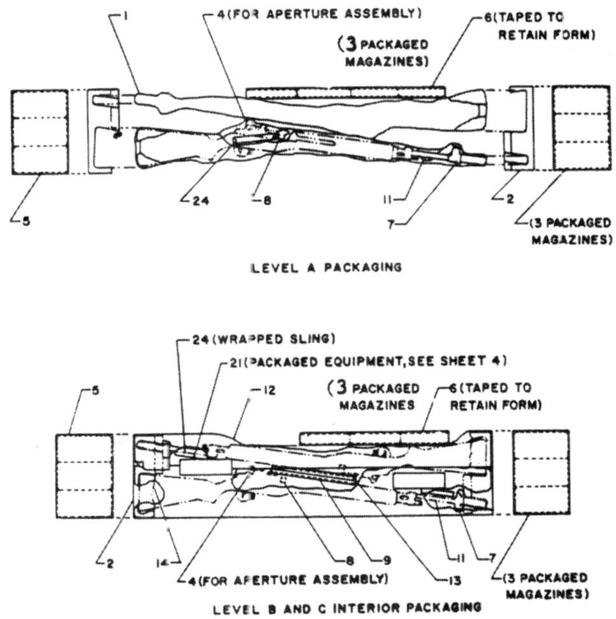

FIGURE 800-1. PACKAGING OF RIFLES

Thompson Ramo Wooldridge, Inc. was a very diversified major corporation. The Thompson Products company began producing parts for the automotive industry in 1928. The company grew by creating products and services providing maximum value and meeting the needs of their customers. Realizing the need to become more diversified, the Thompson Products merged with the Ramo-Wooldridge Corporation to create TRW Inc.The Electro-Mechanical Group of TRW was the industry's largest supplier of jet aircraft engine components. The manufacture of aircraft parts by nature requires the utmost attention to the

Marking on a TRW receiver. The TRW company officials felt that although they would lose money on their first M14 contract, it was simply the entry fee into the firearms manufacturing business. TRW expected more contracts to manufacture the rifles would be forthcoming, allowing them to recover their initial investment and begin making a fair profit.

smallest detail, close tolerances and impeccable quality control. The Division specialized in high precision while controlling production costs through innovative manufacturing methods. While highly regarded in the aircraft industry, the company realized the state of the declining aircraft market and the need to expand into another endeavor.

At first TRW was somewhat reluctant to bid on something as unfamiliar as a military rifle contract, particularly when the other firms that were currently producing the weapon both were small arms manufacturers with years of know-how and experience. However as the TRW engineers studied the requirements and details of the manufacturing process, they realized that their resources were well suited for such a venture.

Although the Thompson Ramo Wooldridge had little experience with small arms they were very well versed in manufacturing operations. The company's painstaking preparation and study of a manufacturing process for the M14 rifle would make them the most successful of the three commercial contractors involved. Their primary goal was a high-quality product at the lowest possible cost per piece. The company expended millions of their own dollars purchasing the latest state of the art machines and equipment for producing the rifles. The company officials felt that although they would lose money on their first M14 contract, it was simply the entry fee into the firearms manufacturing business. TRW expected more contracts to manufacture the rifles would be forthcoming, allowing them to recover their initial investment and begin making a fair profit.

After submitting their bid of $85.54 per rifle, TRW Inc. was awarded the contract on 2 October 1961. The company was given until November 1962 to begin delivering the weapon in quantity. The company planned to manufacture the rifle in Cleveland, Ohio at their newly organized Ordnance Works Division. Like all of the other companies involved in M14 manufacture, TRW used many subcontractors to produce parts. The company concentrated their efforts on the manufacture of the eleven most complex pieces, which included; the receiver, bolt, operating rod, barrel, gas cylinder and piston, flash suppressor, trigger housing, connector, hammer and rear sight base. Parts produced at the company's plant were marked TRW to identify the manufacturer. All parts that were manufactured by subcontractors were thoroughly inspected and gaged to insure at they met the company's high standards. Even though subcontractors were used for many component parts, TRW was responsible for the assembly, testing and delivery of the finished product. The first TRW manufactured M14 was assembled and tested in August of 1962. The first substantial M14 rifle delivery was made in October 1962, a month ahead of schedule.

On 9 October 1962 TRW Inc. was awarded a second contract for an additional 219,691 M14 rifles at an agreed upon price of $79.45 per piece. At that time, the new contract temporarily made Thompson Ramo Wooldridge Inc. the largest producer of the U.S. M14 rifle.

By the end of 1963 M14 production at the Springfield Armory had ended, and most of the tooling was being prepared for storage. At this time a commercial contractor was being considered to produce M14 match-grade rifles for the 1964 Matches at Camp Perry, Ohio. Initially the commercial contractors showed no interest in making close tolerance match-standard weapons. Such an endeavor would be considered too time consuming for a production-oriented firm. The Springfield Armory was no longer in a position to produce match-grade barrels as most of the tooling was disassembled. In January, 1964 the Headquarters, Army Weapons Command announced that Thompson Ramo Wooldridge would be producing National Match M14

M14 machine operation at the TRW plant. (TRW)

During October 1961 an Ordnance Integration Committee was formed for the M14 rifle. The first meeting was held at the Headquarters Ordnance Weapon Command on 25 October 1961. The chairman of the committee was U.S. Army General E. J. Gibson, seated at the rear of the table. (NARA)

rifles, with assistance from the Springfield Armory. This would mark the first time that match weapons were made by a commercial contractor.

When M14 production was abruptly canceled in 1963 the firm of Thompson Ramo Wooldridge, Inc. had produced a total of 319,163 M14 rifles from 1961 through 1963.

The M14 Rifle Integration Committee

During October 1961 an Ordnance Integration Committee was formed for the M14 rifle. The first meeting was held at the Headquarters Ordnance Weapon Command on 25 October 1961. The chairman of the committee was U.S. Army General E. J. Gibson. Attending the meeting were representatives from the

Office of the Secretary of Defense, Office of the Chief of Ordnance, Ordnance Weapons Command, Springfield Armory, The Boston and Cleveland Ordnance Districts, Harrington and Richardson, Winchester and Thompson Ramo Wooldridge Inc., as well as several subcontractors. The primary purpose of the committee was for collecting and analyzing information, and making recommendations in order to integrate all manufacturing activities to increase production and lower costs.

During January 1962 a Technical Subcommittee of the M14 Rifle Ordnance Integration Committee was formed. The purpose of the subcommittee was to address problems of a technical nature that could affect M14 production. Such problems as; inspection, testing, material as well as testing and inspection

During the post World War II era Berlin was considered the hot spot of the world. Secretary of Defense McNamara discovered that the troops there were still being issued the M1 rifle. He immediately ordered new M14 rifles be sent to Germany. (NARA)

Robert S. McNamara reviews the U.S. Army troops in Germany. (NARA)

of completed M14 rifles were addressed. The first meeting was held on 24 January 1962 and was attended by all of the current suppliers to the M14 production program. Subjects addressed at the meeting were the general operating plan, and legal aspects, which were defined and discussed. Other topics that were addressed were the handling of requests for technical assistance and the use of the new magnetic analyzer.

As early as 1961 problems, both real and imagined, began to be appear as articles in many large influential newspapers and magazines. The sources of the largely negative articles originated from the M14 contractors that had filed complaints with their district congressmen. The publicity invoked a Senate investigation of the entire M14 rifle development and production program. The Army didn't panic immediately. Such criticisms had been raised in the past with other weapon's development programs, like the M1 Garand / Johnson rifle controversy prior to the U.S. entry into World War II.

The press was quite ruthless in their criticism of the M14 rifle, calling it a "major blunder" on the part of the Army. Secretary of Defense Robert McNamara was an outspoken critic of the program. The U.S. Airforce was not interested in adopting the M14, instead they wanted to procure the new 5.56 mm AR-15 rifle. The Army protested the Air Force's choice arguing that the AR-15 ammunition was not NATO standard. The Airforce's desire to adopt a different rifle than the Army added to the growing doubts regarding the M14 rifle.

Rebuild Programs

As early as July 1963, 175,822 M14 rifles were deemed ready for an arsenal rebuild. The Springfield Armory accepted the order to undertake the task. Due to a lack of spare parts at the time the Armory was forced to cannibalize and repair components. This lowered the anticipated cost per rifle overhaul from $41.00 to $21.71. Many problems were encountered during the initial phases of the program due to lack of skilled personnel. As many as 30 percent of the rifles that were rebuilt at the Armory failed the targeting requirements. An investigation followed that traced the problems to bent barrels; oversize gas cylinders; undersized gas pistons and loose clamping of the stock. By the end of 1963 most of the problems had been corrected, and the rebuild procedures being used were closely scrutinized. The direct labor involved to overhaul an M14 rifle had been reduced to 1.5 hours per piece.

Secretary of Defense Robert McNamara was an outspoken critic of the M14 program. He is shown here looking over into Soviet controlled East Berlin, Germany. Ironically his body guard carries an M14 rifle. (NARA)

During 1965 a second order was accepted by the Springfield Armory for the rebuilding of 24,000 M14 rifles. A separate contract was given to Harrington and Richardson for the targeting of the completed rifles. The order was completed by December 1967. During the same time period the Springfield Armory was given an order to manufacture 64,000 spare bolts; 35,000 operating rods and 12,000 stocks.

On 21 January 1963 the Secretary of Defense Robert Strange McNamara announced that production of the U.S. Rifle M14 would be terminated after the fiscal year's contracts were completed. On June 30, 1964 M14 production was officially ended after Thompson Ramo Wooldridge delivered the last 200 rifles to the United States Army. The untimely end of M14 rifle production proved to be a financial disaster to all of the commercial companies involved, who had invested millions of dollars in anticipation of a long production run of the M14. Winchester suffered a great financial set back, TRW would never manufacture small arms again, and the government soon closed the Springfield Armory. Harrington and Richardson Arms Company received a contract to manufacture M16A1 rifles during the Vietnam War, and returned to the production of sporting arms. The company eventually closed their doors in 1986. There were a very small number of new and like-new M14 and M16A1 rifles sold at the company's bankruptcy auction in 1986. Fully transferable, the rifles quickly disappeared into private collections.

Presentation Grade M14 Rifles

The very first M14 to come off of the Springfield Armory production line was serial number 2000. This rifle is currently held at the Springfield Armory Museum. Other known presentation M14 rifles are listed here.

Springfield Armory M14 Serial Number One

The Secretary of the Army Wilber M. Brucker awarded a presentation-grade Springfield Armory M14 rifle marked with serial number 0001 to the Infantry Museum at Fort Benning, GA during 1959.

The Eisenhower Presentation M14

As was customary the U.S. Army presented the President of the United States a special rifle made at the Springfield Armory. An M14 rifle built by the Springfield Armory was cho-

The President Eisenhower Presentation M14. (Photograph courtesy of the Springfield Armory Museum NHS)

sen for presentation to President Dwight David Eisenhower. The presentation rifles were highly polished and blued, the wood was specially selected and hand finished.

Before presenting the specially prepared rifle to the president the piece underwent a detailed examination. To circumvent any unforeseen problems the Armory had built two rifles, one to serve as a backup. The first rifle was serial numbered D.D.E. 1 and the second rifle D.D.E. 2. After discovering flaws in the first rifle, the second rifle D.D.E. 2 was presented to President Eisenhower, the other rifle was retained at the Springfield Armory. Interestingly the rifle that was not presented to the President had been an endurance test rifle and had an unserviceable barrel with a rough bore.

Presentation grade M14 rifle serial number 0006 was presented to the National Rifle Association at their annual meeting in 1960. (Courtesy of the NRA Museum, Fairfax, Virginia)

Receiver markings of the NRA's M14 serial number 0006. (Courtesy of the NRA)

NRA Presentation M14 Rifle

The development and progress of the U.S. M14 rifle was closely followed by the National Rifle Association's magazine American Rifleman. During the production of M14 rifle there was a lot of controversy in the press regarding insurmountable problems being encountered with production of the weapon. During the 1960 era the NRA staff visited all of the factories engaged in M14 production and published unbiased reports in several issues of American Rifleman. As it turned out the problems were more with inadequate Con-

Side view of M14 serial number 0006. There were a number of presentation-grade M14 rifles made by the Springfield Armory. Their whereabouts for the most part are unknown. (Courtesy of the NRA museum)

gressional funding, not the design of the M14 or the plants engaged in its manufacture.

The guest speaker at the 89th annual NRA Banquet held on 23 March 1960 was the United States Army Chief of Staff General Lyman L. Lemnitzer. General Lemnitzer delivered a speech commending the National Rifle Association's contributions to marksmanship training and rifle competition. In honor of the contributions made by the NRA General Lemnitzer presented the NRA museum a presentation grade M14 rifle made by the Springfield Armory serial number 0006. The rifle featured a highly polished blue finish and a hand rubbed walnut buttstock. The rifle has early production M14 features such as the M1 buttplate, early gas cylinder and wooden hand guard.

There were several other presentation-grade M14 rifles manufactured by the Springfield Armory. Serial numbers 0013 and 0017 are at the Rock Island Arsenal Museum. The serial numbers and providence of other presentation-grade M14 rifles is largely unknown.

The M14 in Service

The replacement of the old M1 Garand rifle with the M14 was a slow process due in part to budget cuts, and the idea that the M1 was

A U.S. Marine readies for combat in Vietnam. His M14 rifle is fitted with a selector switch and a sling from an AK47 rifle. (NARA)

still a capable Infantry weapon. One of the first Army units to be issued the M14 rifle was the famous 101st Airborne Division. Initial dispersion of the new weapon to other units was not performed with any sense of urgency at least not until Secretary of Defense Robert McNamara learned the troops posted in West Germany were still armed with the M1 rifle. At the time U.S. check points dividing East and West Germany were the considered the hot spots of the world. The United States had expected the next war to begin with a Soviet invasion through Eastern Europe.

By the time of the Cuban Missile Crisis in 1962 the M14 was being issued in substantial numbers to both the Army and Marine Corps. The first Marine Corps units to land in-force on the shores of Vietnam on 8 March 1965 were armed with the 7.62 mm M14. Vietnam would be the first large-scale baptism of fire for the M14. The M14 weapon, like its ancestor the M1 rifle, proved to be hard hitting and reliable under the worst of conditions.

The U.S. M14 Rifle in the Middle East

U.S. Navy SEALs armed with M14 rifles. Note their use of duct tape to secure the handguards on their M14s. The have also used tape to secure two magazines together for extra firepower. (NARA)

With an M60 general purpose machine gun in front, U.S. Marines carrying M14 rifles stand ready. (Courtesy of Ronald Maxson, USMC retired)

A U.S. Marine with Viet Cong prisoners. (NARA)

Long after the last M14 was manufactured it has found a new lease on life in 2001. Unfortunately the majority of them have already been destroyed or given away as military aid. Of the M14 rifles that remain in government stores, they were rebuilt and upgraded by the military as well as under contract by several commercial entities.

Often hailed as "too powerful" firing the 7.62 mm cartridge, the M14 is proving useful for U.S. troops fighting in the ongoing conflicts in Iraq and Afghanistan. The 5.56 mm M4 and

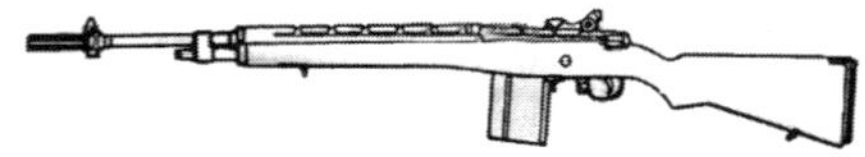

U.S. Marines land in-force on the shores of Vietnam 8 March 1965. The Marines are armed with 7.62 mm M14 rifles. Vietnam would be the first large-scale baptism of fire for the M14. (NARA)

M16A2 weapons while suitable for close engagements in the streets and villages, have often proved wanting in the vast open terrain of the desert. This is where the M14 excels, engaging the enemy at well over 500 meters while remaining well out of range of the 7.62x39 AK weapons often used by the insurgents.

Designated Marksmen Rifle (DMR)

Designated Marksman

During the 1990s the U.S. Marine Corps conceived the idea of having an individual Marine with above average shooting skills be a Designated Marksman. The concept was to have an individual with special training that could engage and dispatch targets at ranges beyond that normally used by the infantryman. The training would not be as intense as that of a sniper, but concentrated more on enhancing long range shooting skills.

The idea of having a sharp shooter as part of a rifle squad is not a new concept. During World War II and Korea a sniper rifle was often part of the squad's Table of Organization and Equipment (TOE). After 1960 sniper rifles were no longer listed on the TOE.

The Marine Corps choose the M14 rifle as its preferred DMR or Designated Marksman Rifle. The M14 DMR is a highly tuned match weapon built by Marine armorers at the Precision Weapons Shop located at Quantico, Virginia. The DMR is equipped with a Harris Bipod, McMillian fiberglass stock, special scope mount and rings and a Unertl or Leupold & Stevens MK4 10-power telescope. A commercial match-grade barrel was fitted to the receiver. Also available for special operations were sound suppressors and AN/PVS night vision scopes.

The Designated Riflemen are issued special M118LR match-grade long range ammunition. The M118LR cartridges are loaded with 175-grain projectiles to give the rifleman greater effective range.

U.S. Navy Mark 86 line-throwing rifle.

11 Aug 2002. At sea aboard the USS Carney, Gunner's Mate 2nd Class Michael Smith uses a modified M14 to fire a shot line to the USNS Walter Diehl in preparation for underway replenishment. US Navy photo by Chief Journalist John Williams/Courtesy Robert Bruce Military Photo Features

Configuration Document
for
MK 14 MOD 0 Enhanced Battle Rifle (EBR) 7.62mm NATO

SMALL ARMS WEAPONS BRANCH
ORDNANCE ENGINEERING DIRECTORATE
NAVAL SURFACE WARFACE CENTER, CRANE DIVISION

NAVSEA WARFARE CENTERS CRANE

UNITED STATES NAVAL SPECIAL WARFARE COMMAND

The U.S. Navy MK14 Model 0 EBR

The U.S. Navy began the refurbishing and upgrading of the 7.62 mm M14 rifle into what are designated as the MK14 Model 0 Enhanced Battle Rifle or EBR. The rifles were developed for use by the Navy Seal Teams and Special Marine Units.

The Navy's Enhanced Battle Rifle rebuilding procedures were developed by the Small Arms Weapons Branch, Ordnance Engineering Directorate at the Naval Surface Warfare Center at Crane, Indiana. The procedure begins with M14 rifles in "F" condition, either NSN 1005-589-1271 or 1005-072-5011. The rifles were completely disassembled and inspected. Parts selected for reuse or disposal was to be according to the MK14 Model 0 build requirements. Several standard M14 parts were no longer used in the MK14 rifle, these parts were retained to service unaltered M14 rifles. Standard M14 parts not used in the MK14 EBR include; Stripper clip guide and retaining pin, rear sight aperture, operating rod guide, bolt stop, front band, barrel and flash suppressor.

The MK14 EBR was built using the basic action of the M14 rifle. Improvements include an eighteen-inch barrel to replace the standard twenty-two-inch M14 barrel. A commercial Chassis Stock System with pistol grip from Sage International, with various Mil Std 1913 rails and ergonomic features, improved sights, tailored see-through scope rings, tactical sling, special tools, cleaning kit, Harris bipod and a discreet Eagle brand soft carry case. Basic issue items include two standard twenty-round M14 steel magazines.

The MK14 EBR was built using the basic action of the M14 rifle. Improvements include an eighteen-inch barrel to replace the standard twenty-two-inch M14 barrel. A commercial Chassis Stock System with pistol grip from Sage International, with various Mil Std 1913 rails and ergonomic features, improved sights, see-through scope rings, tactical sling and a Harris bipod. (Courtesy of Smith Enterprises Inc.)

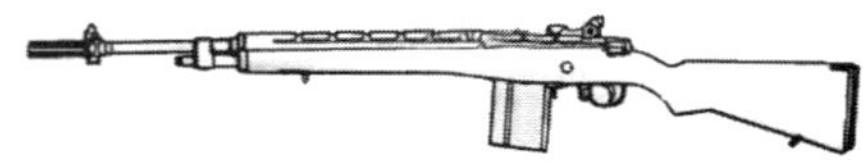

The U.S. Navy MK14 Model 0 EBR The U.S. Navy began the refurbishing and upgrading of the 7.62 mm M14 rifle into what are designated as the MK14 Model 0 Enhanced Battle Rifle or EBR. The rifles were originally developed for use by the Navy Seal Teams and Special Marine Units. The Navy's Enhanced Battle Rifle rebuilding procedures were developed by the Small Arms Weapons Branch, Ordnance Engineering Directorate at the Naval Surface Warfare Center at Crane, Indiana. (Photo courtesy of Robert Silvers)

All used GI and new commercial parts used in the construction of the MK14 rifle were inspected and gaged to insure serviceability. Original parts that were to be reused were bead-blasted and manganese phosphate coated. The gas cylinder was black oxide coated. All used springs were replaced with commercial springs manufactured by the Wolff Spring Company. GI springs could be used as an alternative.

The trigger group was reassembled using gaged, inspected, refinished parts. A Wolff

Laser Engraving of Bolt (MK 14 MOD 0 and anchor)

hammer spring was installed, use of a GI spring was considered acceptable if no Wolff springs were available. Trigger pull was adjusted to fall within a 4.5 to 7.5 pound range. A new commercial extended bolt release and roll pin manufactured by Smith Enterprises were installed on the receiver assembly.

The breech bolt was completely disassembled cleaned and inspected. The bolt was then magnetic particle inspection, paying particular attention to the locking lugs. The bolt was also checked for proper firing pin protrusion and for excess wear of the firing pin hole.

After passing inspection the bolt was laser engraved with the designation; MK 14 MOD 0 and the Navy anchor logo. A new eighteen-inch barrel manufactured by the commercial Springfield Armory was installed on the MK 14 receiver, taking care to obtain proper alignment.

An operating rod guide made by Sage International was fitted. The gas cylinder was secured by a gas lock with dovetail provided by Smith Enterprises. A new commercial flash suppressor also from Smith Enterprises was fitted to the barrel.

The bolt was headspaced to a barreled receiver. The last four digits of the receiver's serial number were marked on the underside of the bolt. The barrel was proof tested using a M60 High Pressure Test Round. After a magnetic particle inspection of the barrel and bolt, the barrel was marked with a letter P on the top right side near the receiver. Headspace was rechecked to insure that it was still within specifications.

A two-slot stripper guide rail manufactured by Sage International was installed in the slot formally used for the stripper clip guide. The Sage top rail/handguard was attached by screws to the top of the main stock. A front vertical grip was installed in the two holes provided on the underside of the main stock.

A new rear sight aperture manufactured by XS Sight Systems was used to replace the original aperture. The replacement sight has a larger .125-inch diameter hole. The hooded front sight also was produced by XS Sight Systems.

The MK14 MOD 0 rifle was test fired using two full 20-round magazines of M80 and M118LR ammunition. Point of aim / point of impact were specified to be calibrated for a range of 100 meters.

Commercial Parts List of items used in the Crane MK14 Model 0 EBR System

Carrying soft case, Eagle Industries

Sling Assembly, Eagle Industries

Return to duty. After a long period in storage, a box full of U.S. 7.62 mm M14 rifles, rebuilt and modified by Smith Enterprises, are being prepared for return to the Army. (Courtesy of Smith Enterprises)

Sight Rings, Badger Ordnance

MK14 EBR Tool, Bondhus/MSC

Bipod, Harris Engineering

Bipod Mount, Knight's Armament

Cleaning Kit, Otis

Bore Snake, Hoppes

U.S. Army M14SE

A substantial number of old, and for the most part forgotten, U.S. M14 rifles have been reposing in government storage for many years. The rifles have been awaiting their unfitting destiny of being turned into scrap metal. However, recent world events have presented these old warriors with a new opportunity.

Smith Enterprises of Arizona has established a program for modifying the old GI M14 rifles into a capable military rifle for the 21st Century. This task has been accomplished by utilizing modern technology and modern precision machining capabilities that did not exist when the M14 was originally designed and manufactured.

A Smith Enterprises government contract M14 EBR is test fired for function and accuracy on the range (Smith Enterprises Inc.)

The Smith upgraded M14 is not designed to be a precision target rifle, but an accurate and rugged military weapon. The original GI synthetic stocks are not glass-bedded or the fore-end of the military fiberglass stocks modified. The newly configured rifle was developed by Smith Enterprises to provide a reliable, proven and cost-effective modernization program for a modern 7.62 mm semiautomatic sniper system or SASS. The modified M14 addresses all of the shortcomings documented in the original M14, M21 and M25 designs. Smith Enterprises has been supplying the U.S. military, as well as many other government entities with the modified M14 "Crazy Horse" rifles.

The M14SE rifles has been designated as the Crazy Horse Sniper System. The basic weapon features a medium-weight Chrome Moly steel 22-inch barrel with a 1:10 twist 4-lands and grooves chambered for the U.S. M118 Long Range 7.62 mm ammunition. The original gas system is modified with a Smith Enterprises special unitized design gas plug, lock, front band and gas piston. The Smith Enterprises Direct Connect Vortex flash eliminator is used in place of the original M14 flash suppressor. The Smith Enterprises rifle features a laser alignment process for barrel installation. The Smith barrels are designed for the M118 LR rounds, not standard M80 ball ammunition

The Leupold Mark 4 3.5-10x40 LR/T M3 Telescopic sight is mounted with Smith's melonite heat-treated scope mount and rings. A heavy-duty M1913 Picatinny ring set; extended bolt stop; 4.5 pound tuned trigger assembly, with special S-7 tool steel trigger and hammer pins complete the upgrades.

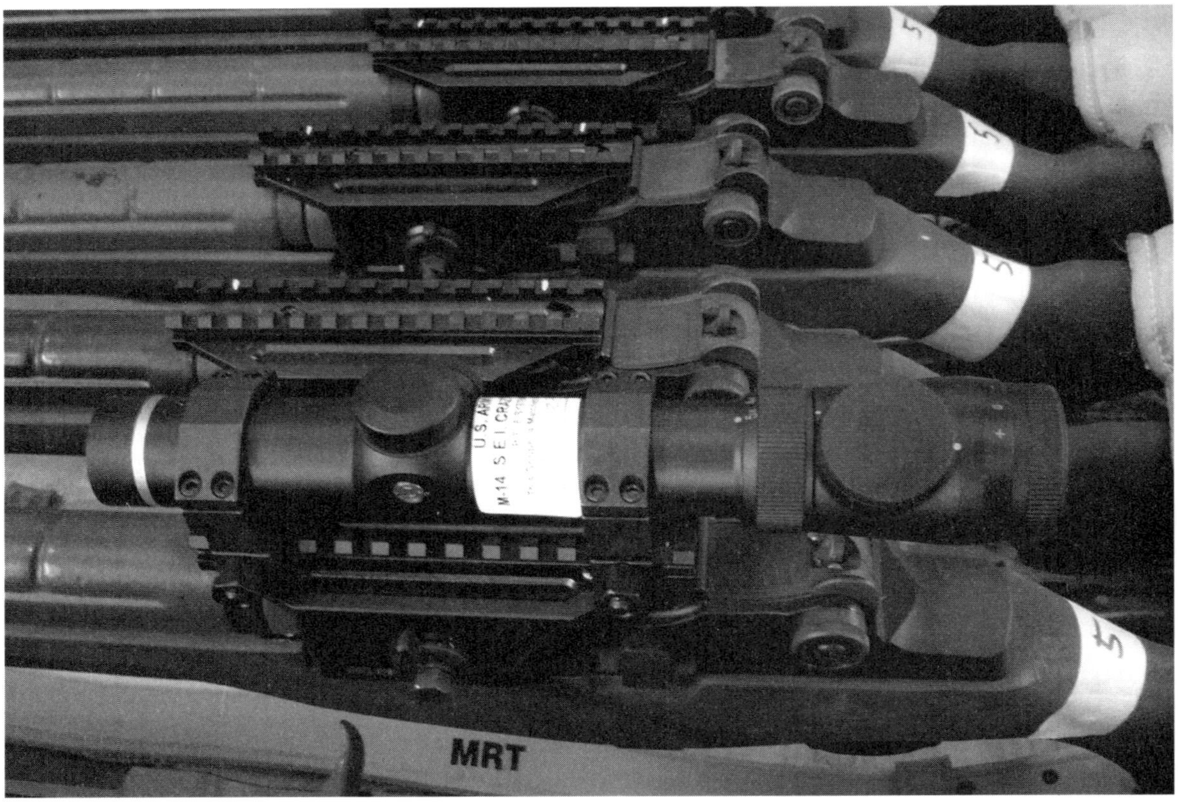

A shipping crate of upgrade Smith Enterprises Inc. "Crazy Horse" M14 rifles await shipment to the U.S. Army. (Courtesy of Smith Enterprises Inc.)

Chapter 6
Rifle Training

Recruits learn how to fieldstrip and maintain the M14 Rifle (NARA)

The U.S. soldier and Marine were introduced to the infantry rifle during their basic training. Many were trained on the M1 and eventually issued the M14 rifle. Later others trained on the M14, but when deployed to Vietnam were issued M16A1

An old seasoned sergeant gives a few pointers on aiming the infantry rifle. (U.S. Army)

weapons. This practice usually occurred because the first-line weapons were generally not issued for training purposes.

The training objectives taught were to develop confidence, will, know-how and skills required to fire a rifle to cause enemy casualties on the battlefield; to apply proper techniques of marksmanship when functioning as an individual or as part of a unit engaged in combat; maintain a continuing degree of proficiency in combat rifle training consistent with the mission of the individual's unit; in times of peace to develop potential marksmen and provide them further training; in times of war provide an instructor base or cadre for sniper training, if required; insure that every individual is able to properly maintain his weapon.

The individual infantryman is trained on the mechanical aspects of his weapon, how it functions, and the nomenclature of all its components. The infantryman is also trained on how to disassemble and assemble the piece. He is also instructed on proper preventive maintenance and cleaning.

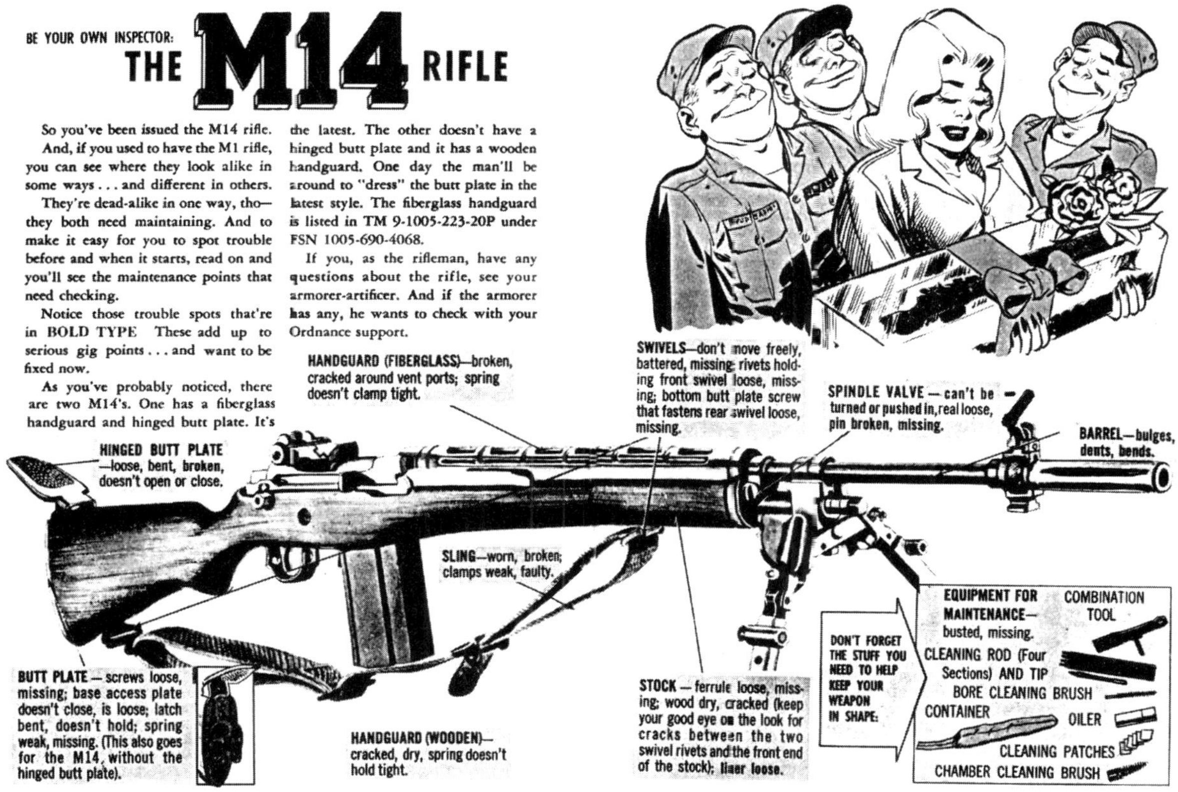

BE YOUR OWN INSPECTOR:

THE M14 RIFLE

So you've been issued the M14 rifle.

And, if you used to have the M1 rifle, you can see where they look alike in some ways . . . and different in others.

They're dead-alike in one way, tho— they both need maintaining. And to make it easy for you to spot trouble before and when it starts, read on and you'll see the maintenance points that need checking.

Notice those trouble spots that're in BOLD TYPE These add up to serious gig points . . . and want to be fixed now.

As you've probably noticed, there are two M14's. One has a fiberglass handguard and hinged butt plate. It's the latest. The other doesn't have a hinged butt plate and it has a wooden handguard. One day the man'll be around to "dress" the butt plate in the latest style. The fiberglass handguard is listed in TM 9-1005-223-20P under FSN 1005-690-4068.

If you, as the rifleman, have any questions about the rifle, see your armorer-artificer. And if the armorer has any, he wants to check with your Ordnance support.

More information and tips from Ms Connie Rodd on maintaining the M14 rifle. (John M. Miller)

Conditions that affect marksmanship in combat include that the enemy is seldom visible. Most targets are identified by smoke, muzzle flash, noise, or movement and are only visible for a brief moment. Most combat targets are linear in nature and will generally consist of troops irregularly spaced in covered or concealed areas. Targets are engaged by using objects near the enemy as reference points. The distance that personnel targets can be detected and engaged will seldom exceed 300-meters. In a defensive position the rifleman will usually be firing from a prepared position such as a foxhole or other type of emplacement. Selecting a point of aim in elevation is difficult because of the low outline and obscurity of enemy personnel targets. Mechanical adjustments of the rifle's sights are

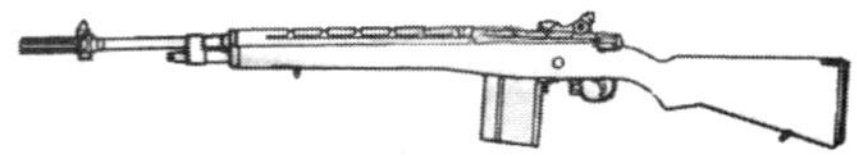

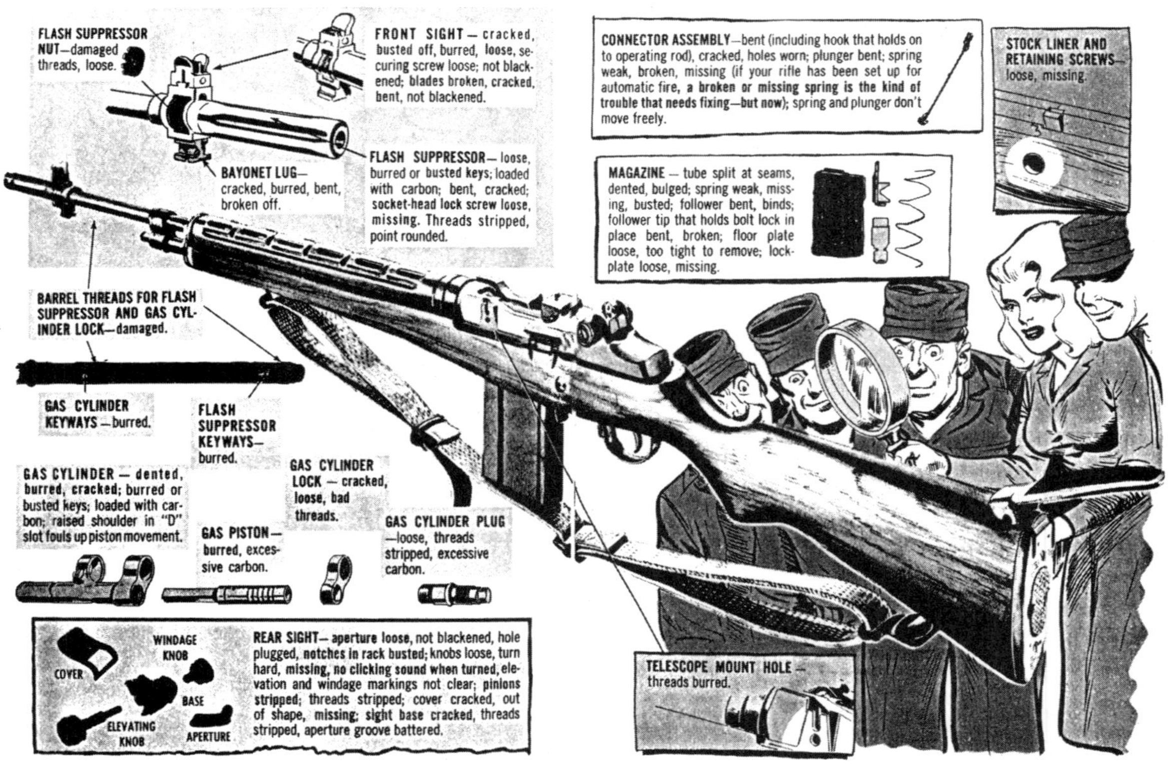

rarely required in combat. Targets in combat requiring time-pressure fire are basically of two types. The first is a single running individual that must be engaged within an unknown minimum time frame. The second type of combat target consists of a number of enemy in scattered positions. This type of fire allows the rifleman more time to select targets.

Rifle Marksmanship Fundamentals

A proficient combat rifleman should be able to detect targets, determine the range of the target and hit the targets that he engages. There are many factors that can affect an individual's ability to detect targets and distance. However the conditions that affect his abilities are constant. A rifleman must be able to assume a number of shooting positions that enable him and his weapon to form a single steady unit.

One of the fears of new recruits is of rifle recoil, particularly to those with little or no previous firearms experience. A preliminary firing demonstration was usually performed by a well-trained rifleman in several positions, sometimes including firing while the rifle's butt is resting on his chin! The primary purpose of the demonstration is to show the troops that the recoil is harmless, and that the key to accurate shooting is holding the weapon firmly against the body. After a brief familiarization, a firing exercise allowing each individual to fire a minimum number of rounds at a distance of 25 meters was performed.

One of the first lessons in military marksmanship training was to teach proper use of the sights, including sight alignment, the proper aiming point and trigger control. Also stressed was proper shouldering of the rifle. Six standard rifle-firing positions were taught to the recruit; prone, prone supported, kneeling, kneeling supported, standing and foxhole.

The rear sight of the M14 rifle was very well designed and provides an excellent sighting system for a battle rifle. The elevation knob of the sight is calibrated and marked 2, 4, 6, 8, 10 and 12 representing hundreds of meters,

A young infantryman aims his weapon while his counterpart observes. The soldier aiming the M14 has the M1956 universal magazine pouch attached to his belt.

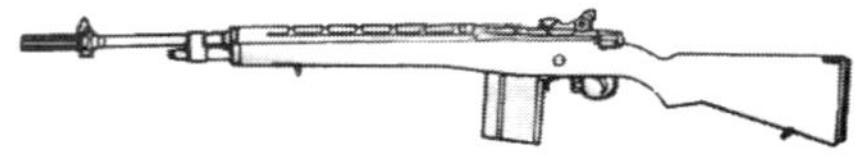

During a VIP tour of the rifle range an Army instructor points out features of the pop-up silhouette target. (U.S. Army)

when marked with a letter "M". If no letter "M" is present, (configuration of the M1 rifle sight) the rear sight is calibrated in yards not meters.

On the standard issue M14 rear sight, each audible click changes the point of impact 1-inch at 100 yards (1 minute of angle). Elevation is adjustable from 0 to 72 clicks. Windage can be adjusted 16 clicks to the right or left of center. National Match rear sights offer elevation and windage adjustments in 1/2 minute of angle increments for more precise target shooting.

Zeroing the M14 Rifle

The definition of zeroing a military rifle is having the rear sight elevation and windage set so that the projectile strikes the center of the intended target at a predetermined distance. There are two types of zero adjustments of interest to civilian shooters. One is target zero, the other battle sight zero.

Using a typical six-o'clock sight picture, the projectile strike will be in the center of the target or "bullseye" using a target-zero. A battle sight zero is a point of aim setting. With a battle zero an infantryman can engage a man-size target at ranges from 0 to 400 meters by aiming at the target's center of mass.

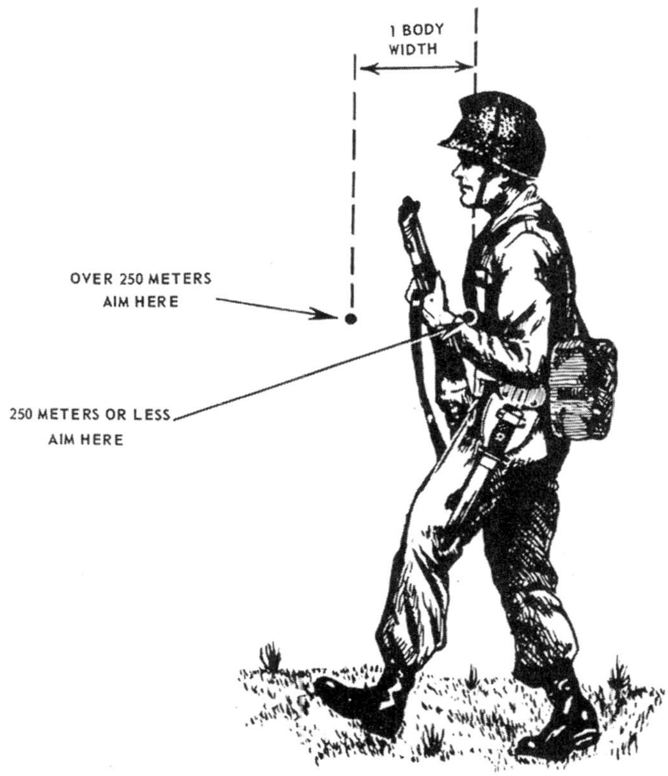

Aiming points, from the manual. (Donald G. Thomas)

To establish either zero setting, set the windage adjustment of the rear sight so that the index line is aligned with the line inscribed on the receiver. This position is known as a mechanical zero setting. After making the initial setting described, fire 3 to 5 rounds at a target positioned at a 100 meter distance. Examine the group of projectile strikes on the target. To center your group move the front sight in the opposite direction that you need to center your group onto center of the target. To adjust the front sight the retaining set-screw must be loosened. Drifting the front sight .008-inch will move the group 1-inch at 100-meters. After adjustment retighten the set-screw.

To establish an elevation target zero, turn the elevation knob on the rear sight counter-clock wise until the aperture bottoms out. Next raise the aperture by turning the elevation knob clockwise eight to twelve clicks, this preliminary adjustment will place the shot group close to the aiming point at 100-meters. Fire additional groups, making adjustments as needed to center the strikes onto the center of the target.

Battle Sight Zero

The standard 25-meter target was used for the battle sight zero exercise. The vertical and

U.S. Army one on one training with the 7.62 mm M14 rifle. (NARA)

horizontal lines printed on the target form 1.4-centimeter squares. One click of elevation would move the strike of the projectile .7-centimeter at a range of 25-meters, thus 2-clicks would move the projectile strike 1-square on the target. The 250-meter battle sight zero is determined by firing a series of 3-rounds at the 25-meter target. The shooter fires at the bottom center of the black rectangle on the target and adjusts his sights until the center of his shot group is located 4.6-centimeters direct above the point of aim. With this setting an aiming point at 250 meters will coincide with the projectiles point of impact. Once the zero had been established no further adjustment was needed unless engaging targets beyond a 250-meter range.

Calibration of the M14 Rifle Rear Sight

After obtaining the battle sight zero for his weapon the infantryman must calibrate the rear sight. The procedure was necessary to insure that the sight settings remained the same, and could be easily re-adjusted if needed. The procedure was to turn the elevation knob for-

ward until the rear sight aperture was at its lowest possible position. While turning the elevation knob the infantryman counts the number of clicks noted until the sight is adjusted to the lowest setting. The elevation knob is then turned forward until the 250-meter index line (the line between the 2 and 4 positions on the elevation knob.) is opposite the index line on the receiver. From this point the elevation knob was turned forward the number of clicks to the 250 meter battlesight zero position. Holding the elevation knob in this position tighten the center screw. Next turn the elevation knob to the rear until it is at its highest possible setting and tighten the screw. To check for proper adjustment set the 250-meter index line on the elevation opposite the index line on the receiver. Then turning the knob forward count the number of clicks, they should be equal to the battlesight setting.

Effects of Wind

Troops were taught the effects that the wind has on a projectile moving towards the target. The effects varied depending on velocity and direction of the wind and the distance to the target. To compensate for the wind "hold-off" was taught.

Rapid Reloading

Techniques were taught on how to quickly reload the M14. During several firing exercises a time limit was imposed to engage the target and reload. During firing exercises in which time limits have been imposed the recruit is closely observed to insure that he is using correct trigger techniques and not jerking the trigger. During training a dummy round was loaded into the magazine to test the infantryman's ability to apply immediate action to clear the round from the rifle, resume his position and continue firing at the target. The infantryman was trained until he could act instinctively to clear stoppages and malfunctions of his weapon.

Range Estimation

Determining the range to a target is the process of finding the distance between two points. One point is the infantryman's position; the other is that target's position. Being able to accurately determine range is an extremely important skill that must be mastered by the infantryman.

Tom York, great-grand-nephew of World War I hero Alvin York, fires the M14 rifle. (NARA)

An Army rangemaster points out to a young recruit how to correct his aim at the target. (U.S. Army)

There are a number of ways to determine the distance between two points. One is by using a map, pacing the distance or using an optical range finder and firing a round at the point in question. All of the aforementioned methods are not always practical for to the combat infantryman to use. There are two other methods that are taught to determine distance. One is the 100-meter unit of measure. The infantryman must be able to visualize a distance of 100-meters on the ground. For longer distances, like 500-meters he would need to determine how many 100-meter increments there are to the target (five). Beyond ranges beyond 500 meters the infantryman determines a distance halfway to the target, determines the number of 100-meter increments there are to the halfway point and multiples it by two.

Another method employed by infantry troops is the appearance of objects method. This method determines range by determining the size and other characteristic details of the object. To apply this method the infantryman must be familiar with the characteristics of enemy personnel at known ranges, then he can mentally compare the same characteristics at unknown ranges. To sharpen his skills the infantryman must study the appearance of a man-size target in varied positions at ranges from 100 to 500 meters. This method is severely handicapped in conditions of poor visibility such as smoke, fog or darkness.

Crack and Thump

The term crack is used to describe the sound of the projectile exceeding the speed of sound as it passes near the infantryman. The term thump is used to describe the sound of the expanding gases escaping into the atmosphere when a weapon is fired. The object of "crack and thump" training is to determine the location and range to a concealed target firing at the infantryman's position. The first sound that an infantryman will hear when being fired on is the crack of the projectile as it passes, the second sound will be the thump. By using a

Vietnam 1966, troops fire both M16 and M14 rifles during target practice at a base range. (NARA)

A U.S. Army presentation familiarizes newly arrived officers with a well-equipped infantryman.

technique to mentally align the cracks with the thumps, an infantryman can determine in which direction the enemy fire is coming from. To determine the range of the enemy the rapid count method is employed. The count starts when the first projectile crack is heard and continues until he hears the thump. The count is conducted at a fast 5-counts per second. During automatic fire the count starts with the first crack and ends with the last thump. The final count is multiplied by 100 to obtain the range estimate in hundreds of meters.

Live ammunition was fired over the heads of recruits to practice the technique. Training was conducted under all types of weather to insure effective target detection under any conditions that may be encountered in combat.

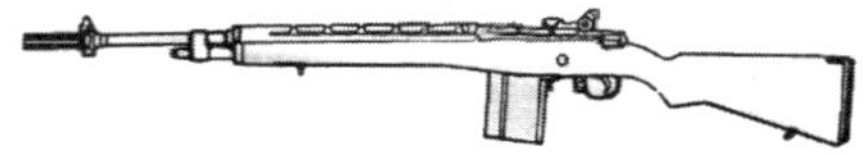

After training the infantryman should be able to determine the location of a concealed target within 10-meters 60-percent of the time. Special overhead fire ammunition was used for the training exercise.

Engaging Targets

U.S. Army soldiers train in West Germany during the cold war era. For many years Germany was where the next world conflict was expected to begin. (NARA)

Unless receiving orders to the contrary, combat targets are to be engaged immediately upon detection.

In the case of enemy personnel there are three types of targets; a stationary target, a slowly moving target and a rapidly moving target. A stationary target can be engaged by using reference or aiming points. Stationary targets are usually well concealed making engagement as much of a detection problem as it is of marksmanship.

Although moving targets are easily detectable, the movement complicates the selection of an aiming point. Unless the enemy is totally unaware of the infantryman's presence he will change positions by running from one concealed position to another. While moving the enemy presents a rapidly moving target. However, for a brief moment as he begins and ends his run to a new position, his movement is relatively slow. These two moments offer the most vulnerable exposure to aimed fire from the infantryman.

An enemy target moving towards the infantryman's position is engaged in the same manner as a stationary target. However, to successfully engage a target moving laterally across his front the rifleman must lead the target. To engage a walking target at a range of 200-meters or less the rifleman should aim at the forward edge of the enemy's body. For

An Army instructor enlightens new recruits preparing to fire their weapons. (NARA)

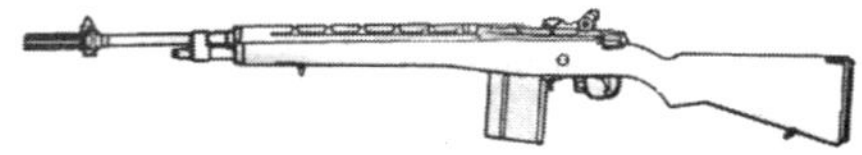

ranges over 200-meters the rifleman should select a point of aim approximating one body-width in front of the target. If the enemy target is running the target leads are doubled, for example a running target at less than 200-meters the aiming lead would be one-body width in front of the target. For distances beyond 200-meters the lead of a running target would be two-body widths in front of the target.

Automatic Rifle Marksmanship

Automatic fire is the firing of two or more consecutive rounds without releasing the trigger. Bursts of two or three rounds are usually fired to minimize dispersion of fire. Automatic fire is employed when engaging enemy positions out to 460 meters, when engaging large targets such as crew served weapons, unarmored vehicles and openings in buildings. Automatic fire is also employed when fire superiority is warranted by the tactical situation.

Decreased accuracy associated with automatic

fire is compensated by the delivery of a heavy volume of fire. Maintaining a high volume of fire depends on the automatic rifleman's ability to load and change magazines. The automatic rifleman should be able to perform magazine changes in four to five seconds. This level of proficiency is only accomplished through intensive training and the fundamentals of automatic fire.

For infantryman issued the M14A1 rifle or the M14 with the M2 bipod and selector (M14 modified) slightly different fundamentals are employed. The automatic rifleman is required to further develop his marksmanship skills to successfully employ the M14A1 or M14 (M) weapons.

The infantryman must become proficient in establishing a stable body position when employing the M14 in the automatic role; become skilled at rapid and systematic magazine

Anyone who has served in the military knows this drill. (NARA)

changes; controlled disposition of fire and additional knowledge on operation of the weapon's rear sight.

Infantrymen were taught to engage targets at 200-meters or less by aiming at the bottom center of mass. At targets beyond 200-meters the infantryman was trained to fire at the center of the mass. The maximum effective range of the M14A1 and M14 Modified with bipod support is 700-meters. Since the rear sight 250-meter battle sight zero setting is only effective out to 500 meters, the method of engaging targets between 500 and 700-meters is accomplished by first determining the approximate distance to the target. The elevation of the rear sight is then adjusted to the estimated range. By then employing the adjusted aiming point method the automatic rifleman should be able to obtain kills at that range

Firing the M14 from the kneeling position. (NARA)

The M14A1 automatic rifle should be fired from the right shoulder. The primary reason for this is that the muzzle stabilizer, attached over the flash suppressor, is designed to compensate for dispersion characteristics of right-handed shooters. For right-handed shooters the projectile dispersion pattern is generally high and to the left, the dispersion pattern for left-handed shooters is high and to the right.

Aiming the automatic rifle is the same as taught during semiautomatic fire. A firm steady hold while aligning the sights and firing the weapon is paramount in successful target engagement.

In unsupported kneeling or standing firing positions the techniques are the same as semiautomatic except for the grip of the hands and the spot-weld of the cheek. On the M14A1 rifle the front handgrip is grasped with the left hand while exerting a strong rearward pressure against the shoulder. The foregrip should be adjusted so that it cants forward at about a 20-degree angle, perpendicular to the stock. The adjustment should be made so that the sling absorbs the tension instead of the grip itself, to prevent breakage of the foregrip or stock. When firing from bipod-supported position the hinged shoulder rest of the buttplate should be used. The weapon should be positioned against the shoulder at the point where the neck and shoulder join so that the recoil pad is against the collarbone. The tighter the weapon is held against the body the less dispersion of fire will be. Care should be taken not to buck the shoulder into the weapon as this will cause fire to be displaced down and to the left. The pistol grip of the M14A1 stock should be grasped so that the rear of the grip is positioned between the V formed between the thumb and forefinger of the right-hand. The fingers are wrapped tightly around the rear pistol grip exerting only

The M14A1 rifle.

slight rearward pressure. The tip of the forefinger is placed on the trigger so that there is no contact between the finger and the stock. This allows the trigger to be pressed straight to the rear without disturbing the lay of the weapon.

The overtightening of the gas cylinder plug was an ongoing problem with the M14 rifle. Many PS articles were written attempting to address the problem. (Bill Ricca)

Concentrated Fire

The automatic rifleman was thoroughly trained to deliver fire at targets, which have one or more aiming points. When fire is delivered at one aiming point it is called concentrated fire. The ability to deliver concentrated fire requires a high degree of accuracy. Basic rifle marksmanship training teaches the infantryman to think principally in terms of concentrated fire. The automatic rifleman is trained to apply the integrated act of firing to distribute as well as to concentrate his fire.

Distributed Fire

Distributed fire delivered by the M14A1 and M14 Modified weapons, is fire in depth and width so that a target is effectively covered. The primary object of distributed fire is to place a heavy volume of fire between the known or suspected flanks of a target. The automatic rifleman must attempt to place fire within the area of such a target. The inability to see enemy personnel should not be considered a reason not to fire into an area if there is reason to suspect the presence of a covered or concealed target present

Quick Fire

Infantrymen were instructed in a method of instinctive shooting called quick-fire. The method is based on unaimed fire delivered in a quick reflex action. As a prerequisite to survival on the battlefield, there was no substitute for a working knowledge of quick-fire.

Standard configuration U.S. M14 rifle fitted with the M2 bipod. (U.S. Army)

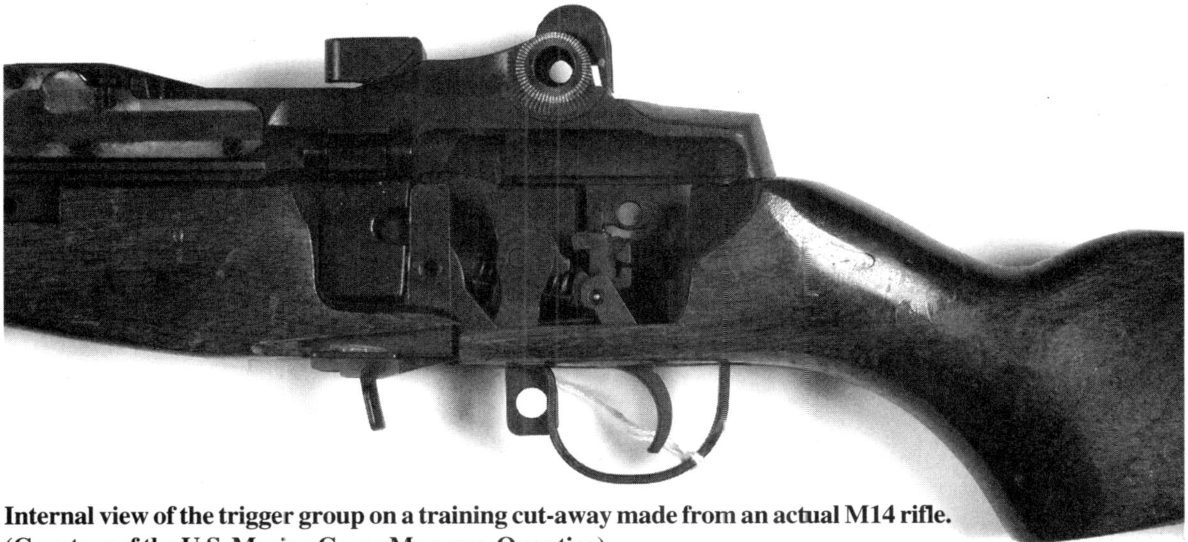

Internal view of the trigger group on a training cut-away made from an actual M14 rifle. (Courtesy of the U.S. Marine Corps Museum, Quantico)

The infantryman effectively trained in quick-fire methods does not consciously align his weapon's barrel when picking up his target, he must be able to relate himself to it.

To become proficient in quick-fire, practice firing with tape placed over the sights was used to reduce the tendency of the trainee to aim using the sights, and at the same time assist the infantryman in obtaining the relationship between his rifle and the target.

The proper position for quick-fire target engagement was for the infantryman to lean

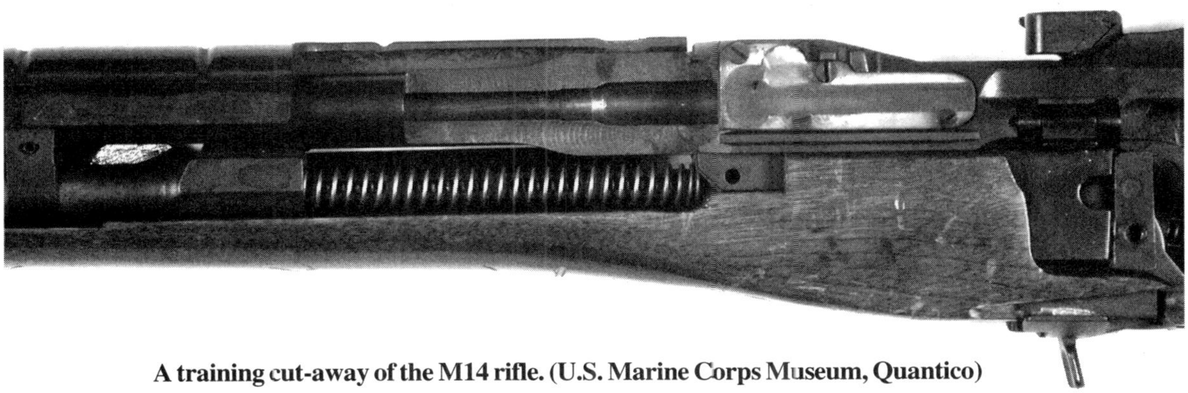

A training cut-away of the M14 rifle. (U.S. Marine Corps Museum, Quantico)

slightly forward, while holding his weapon at the low port position. The feet were spread at a comfortable position, and the shooter's weight was balanced on the balls of the feet so targets could be engaged within a 120-degree arc to the front without having to shift his position. While shouldering his weapon or a quick-fire engagement, both eyes remain open and riveted to the aiming spot at the base of the target.

At initial engagement the infantryman was instructed to look at the base of his intended target. This would prevent the natural tendency to fire over top of a target because of overestimating distances, and a natural urge to draw a comparison between the barrel and the target. Additionally if the infantryman misses with his initial low shot there remains a possibility that a ricochet will hit the target.

Amen, brother, amen.

And here's the latest scoop on blank firing attachments for 7.62 ammo with both the M14 rifle and the M60 machine gun.

On the M14 rifle it's a two-for-one proposition . . . 'cause FSN 1005-893-0902 gets you both the M12 blank firing attachment as well as the M3 breech shield.

And, natch, the blank firing attachment gets shoved into the flash suppressor and snapped over the bayonet lug while the breech shield is secured to the cartridge guide—like it says in TM 9-1005-223-12, May 61, (See PS 114, page 7, "Slow Down—Shoulders Ahead" if you have problems).

With the M60 machine gun, it's a horse of another color, and one FSN gets you one item.

In this case FSN 1005-893-0897 supplies you with only the M13 blank firing attachment. Just the muzzle piece is needed on the MG since its cover acts as the shield.

Tips from a PS article on using the blank adapter and breech shield for training. (John M. Miller)

Ship board practice with the M14 rifle.

During training on the firing range the infantryman was instructed to aim at a spot located at the lower portion placed on a standard E-type silhouette target. Should the initial target be missed the infantryman was trained to engage a different target. The reason was when a trainee missed his intended target he was inclined to seek the strike of the projectile and in the case of a miss he would shoot in the same spot repeating a missed shot. When shifting to a new target it would erase the visual picture from his mind. It was emphasized that the shifting of targets was a training exercise,

U.S. Marines train with M14 rifles. (Courtesy of Ronald Maxson USMC retired)

used to develop proficiency using quick-fire methods. During an actual combat engagement an infantryman would not shift to an alternate target if his initial shot missed the target. He would continue to engage his immediate threat target and continue to fire.

The infantryman was instructed to shoulder his rifle in one smooth fluid movement, not hurriedly jerk it into position. As soon as his weapon was shouldered and locked into position he was to fire immediately. Pausing would increase the natural tendency to aim or attempt to draw a comparison between the target and muzzle.

Training in quick-fire on the range was implemented at a range of 15-meters. The E-type silhouette target had a small two to three inch circle located on the lower third of the target. The infantryman was to focus his vision on the circle. After the trainee was deemed to be 80 percent effective at the 15-meter range, the firing line was moved back to a 30-meter range. The trainee would repeat his attempts to successfully engage the target. After becoming proficient engaging the target at the 30-meter range the tape was removed from the sights of the rifle. With the tape removed the trainee was allowed to engage the targets at the 15-meter and 30-meter ranges. Instructors carefully observed the trainee to insure that he was not using the rifle's sights while engaging the targets using quick-fire methods. If the trainee attempted to aim with the sights, the training was to be repeated.

29 Jan 03. At sea aboard the USS Ardent, Mineman 3rd Class Arthur Cobles fires an M14 rifle during weapons qualification training. The Ardent is performing mine counter measures in the Central Command Area of Operations in support of Operation Enduring Freedom. US Navy photo by PHM2C Jeffrey Lehrberg/Courtesy Robert Bruce Military Photo Features

Chapter 7

U.S. Military Sniper Rifles

8 Jan 02, Kandahar, Afghanistan. USMC Sergeant Jason Capozzoli, a Force Reconnaissance Marine with 26th MEU (SOC), gets acquainted with the Corps' new Designated Marksman Rifle, a specially accurized and equipped M14 rifle built by armorers of the Precision Weapons shop at Quantico. USMC photo by Sergeant Andrew Pomykal/Courtesy Robert Bruce Military Photo

The Military Sniper

The sniper is a soldier or Marine that has special abilities, training and natural skill. He is issued special equipment to enable him to perform his deadly task. His job is to deliver accurate discriminatory fire against enemy personnel. The sniper's primary targets are those which are not normally engaged by the infantryman, because of range or visibility. A successful sniper must be highly proficient in basic infantry skills. The sniper is a force multiplier, his craft requires him to train until he has mastered the skills for him to survive in the field. He must be a highly skilled long-range marksman as well as being well-versed in field craft. He must be in top physical and mental condition. The sniper must also be skilled in observing and collecting intelligence data. A skilled sniper is the best defense against an enemy sniper.

United States' Sniper Rifles

The M1903 Rifle

During the world's many conflicts and wars snipers have often been deployed to keep the enemy on edge and to demoralize him. When the Second World War began the U.S. Army had no trained snipers, having abandoning the use of them after World War I. The Marines on the other hand resumed training snipers just prior to the U.S. entry into the World War II, and thus were able to deploy them during the first U.S. ground action at Guadalcanal Island in 1942. The Marine sharpshooters were equipped with the M1903A1 bolt action rifle fitted with a Lyman 5A telescopic sight, basically the same weapon fielded by the Corps during World War I. The 5-power Lyman scope was a dated design by 1942. After

A World War II soldier armed with the M1903A4 bolt-action sniper rifle. The effective M1903A4 saw action later in both the Korean and the Vietnam Wars (NARA)

M1903A4 Sniper Rifle with the Weaver 2.5-power M73B1 telescope. (Courtesy of Scott Duff)

some research and testing the Marines adopted the 8-power Unertl telescopic sight as Standard for use on the M1903 rifle. The scope was also mounted on a few civilian Winchester Model 54 and Model 70 bolt action rifles for Marine sniper use. The old M1903 continued its sniper duties with the Marines during the 1950s Korean War. In 1951 the Equipment Board declared the M1903 as a Limited Standard sniper rifle and adopted the M1C Garand semiautomatic rifle as the Standard sniper weapon.

As a direct result of the Marine snipers' success during the early months of World War II, the Army issued an urgent requirement for a sniper weapon based on the M1903 rifle. To expedite the fielding of a sniper rifle the commercial Weaver 330-C telescopic sight was chosen to be mounted on specially selected rifles.

Due to a shortage of the M1 Garand service rifles early in World War II, the U.S. Ordnance Department had ordered the M1903 bolt-action rifle back into production. Simplified for easier manufacture, the new standard issue M1903 was designated as the M1903A3. The Ordnance department ordered that a number of the newly manufactured M1903A3 rifles be converted into a sniper's weapon. The sniper version of the '03 was designated as the M1903A4 rifle. Several different scopes were tried on the

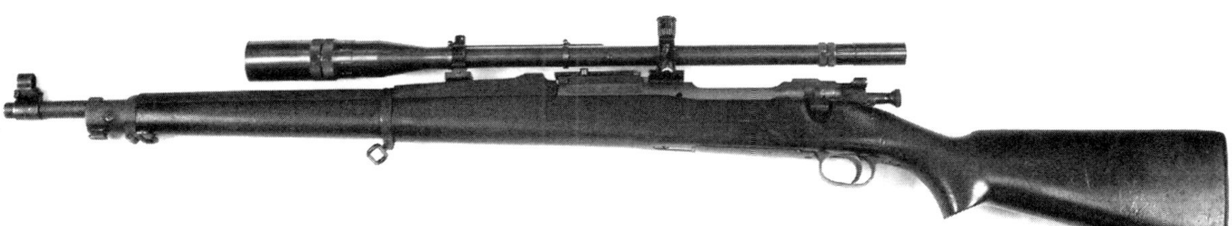

The M1903A1 Sniper Rifle with the 8-power Unertl telescope. This weapon was fielded by the U.S. Marines during World War II and later in the Korean War. (U.S. Marine Corps Museum, Quantico)

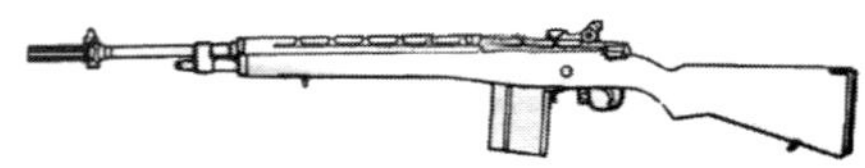

M1C Sniper rifle with a Stith-Kollmorgen Model 4XD telescope, T4 cheek pad and M2 flash hider. (U.S. Marine Corps Museum, Quantico, Virginia)

M1903A4, but virtually all were fitted with the Weaver M73B1. During the Korean War the M84 scope was fitted. The M1903A4 rifle with the M84 scope saw limited use as late as the Vietnam War.

The M1 Garand

After the M1 Garand rifle was being produced and issued in numbers during World War II, a directive was issued to develop a sniper's version of the new semiautomatic rifle. One obstacle that was immediately encountered was the M1 rifle's top-loading (and ejecting) ammunition clip. This problem was addressed by development of an off-set scope mount. An adjustable leather cheek pad was supplied to help align the shooter's eye with the scope. The Springfield Armory developed two experimental sniper weapons based on the M1, the M1E7 and the M1E8. The scope mount for the M1E7 required that the receiver be drilled and tapped to attach the mount. The M1E8 mount used a steel block pinned to the barrel. After extensive testing, the M1E7 fitted with a 2.5 power Lyman Alaskan (M73) telescopic sight was chosen and adopted in June of 1944 as the Rifle, Caliber .30, M1C sniper weapon. An M82 scope was also used for the M1C, the only difference was that the M81 used a cross hair reticle, and the M82 used a tapered post style reticle.

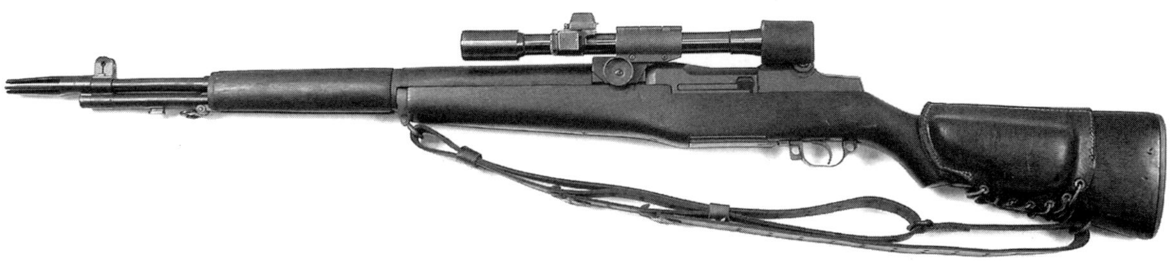

The M1D sniper rifle, this weapon originally developed during World War II, saw action during the Vietnam War. This rifle is fitted with the T37 prong-type flash hider adopted in 1958.

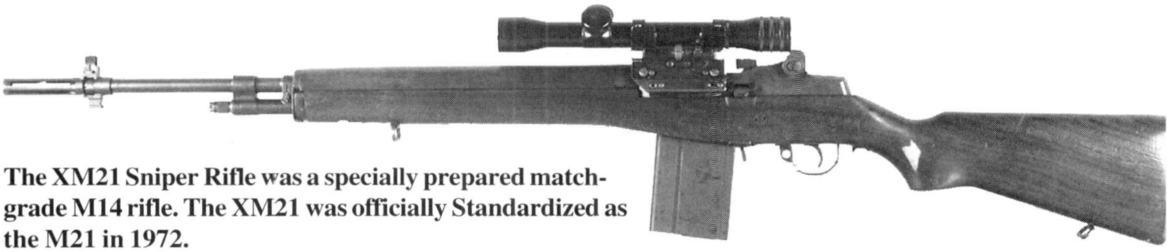

The XM21 Sniper Rifle was a specially prepared match-grade M14 rifle. The XM21 was officially Standardized as the M21 in 1972.

The M1E8 variation was adopted in September of 1944 as the M1D to supplement the M1C rifle. The prototype M1E8 also used the M73 telescopic sight. Both variations were fitted with the funnel-style M2 flash hiding device for dusk or night operations. Because of their late arrival in combat theaters, few M1C sniper rifles saw action. At the war's end only a small number of the M1D variation were available for testing and it is believed that none were fielded during WWII. After the war ended the sniper weapons continued to be produced at the Springfield Armory. The M1C would eventually see more action during the Korean War. During the Korean War the 2.2 power M84 scope was used on Army M1C rifles, while the Marine Corps opted for the 4XD Stith-Kollmorgen scope. The M1D model used a specially designed barrel to accommodate the scope mount. Most M1D rifles were assembled during the rebuild process rather than built as new production rifles, the M1D saw little use during the Korean War. During the 1950s the more effective T-37 prong-type flash hider was adopted to replace the M2 funnel style.

A well-trained sniper using an M1C reportedly could be effective at distances of 400 to 600 yards. The lack of special match-grade ammunition and the relatively low power of the M81, M82 and M84 telescopic sights mounted on the weapons hampered long-range engagements.

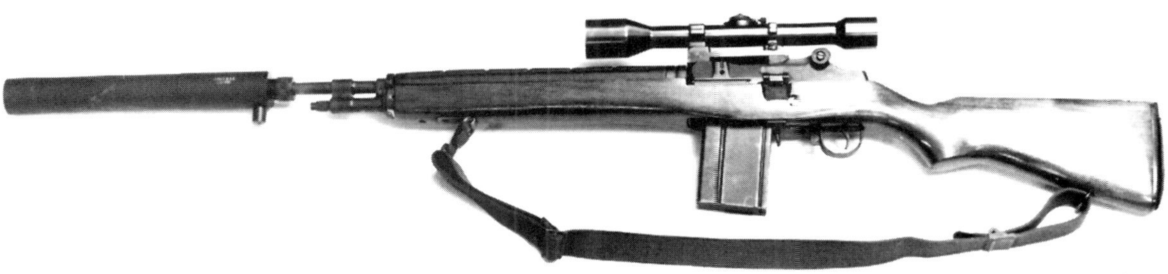

XM21 Sniper rifle with SIONICS suppressor. (Donald G. Thomas)

A group of officers from Paraguay visiting Fort Benning, Georgia inspect a U.S. M14 rifle fitted with an infrared night-vision scope. (U.S. Army Signal Corps)

A number of the M1D rifles were used in the early stages of the Vietnam War. These M1Ds were assembled from M1 rifles that were undergoing the rebuild process. A substantial number of M1D rifles were built and never issued. Highly prized by collectors, a number of these (rebuilt) unissued, M1D Garand rifles were sold through the Civilian Marksmanship Program in the 1990s.

Night Vision Devices

During World War II the U.S. Army developed infrared sights for infantry weapons for engaging targets at night. The problem with the early infrared systems was that the enemy began to use an optical device that could detect the infrared rays being projected. The device enabled the enemy to locate the shooter's exact position. Although limited success was achieved with the I.R. systems, the equipment was bulky and heavy. Another problem that was encountered was the inability to recharge the batteries while operating in the field.

During the war in Vietnam the U.S. made great strides with scopes that allowed the shooter to locate and dispatch targets at night. Infrared systems were fielded again during the early phases of the war.

By the 1960 era Varo Inc of Texas and Polan Industries of West Virginia had developed new improved infrared systems for the Army. However, when the AN/PAS-4 infrared rifle scopes were used in the early stages of Vietnam by Special Forces, they discovered that reflections from the dense jungle foliage limited the

A U.S. soldier peers through the AN/PVS-2 Starlight Scope mounted on an M14 rifle. The weapon has an M2 bipod attached. (U.S. Army)

The U.S. Army night vision device known as the Starlight scope. Shown is the PVS-2 model in its storage case and nylon carry bag. (Lawrence Heiskell, M.D. collection)

effective range of the devices. Additionally the Russians had also developed I.R. technology and were known to be supplying it to Communist satellite nations. Any enemy soldier using infrared equipment could easily pinpoint the source of I.R. devices in use by U.S. troops, making soldiers reluctant to use the equipment.

The next technological step was light intensification devices. Fully aware of the limitations of infrared devices, the Army Warfare Vision Division had begun working on new systems as early as 1955. By 1957 much progress had been made in using light-amplifying tubes, this was a passive system that could not be detected by the enemy. Continuing government studies on limited warfare recognized the need for night-vision devices and funding was granted to accelerate the research. Research for a suitable night vision device was ongoing,

The U.S. AN/PVS-2 Starlight scope weighed nearly six-pounds with an overall length of 18.5 inches, it was powered by small battery. While the scope didn't offer precise, clear vision it was effective in the hands of a trained operator and best of all was undetectable by the enemy when in use. (Lawrence Heiskell, M.D. collection)

and intensified, as the enemy in Vietnam preferred to operate under the cover of darkness. Once developed the new light amplifying devices were much lighter in weight then the earlier I.R. systems. The new systems operated by greatly intensifying existing light from the moon or stars. The new night vision devices were aptly named Starlight Scopes.

By 1961 a suitable prototype light-intensifying, night-vision scope was developed and built. By June of 1962 contracts were awarded to the Bell & Howell Company for engineer design models. During 1964 a production contract for a night vision scope for individual weapons was awarded to the Electro-Optical Systems Company. In the interim, research continued by several government operated research and development laboratories. As U.S. military involvement was escalating in Vietnam additional contracts were awarded for manufacture of the light-intensifying Starlight Scopes, Varo Inc. was awarded a contract during 1967. The small Starlight scopes could be hand-held or mounted on M14 and M16 rifles using adapter brackets. The "small" AN/PVS-2 Starlight scope weighted nearly six-pounds with an overall length of 18.5 inches, it was powered by small battery. While the scope didn't offer precise, night. Starlight scopes were fielded in a variety of configurations including rifle scopes, crew-served weapons and helmet mounted binoculars. The smallest scope was the AN/PVS-2, it had a sight range of 300 meters. Starlight scopes proved especially effective on sniper weapons. The improved second-generation model of the Starlight scope was the AN/PVS-4.

M14 rifle with Starlight scope. (Lawrence Heiskell, M.D. collection)

clear vision it was effective in the hands of a trained operator and best of all was undetectable by the enemy when in use. Usually the first indication of the Starlight scopes presence to the enemy was when the first projectiles began to accurately strike their targets in the dark. One of the few disadvantages was that the device would not work in total darkness encountered during an overcast, starless

The M21 Sniper Rifle

During 1966 an urgent requirement was once again issued by the U.S. Army for a modern sniper rifle, to replace the obsolete M1903A4 and M1D weapons currently available.

All production M14 rifles were manufactured with a threaded hole and grooves on the receiver's left side for mounting an optical sighting device. During 1966 a hinged type mount was designed for the M14 rifle mounting the old World War II M84 telescope. The combination proved little better than the previous, and dated, sniper rifles based on the M1 Garand. However due to the large amount of M84 telescopic sights available, the M14 with the M84 scope was issued until a suitable replacement could be developed, tested and fielded.

Much thought was given to the use of the M14 rifle as an effective sniper weapon resulting in the development of specially prepared and accurized match-grade weapons using match ammunition. The M14 rifles destined for sniper use were rebuilt according to the rigid specifications of the U.S. Army Marksmanship Training Unit. The USAMTU was well versed in building extremely accurate M14 rifles for competition.

After a number of telescopic sights were evaluated and tested, the ART or Auto Ranging Telescope was chosen for the M14. The

The AR TEL scope and aluminum carrying case. This scope was used to good effect on the sniper version of the M14, the XM21 rifle (John M. Miller)

Soldiers take a break while attending the U.S. Army Sniper School in Vietnam. (U.S. Army)

first ART scope was a specially modified commercial Redfield manufactured Accu-Range 3 power to 9 power variable power scope with an exterior cam physically attached to the power adjustment ring, which could be used to raise or lower the eye-piece end of the telescope. This arrangement enabled the correct angle of elevation for firing the M14 rifle with M118 ammunition at ranges from 300 to 900 meters.

The ART scope removed the guesswork from range estimations making long range hits of up to 900 meters common. The development of the ART scope also enabled sniper candidates to engage long range targets with less training and practice. The design was perfected through several pre-production prototype models. The production model was designated as the AR TEL scope. U.S. Army Lieutenant James Leatherwood developed the concept of the Auto-Ranging-Telescope (ART) while assigned

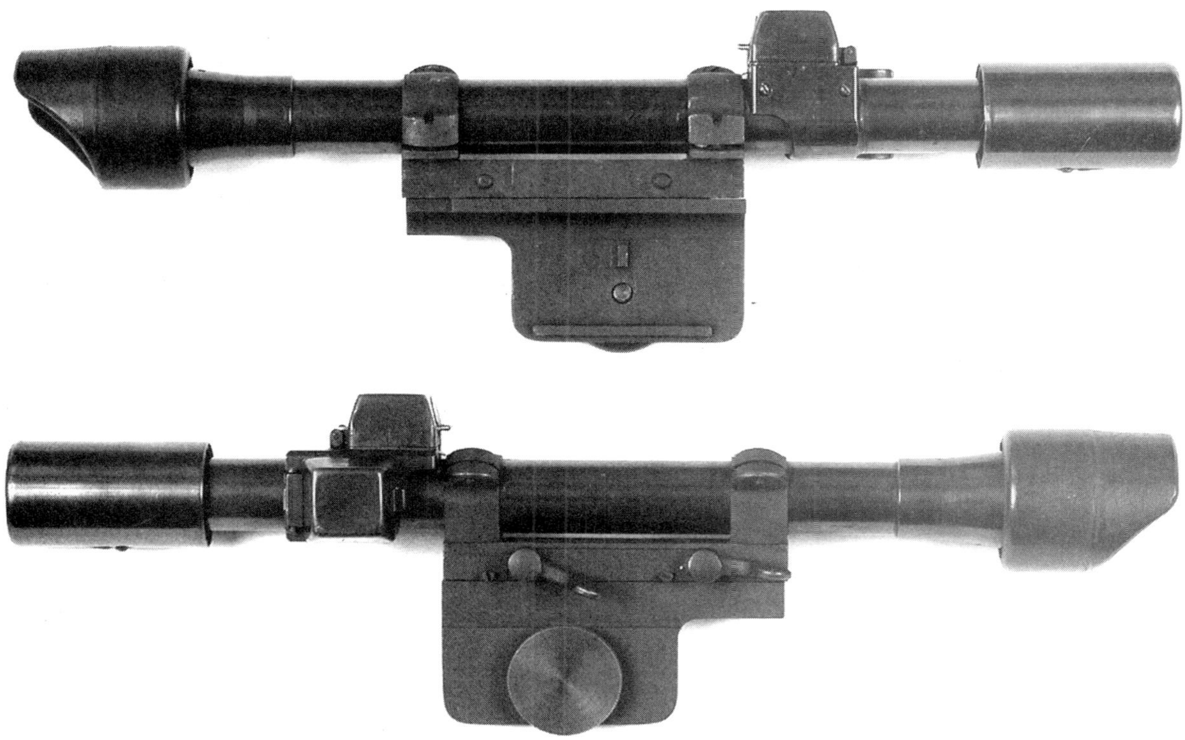

U.S. M84 telescope in a M1C rifle Griffin & Howe mounting bracket, adapted for use with the match grade M14 rifles. The M84 scope was a carry-over from earlier sniper rifles. During the early days of the Vietnam War the M84 was pressed into service with the M14 until a modern alternative could be fielded. (Courtesy of John M. Miller)

to Fort Benning, Georgia. The principle was perfected at the U.S. Army Limited Warfare Laboratory at Aberdeen Proving Center.

The sniper variant of the M14 was designated as the XM21 and announced by the U.S. Army Weapons Command on 18 September 1969. The X designation assigned to the nomenclature of the weapon served to identify any item that is under test or development and not yet a Standard item. A special seven-round magazine was developed and issued with the XM21. The compact magazine with a height of only three-inches allowed the sniper less exposure by allowing him to get closer to the ground when engaging targets in the prone position. Ongoing development and evaluations of the XM21 continued based on reports

The ART scope removed the guesswork from range estimations making long range hits of up to 900 meters common. The design was perfected through several pre-production prototype models. The production model was designated as the AR TEL scope. U.S. Army Lieutenant James Leatherwood developed the concept of the Auto-Ranging-Telescope (ART). The principle was perfected at the U.S. Army Limited Warfare Laboratory at Aberdeen Proving Center. (John M. Miller)

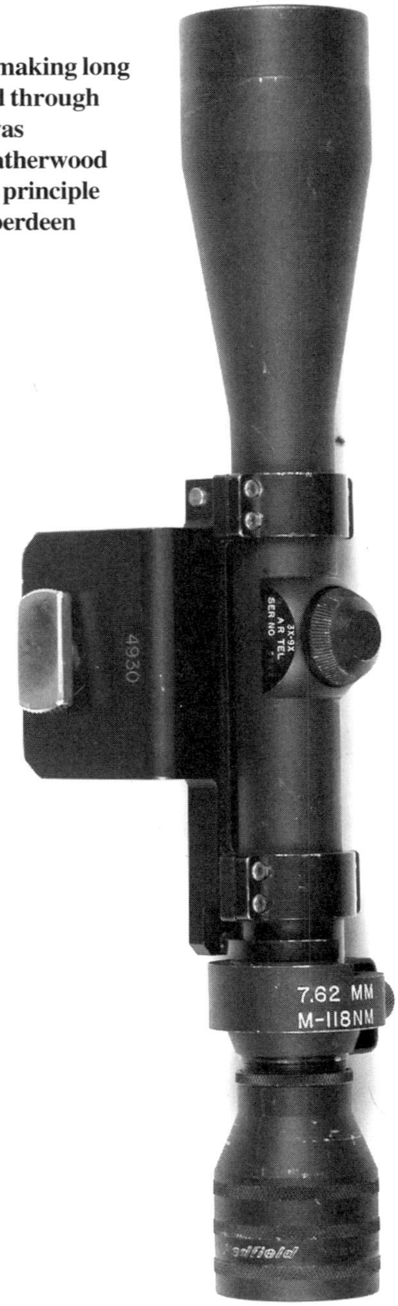

M21 Sniper Rifle

Leave It to the Experts

Your M21 sniper rifle requires special treatment.

Because of the M21's close tolerances and critical alignment, major maintenance—the kind requiring disassembly—must be done by repairmen specially trained at the National Match Rifle and Pistol Maintenance School. If you try to disassemble your M21, you can damage it.

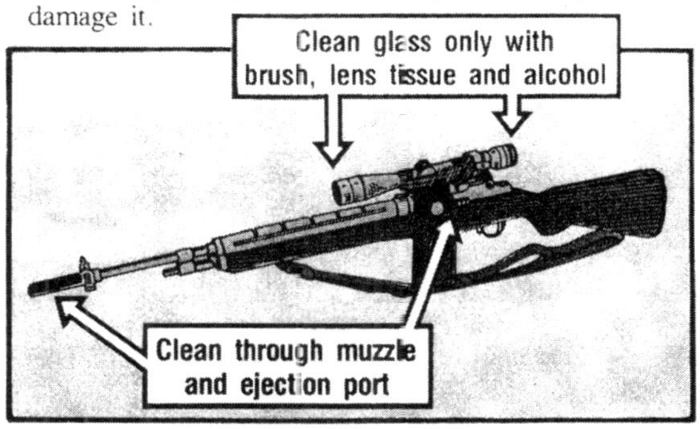

You can clean the M21 without taking it apart.

Run a chamber brush, NSN 1005-00-690-8441, dipped in CLP or Rifle Bore Cleaner (RBC) through the ejection port.

Push a CLP- or RBC-dipped bore brush, NSN 1005-00-556-4174, through the muzzle, taking care not to damage the crown of the barrel with the cleaning rod. That can ruin accuracy.

Dry the bore and chamber with a clean patch. Lubricate the bore with a patch dipped in CLP, LSA, or LAW.

Before firing again, dry the bore and chamber with a clean patch.

Clean the sight's glass only with a clean artist's brush, such as NSN 8020-00-597-4767, lens tissue, NSN 6640-00-597-6745, and alcohol. Cleaning cloths, polishing liquids or pastes, and abrasives will damage the sight.

If you spot problems, tell your armorer. He can send the M21 to the folks trained to give the M21 the special treatment it needs.

DEC 86

from the field. As a result the USAMTU had tested and developed an XM21 rifle fitted with a SIONICS sound suppressor. A small number of the weapons were shipped to the 9th Infantry Division operating in the Mekong Delta area of South Vietnam, who had been the leader in deploying snipers in country. Much of the early sniper activity in Vietnam by the Army was done on an informal basis. Marine Corps units on the other hand had well trained and organized Scout and Sniper Teams.

U.S. Army 9th Division Sniper School Vietnam. An instructor uses a spotting scope to observe the shooter's strikes on the target. (U.S. Army)

A sound suppressor was used for special operations. During the Vietnam War the SIONICS suppressor was used in combination with the XM21 Rifle. The sound suppressor replaced the flash suppressor on the XM21. The

Lubricate the metal parts very lightly. Too much lube can ruin the glass bedding between the metal and wooden parts. Put four drops of CLP, LSA, or LAW on a cleaning patch and rub it over the exterior.

Because the M21 must shoot accurately over very long distances, it has close tolerances and critical alignment. Most maintenance has to be done by specially trained repairmen. But there are some very important things you can do to help your M21 see and shoot straight.

Clean it. After every firing, run a chamber brush dipped in CLP or RBC through the ejection port.

One problem with pushing a cleaning rod through the muzzle is that it's easy to nick or burr the crown of the barrel. That ruins accuracy.

To prevent that, carefully insert the cleaning rod through the muzzle all the way to the chamber. Attach the cleaning rod tip through the chamber. Pull the cleaning rod straight back and out.

Dry the bore and chamber with a clean patch. Lightly lubricate the bore with a patch dipped in CLP, LSA, or LAW. Before firing, dry the chamber and bore again.

Clean the scope lenses only with an artist's brush, NSN 8020-00-597-4767, lens tissue, and alcohol. Anything abrasive—like cloth or polishing liquids—scratches the glass.

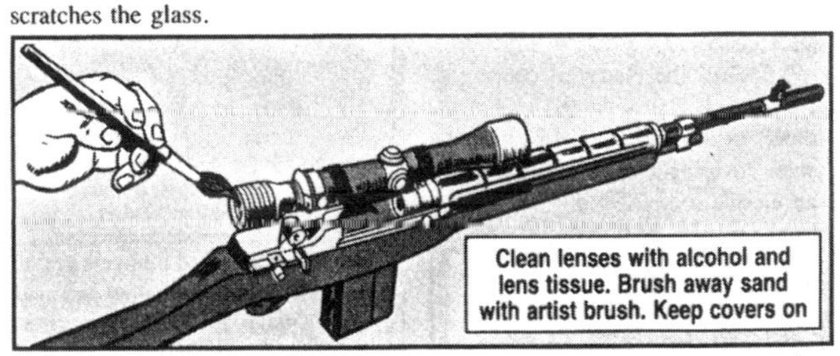

Protect the scope lenses from dirt and sand by keeping their covers on when you're not shooting.

Take the scope adjustment knob covers off only to zero the weapon. Moisture and dirt can get in and damage the adjustment knobs.

Use the artist's brush to clean away sand and dirt from the scope and lenses.

If you run into other problems, don't try to solve them yourself. You've done all you should. Your armorer can send your M21 to the folks who know what to do.

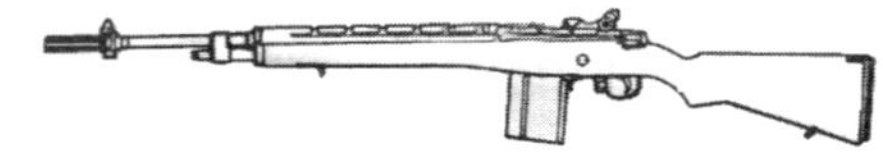

U.S. Army sniper trainees practice long-range shots from a sitting position with and without telescopes. Note that the shooter in the foreground is wearing ear protection. (U.S. Army)

SIONICS suppressor was secured by a nut and split bushing. Suppressing the sound of the sniper rifle made it difficult for the enemy to locate its exact position. Operators often turned the spindle valve to the grenade launching position so that the rifle would not make noise from the cycling of the action. During night operations a seasoned sniper would fire a round into the ground and then quickly tape the muzzle end of the suppressor. This would trap inert gas in the barrel, preventing a muzzle flash when the next round was fired. When fitted with the suppressor, AR TEL scope and 20 round magazine the XM21 weighted 12.5 pounds.

The XM21 was officially adopted by the U.S. Army in 1972 as its Standard sniper rifle, changing the weapon's designation to the M21. At the about the same time a new telescopic sight was introduced. The new ART II model was in its basic design, similar to the earlier AR TEL scope, except that used a two-point mount. One mounting point was the original hole drilled and tapped on the side of the receiver; the second attachment point was a hole

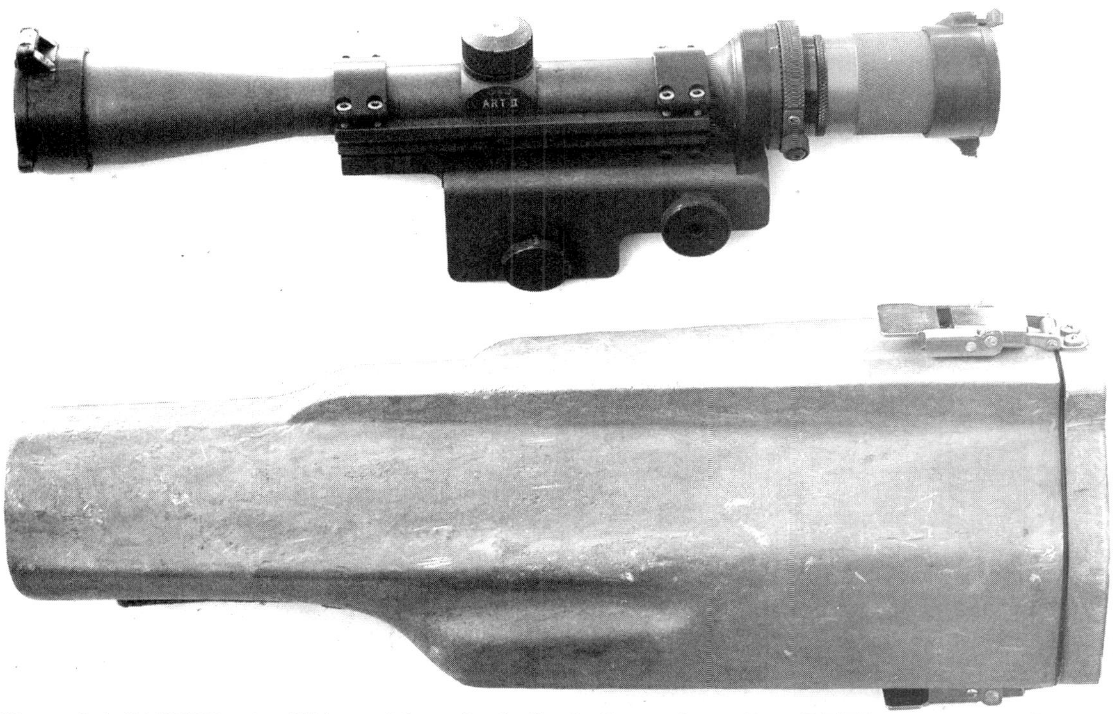

The updated ART II scope. This model was basically similar to the earlier AR TEL scope, except that used a two-point mount. One mounting point was the original hole drilled and tapped on the side of the receiver; the second attachment point was a hole that was drilled into the stripper clip guide. The ART II design proved rugged and accurate regardless of the elements to which it was exposed. (Courtesy of CWO4 James McCollum ALARNG and Scott Duff)

that was drilled into the stripper clip guide. The ART II design proved rugged and accurate regardless of the elements to which it was exposed.

The M25 Rifle

The XM25 and M25 rifles were conceived during the mid-1980s for issue to Special Forces Groups. The idea originated at the 10th Special Forces Group at Fort Devens, Massachusetts. An updated version of the Vietnam era M21 rifle was requested by Special Forces sniper teams. The M25 differed from the earlier M21 weapon in quite a few areas. The M25 weapon was an M14 rifle built to National Match specifications. The barreled action was carefully glass bedded in a com-

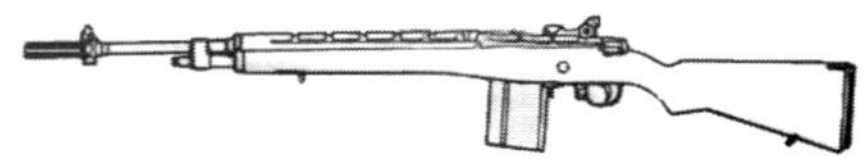

U.S. Army sniper with XM21 rifle.

mercial McMillan fiberglass stock. A special National Match spring guide and an advanced scope mounting system made by Brookfield Precision Tools were also incorporated. The M25 rifles fielded by the U.S. Army generally used the Leupold Ultra MK4 series of telescopic sights. The Navy SEALS also use the Leupold MK4 as well as the VariX-III LR M3. Sound suppressors could be fitted on the M25 for use on special operations. The M25 was also utilized as a weapon for the spotter of a sniper team when the primary weapon is the 7.62 mm M24 bolt action sniper rifle.

The Designated Marksman

A Designated Marksman is a concept that has been recently integrated into infantry units. A Designated Marksman is not as intensely trained as a sniper in field craft, stalking or camouflage. A sniper's mission is usually to achieve long-range fire from a concealed position; the primary role of the Designated Marksman is to provide accurate fire support for line troops. A D.M. is usually an individual that possesses extraordinary marksmanship skills and by using special weapons can be a force multiplier by engaging and dispatching targets at a greater distance then the average

NAVAL SPECIAL WARFARE
SW370-BG-OPI-010
OPERATOR'S MANUAL
FOR
SNIPER RIFLE
M14
NSN 1005-LL-L99-5690

June 1996

DISTRIBUTION STATEMENT D: Distribution authorized to DoD and DoD contractors only; due to Specific Authority, (Public Law 90-629-ARMS EXPORT CONTROL ACT AND DoD 5230.25 WITHHOLDING OF); 30 June 1996. Other requests for this document shall be referred to: Commander, Naval Surface Warfare Center (Code 4081), 300 Highway 361, Crane, IN 47522-5001.

infantry rifleman. During World War II a similar concept was used. The best marksman in a unit was often issued an M1C Garand or M1903A4 scoped rifle to handle special situations.

Designated marksmen are specially chosen for their position. To qualify as a Designated Marksman an individual must be in good physical condition, have a high rifle qualification score and have the potential to excel as a long-range marksman. One of the roles of the Designated Marksman is surveillance. One example is for the D.M. to observe and stop terrorists from planting explosive devices. Using specially prepared M14 rifles with optical sights allow the marksman the ability to neutralize enemy targets without collateral damage.

The old Vietnam-era M14 rifle once considered obsolete, has been recently resurrected as a Designated Marksman Rifle or D.M.R. The powerful 7.62 mm cartridge of the M14 allows the Designated Marksman the ability to engage targets out to a distance of 800

January 2004, Al Fallujah, Iraq. Covering his buddies during an early morning cordon and search raid, a sniper from B Company, 1st Battalion 505th Parachute Infantry Regiment, 82nd Airborne Division, uses the improvised armor plate in the back of a HummVee to steady his M25 rifle, a Vietnam-era 7.62mm M14 equipped with telescopic sight. US Army photo by Staff Sergeant Charles Johnson/Courtesy Robert Bruce Military Photo Features

meters. The insurgents, when realizing the killing zone for an infantry unit has been greatly extended, are far less likely to close with a unit. Thus preventing them from getting into a 400 to 500 meter range where they can effectively employ rocket propelled grenades or light machine guns. While the newer sniper weapons like the bolt-action M24 are more accurate for use in a sniper role, the semiautomatic M14 rifle allows the Designated Marksman to engage several targets in a short period of time. The official view of using the M14 rifle as a DMR is that the rifle is simply a stop-gap effort and not supportable due to lack of parts and qualified support personnel to maintain the weapons.

The Designated Marksman Rifle

As mentioned earlier, U.S. military forces are pulling the old warhorse M14 rifles from storage, modernizing them and sending them out into the field. The M14 rifles currently being issued were manufactured before most of the Designated Marksmen using the weapons were born. All of the M14 rifles destined for fielding as DMR weapons are being completely rebuilt, in most cases by civilian contractors. Special upgrades are being incorporated into the M14, which is an inherently accurate weapon to begin with. The original M14 stock was replaced by a premium McMillan M2A fiberglass stock, for stability a Harris bipod

7 Oct 2004, Gayan District, Afghanistan. Armed with the long-reaching and hard hitting M25 rifle, Specialist Ross Henderson of 2nd Battalion, 27th Infantry Regiment, 25th Infantry Division, watches out for enemy forces during a convoy halt. US Army photo by Specialist Jerry Combes/Courtesy Robert Bruce Military Photo Features

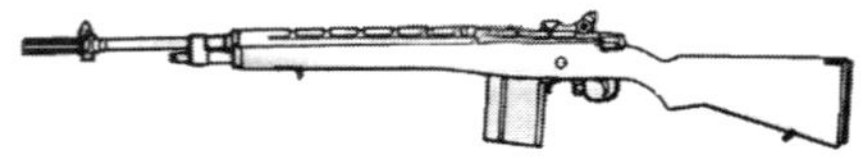

A shipping crate of modernized U.S. M14 rifles boxed up for shipment to Iraq and Afghanistan. The rifles have been completely rebuilt and upgraded by Smith Enterprises of Arizona under contract with the U.S. Government. The rifles feature a medium heavy chrome-moly steel barrel with a 1:10 twist 4-lands and grooves chambered for the U.S. M118 Long Range 7.62 mm ammunition; a Smith Enterprises gas system with special design gas plug, lock, front band and gas piston; Smith melonite heat-treated scope mount and rings; heavy duty M1913 Picatinny rail; Smith Enterprises Direct Connect Vortex flash eliminator, extended bolt stop and a 4.5- pound tuned trigger assembly. (Photo courtesy of Smith Enterprises)

was mounted on the stock's forearm. A Leupold MK4 10-power telescopic sight was mounted using a GG&G scope mount and rings. A match-grade barrel is fitted to the DMR weapon. Specially developed long range M118LR match ammunition was issued to extend the range of the DMR. M118LR 7.62 mm ammunition features a 175-grain match grade HPBT projectile with a velocity of 2,580 feet per second.

The M14 rifle in its sniper guise has been resurrected for duty in Iraq and Afghanistan. The long effective range of the 7.62 mm cartridge is well suited to the vast expanses encountered in those countries. (DOD photo)

Squad Designated Marksman (SDM)

The U.S. Army developed new marksmen techniques at their Fort Bragg sniper school. The course was designed teach selected soldiers to accurately engage targets beyond a 300 meter range. The Army's goal is to assign a precision marksman to each squad to eliminate threats not normally engaged by the infantryman in the field. The SDM training includes range estimation, target detection and engaging moving targets at over 300 meters. The Army uses the M16A4 rifle fitted with the advanced combat optical sight as well as the modernized EBR M14 rifles.

The latest M14 upgrade; the Enhanced Battle Rifle or EBR (Courtesy of Walt Kuleck Fulton Armory)

Chapter 8

The M14A1(E2) Rifle

The M14A1 Rifle

Anyone that has fired the U.S. M14 7.62 mm rifle in its standard configuration will attest that the weapon is very difficult, if not impossible, to control during the automatic mode of fire. The disappointing full-automatic performance of the M14 rifle was simply a matter of basic physics. The rifle was too light, the cyclic rate too high and lastly the 7.62 x 51 mm cartridge was simply too powerful for manageable automatic fire. The Army had insisted on using the T65E3 cartridges (later adopted as the 7.62 mm NATO round) in a lightweight shoulder fired weapon.

Although the U.S. M14 rifle possessed se-

Firing the 7.62 mm M14 rifle full-automatic in the A1 configuration.

lect-fire capabilities (ability to fire either semi-automatic or full automatic) most of the rifles were issued with a nonfunctional round knob or lock in place of the select-fire lever. The locks could be easily removed to restore select-fire capability to the rifles. This decision was generally left up to the unit commander, as a result the selector locks usually remained in place. By 1964 the Army had a Standard infantry weapon that filled few, if any, of the desired requirements originally specified.

A few early attempts at enhancing the weapon's automatic capabilities were the addition of the hinged "hook" buttplate, 1.75-pound M2 bipod and ventilated fiberglass top handguard. These additions were originally slated for the use on the M15 rifle, a heavy-barreled M14 variation that was dropped from consideration in December 1959. M14's in the aforementioned configuration were designated as the M14 (M) or Modified.

Infantry Board testing at Fort Benning, Georgia further confirmed that the M14 rifle as it existed, simply was not effective as an automatic weapon able to replace the BAR in the assault role (AR). Without a useful automatic

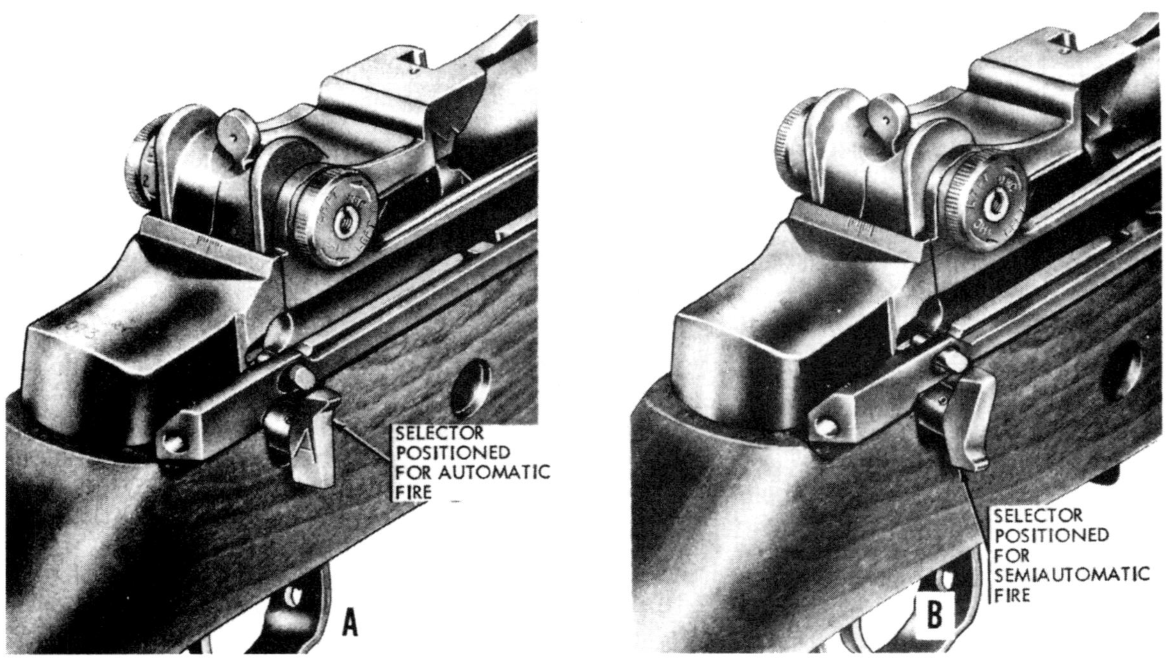

Positions of the mode of fire selector on the M14 rifle.

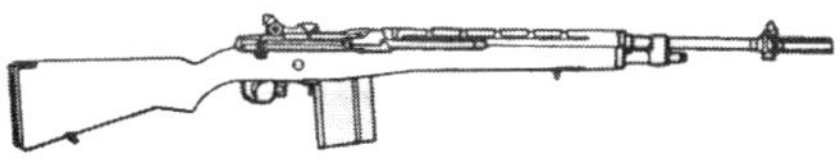

M14 (M) Modified Rifle featured a flip-up buttplate, M2 bipod and a ventilated plastic handguard. All features that originally appeared on the short-lived M15 rifle.

mode of fire the M14 was basically a product improved World War II M1 Garand with a twenty-round detachable magazine. In an attempt to exploit the M14 rifle's design capabilities fully, efforts were made at Fort Benning to modify the weapon to enhance its performance. In 1959 U.S. Army Captain Durward Dean Gosney was serving as a weapons technical advisor of the Infantry Board evaluating and testing the new M14 rifle. Captain Gosney gave some thought as to just how the M14 could be tamed. He envisioned a new stock that would place the centerline of the barrel in-line with the stock's butt. A plastic rear pistol grip, a folding vertical foregrip and a rubber recoil pad attached to the buttplate would complete the new stock. A rudimentary prototype M14 inline stock was built at Fort Benning based on Captain Gosney's idea.

Muzzle climb of the M14, because of the bullet's torque and right-hand twist would violently pull the weapon's barrel upward and to the right during automatic fire. In addition to the inline stock, a muzzle compensator or stabilizer was needed. An experimental muzzle brake was fashioned and installed in place of the flash suppressor. The M2 bipod was used

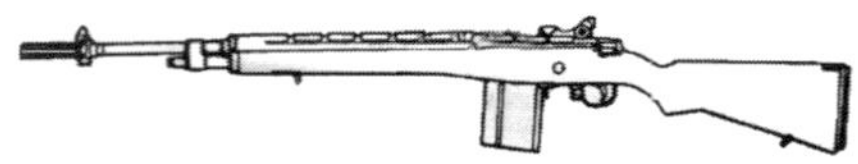

Early experimental M14 rifle tested at Fort Benning in an attempt to control the weapon during automatic fire. A slightly modified version was adopted as the M14A1 Rifle. (Springfield Armory Museum)

to aid in controlling the weapon's fire, and the added weight also helped in keeping the barrel from rising upward. Results of the initial test firing of the M14 with the new stock were encouraging. The Infantry Board shipped their prototype M14 rifle to the Springfield Armory in the spring of 1962 for further assessment.

The U.S. Ordnance Weapons Command, located the Rock Island Arsenal Illinois, was sent test reports of the Infantry Board's favorable test firing of the M14 rifle with the experiential stock and muzzle stabilizer. Based on the Infantry Board's recommendations, an order was issued to develop a production model of

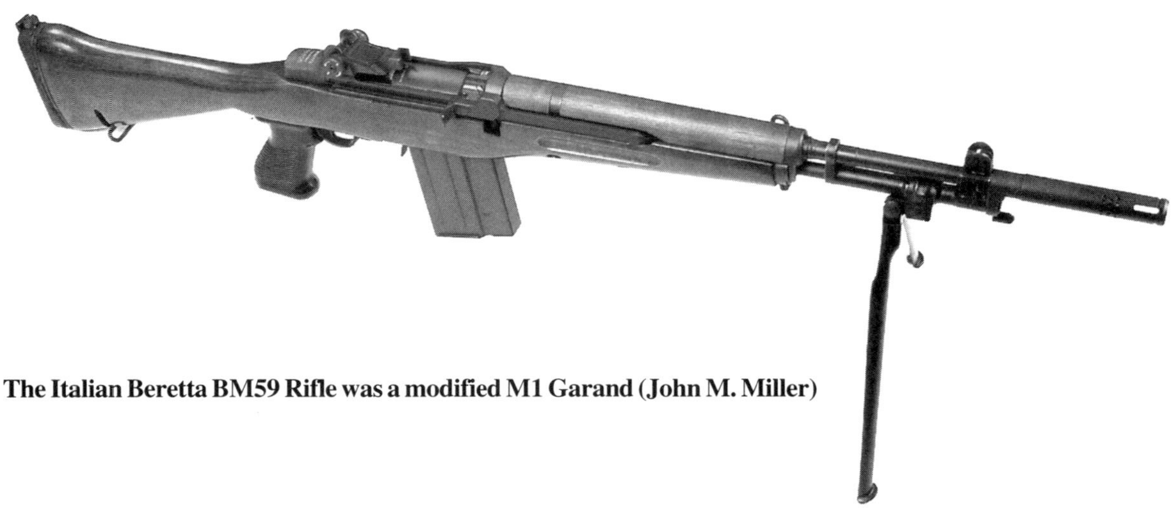

The Italian Beretta BM59 Rifle was a modified M1 Garand (John M. Miller)

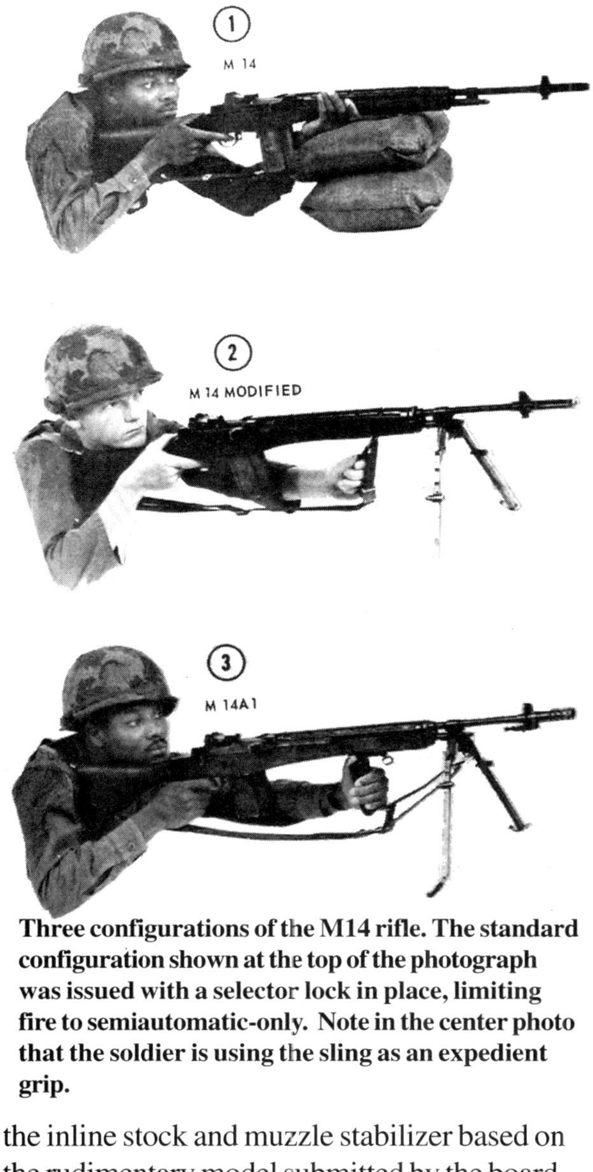

Three configurations of the M14 rifle. The standard configuration shown at the top of the photograph was issued with a selector lock in place, limiting fire to semiautomatic-only. Note in the center photo that the soldier is using the sling as an expedient grip.

the inline stock and muzzle stabilizer based on the rudimentary model submitted by the board. The project was turned over to the Government Springfield Armory in Massachusetts.

After receiving the prototype M14 from the Infantry Board, the Springfield Armory began their own evaluation of the weapon. The Armory concluded that the Infantry Board's prototype resembled the Italian Beretta Company's BM59, a rifle based on the U.S. M1 Garand. The Armory stated that each of the modifications of the Infantry Board's prototype M14 rifle would be thoroughly evaluated as to their percentage of contribution to the overall performance, and additional factors would be considered to result in a better overall design and ease of manufacture. After incorporating several improvements, five prototype M14 rifles were fabricated and assembled at the Armory in October of 1963. In November of 1963 the rifles were shipped to the Infantry Board for evaluation. Eight M14E2 configured rifles fitted with winter triggers were shipped to the Arctic Board in Alaska for testing.

After the initial testing was completed and reports compiled and studied, several improvements were suggested. One was to strengthen the muzzle stabilizer and the forward section of the stock in the area the foregrip was attached. The brief, but largely successful, trial period held by the Springfield Armory, Arctic Board, Aberdeen Test Center and the Infantry Board determined that the M14 in the E2 configuration was able to consistently meet the accuracy required of a squad automatic weapon. That requirement was defined as

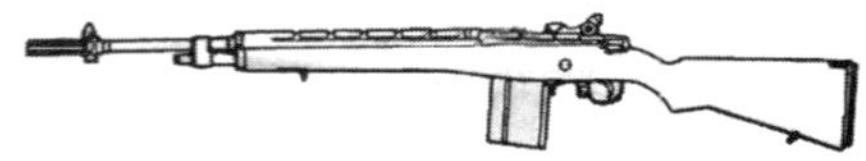

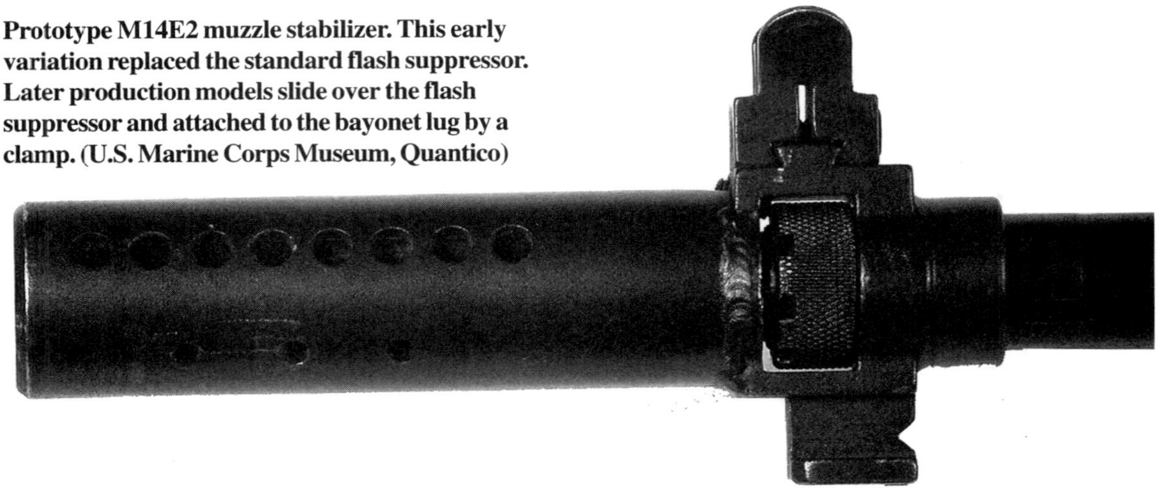

Prototype M14E2 muzzle stabilizer. This early variation replaced the standard flash suppressor. Later production models slide over the flash suppressor and attached to the bayonet lug by a clamp. (U.S. Marine Corps Museum, Quantico)

being able to place 80% of the hits to fall within a 40-inch circle at a distance of 200 meters, firing the weapon in short 2 to 3 round bursts.

After the suggested improvements had been implemented, the authority for the building of 8,350 standard M14 rifles into the E2 configuration was granted. The Springfield Armory was requested to fabricate the new E2 stock assemblies and to ship them to either Winchester or TRW to be assembled onto new M14 rifles. These early SA manufactured stocks were made of walnut. All of the other M14E2 conversion parts except for the wood and sling swivels were subcontracted out to commercial suppliers.

Assembly of the components onto the new stocks was to be performed at the Springfield

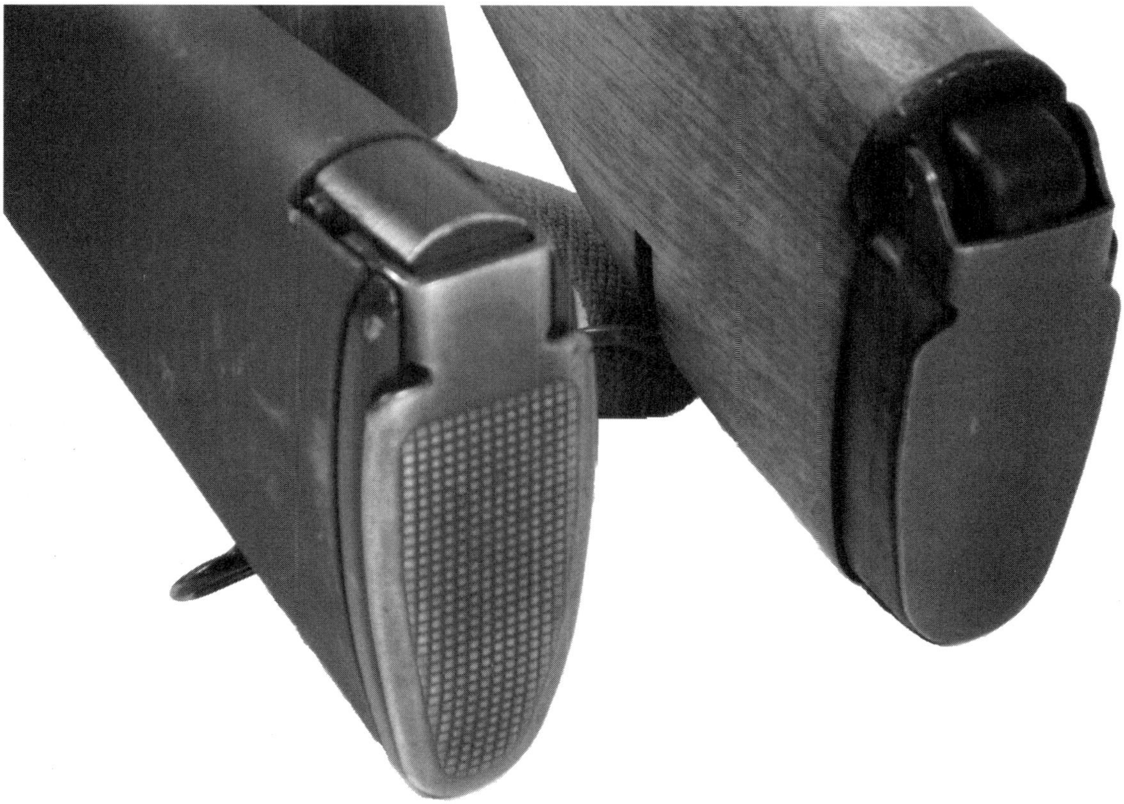

Standard M14 rifle buttplate (left) is compared to the production M14A1 butt (right). The M14A1 buttplate featured a rubber recoil pad, but lacked a trapdoor to contain the cleaning kit. A separate belt pouch was issued to the M14A1 automatic rifleman to carry the items normally stored in the M14's stock.

Armory and then delivered to the prime contractors in April of 1964, to be assembled onto new rifles. The late arrival of some of the parts necessitated a change in the original plan. By the time the E2 stocks were ready, production of the M14 rifle had been terminated. Fifty standard M14 rifles were obtained from Field Service stores, and shipped to the Springfield Armory for installation of the new stocks and other modifications. Personnel at Springfield were then instructed to function fire the rifles in the E2 configuration with the bipod attached.

An order was received in June 1964 to assemble and ship 224 M14E2 rifles to Southeast Asia. However there was a delay due to a problem obtaining the front folding grip handles. General Tire, the contractor, had only a single cavity mold for fabricating the foregrip and thus did not have the capacity to fill

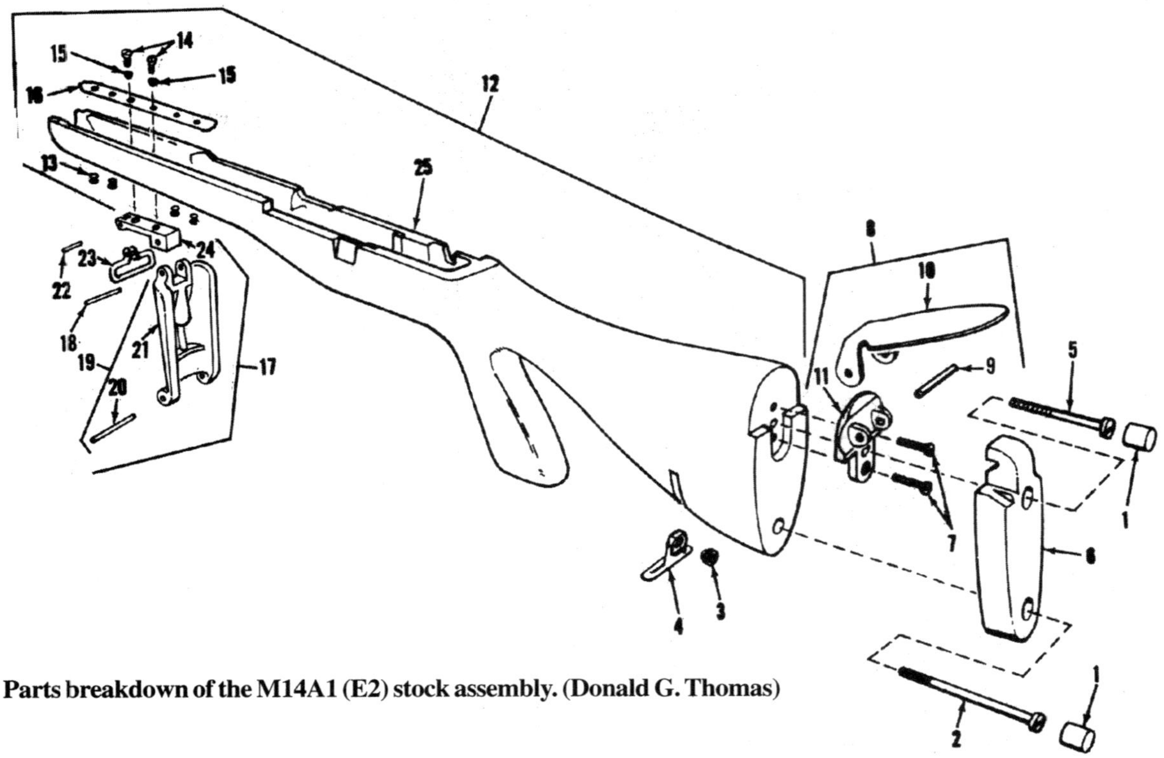

Parts breakdown of the M14A1 (E2) stock assembly. (Donald G. Thomas)

the order in time. The grips were delivered at the end of June and the rifles were completed and shipped in early July. By December 1964 all of the necessary parts and rifles had been obtained and the entire order consisting of 8,350 M14 rifles were converted into the E2 configuration.

After the M14E2 was being issued to the troops in substantial numbers, reports began coming in from the field indicated that several problems were being encountered. The muzzle stabilizers were coming loose during firing; the front handgrips were failing and the stocks were breaking in the area were the foregrip was attached. Additionally the M2 bipods were also coming loose and falling off of the rifles during firing.

A study was immediately launched to eliminate the problems identified with the M14E2 stock, stabilizer and bipod. Assigned to investigate and find a solution to the discrepancies was Springfield Armory project engineer P.W. Marshall. The results of his efforts were; a redesigned muzzle stabilizer assembly that

M2 Bipod on the M14A1 rifle.

The M14A1 buttstock lacked a storage compartment for a cleaning kit, requiring the operator to carry his cleaning equipment in a separate pouch. (Bill Ricca)

featured a more positive locking mechanism; improved foregrip design and new bipod locking jaws to provide a secure grip on the gas cylinder. In addition a new recoil pad made of synthetic rubber was now being installed on the buttstock. Cleaning agents and lube oils were attacking the earlier rubber butt-pads causing them to disintegrate.

A sling swivel was added to the M2 Bipod for use in conjunction with the sling attachment

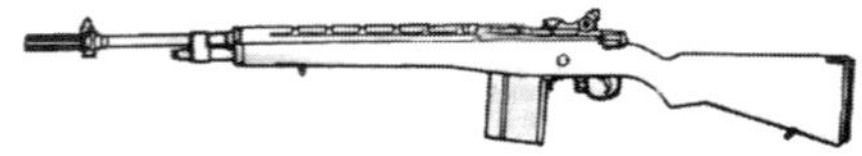

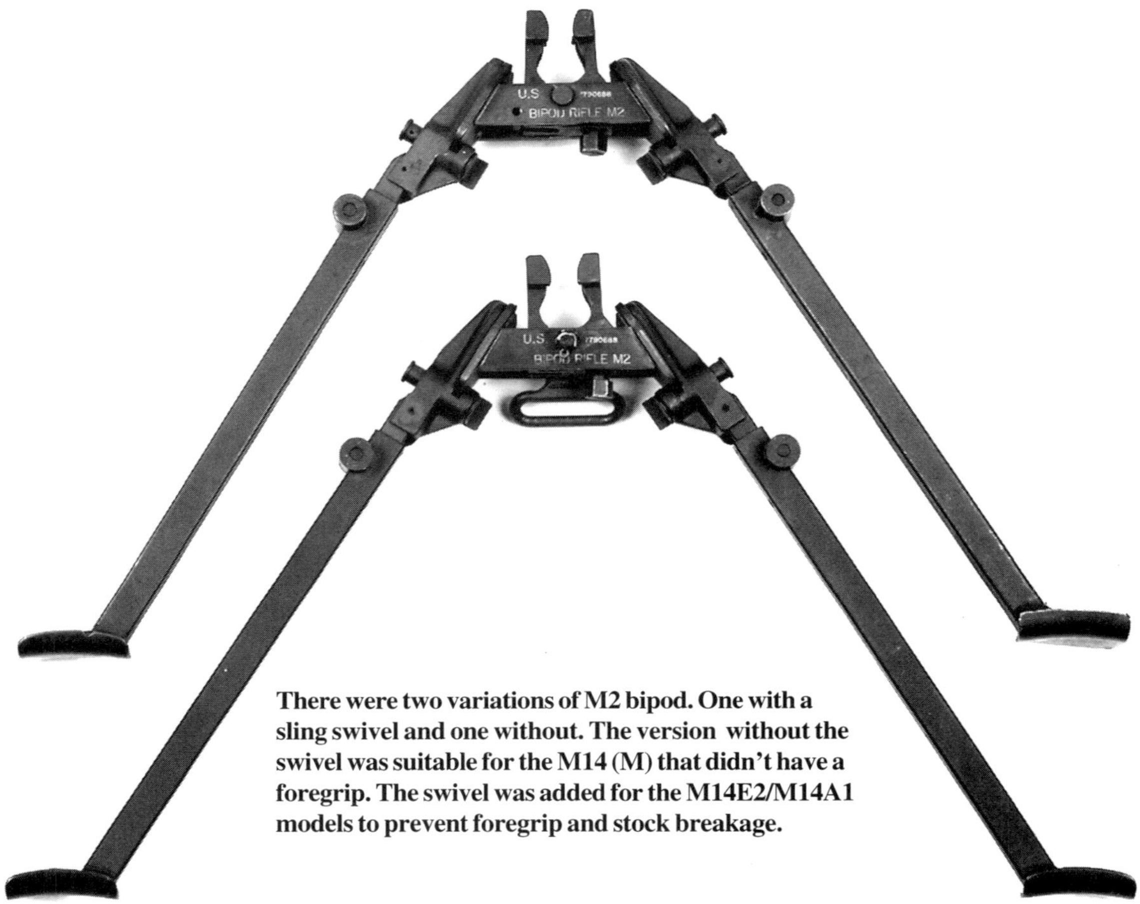

There were two variations of M2 bipod. One with a sling swivel and one without. The version without the swivel was suitable for the M14 (M) that didn't have a foregrip. The swivel was added for the M14E2/M14A1 models to prevent foregrip and stock breakage.

M14A1 rifle with bipod and foregrip in the folded position.

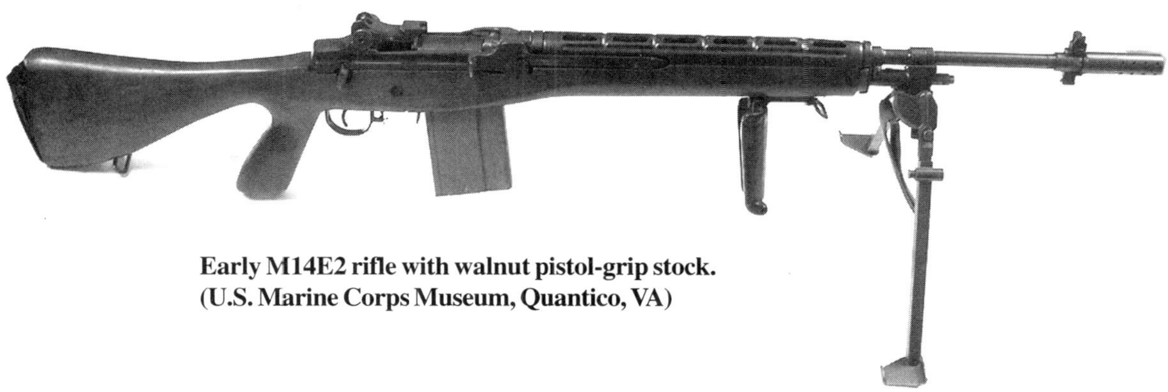

Early M14E2 rifle with walnut pistol-grip stock. (U.S. Marine Corps Museum, Quantico, VA)

pin located on the bottom of the foregrip assembly. When properly adjusted, the sling provided tension between the bipod and foregrip, relieving pressure from the stock forearm. The rearward pressure from operator's hand was absorbed by the bipod sling swivel, preventing the breaking of the grip or stock where the grip-mounting bracket was attached. A second advantage of the bipod sling swivel was that downward pressure was exerted on the bipod, resulting in a more controllable weapon. A longer 64.5 -inch sling, part number 11010038, (FSN-1005-072-5376) with a second attachment clip was issued for the M14E2 and M14A1 rifles with the bipod sling swivel.

The later production M14A1 buttstocks were all made from birch wood. The rear pistol grip

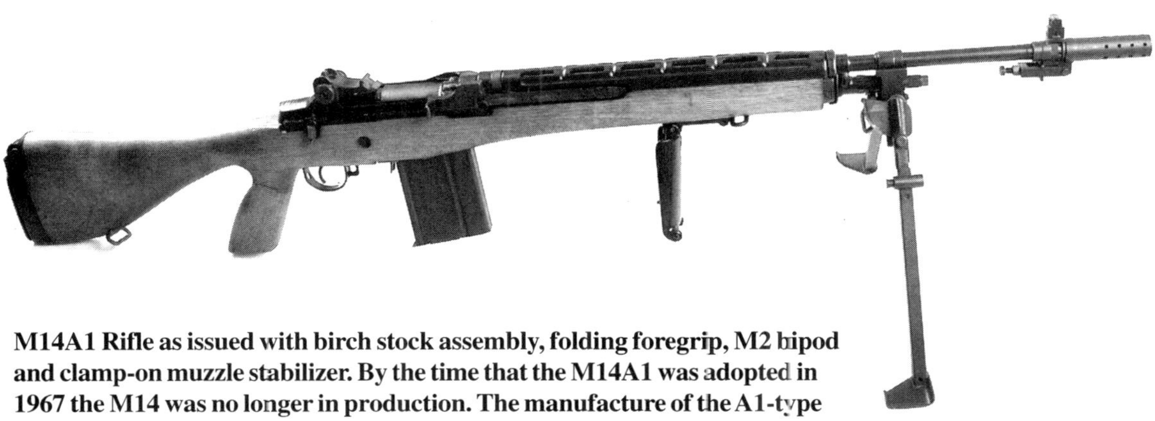

M14A1 Rifle as issued with birch stock assembly, folding foregrip, M2 bipod and clamp-on muzzle stabilizer. By the time that the M14A1 was adopted in 1967 the M14 was no longer in production. The manufacture of the A1-type stocks were contracted out to Canadian Arsenals Limited of Canada.

U.S. Army infantryman trains with the M14A1 rifle. The M14A1 was issued with the fire selector lever in place, allowing either semiautomatic or full-automatic operation. (U.S. Army)

M14E2 with walnut stock, M2 bipod and prototype stabilizer. (U.S. Marine Corps Museum, Quantico, VA)

was fabricated from a separate piece of wood that was doweled and glued to the stock. The front part of the stock under the handguard was designed with a greater cross-section for additional strength in the area where the front handgrip was attached. The folding front handgrip was made from hard rubber. A spring-loaded T shaped latch was provided to secure the grip in both the folded and extended positions. To accommodate a variety of shooters the grip assembly could be moved longitudinally over a range of four inches.

By the time of adoption of the M14A1 version, the M14 had been out of production for nearly four years. Having no U.S. contractor available to produce the inline A1 stocks, a contract was negotiated with Canadian Arsenals Limited of Canada. This company was formally known as the famous Long Branch Arsenal. All of the Canadian manufactured A1 style in-line stocks were made from birch wood. A number of the earlier A1 stocks made at the Springfield Armory were made of walnut. All of the stock's hardware, except for the sling swivels, was primarily made by commercial subcontractors.

The result of both the Infantry Board and the Springfield Armory's efforts were the modified M14 rifles being officially designated as U.S. Rifle M14E2, 7.62mm (the letter E stood for experimental status). During April of 1966, after extensive testing in the field, the M14E2

The production muzzle stabilizer for the M14A1 rifle fit over the flash-suppressor and clamped onto the bayonet lug.

RIGHT WAY IS BEST WAY

Sure, it'd be nice to have a loss-proof muzzle stabilizer on your M14A1 rifle. But till they come up with one, the best you can do is to make sure your present one won't work loose. This means putting the stabilizer on the right way.

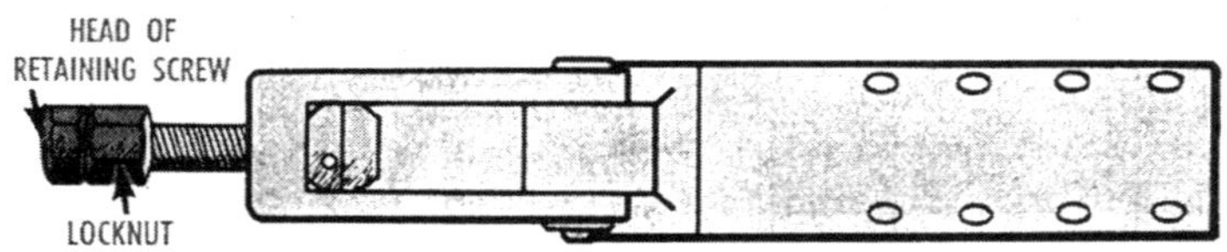

First . . . run the lock nut up against the head of the retaining screw.

Then put the stabilizer over the flash suppressor and slip the yoke assembly over the bayonet stud.

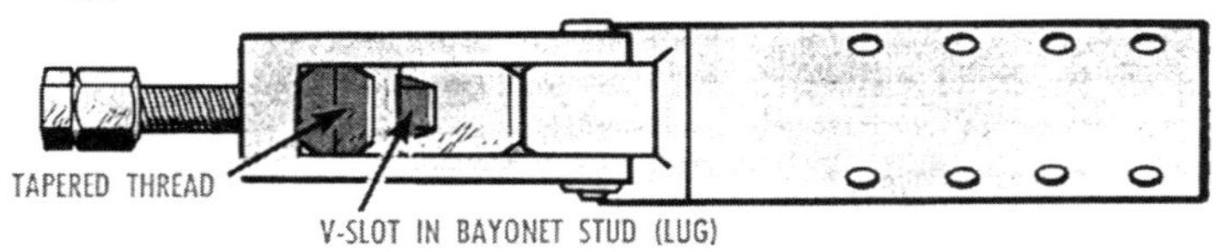

Next . . . use your fingers to turn the retaining screw so that the tapered stud at the end of it goes into the V-slot in the bayonet stud. Tighten the retaining screw with your combination tool.

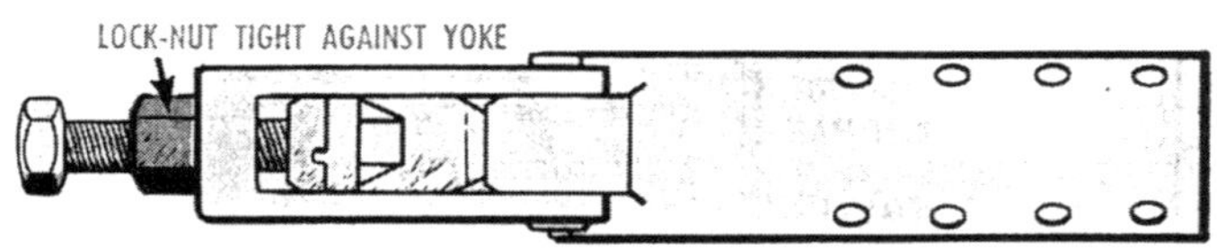

Push up and down on the yoke assembly to make sure it has no give. If it's not loose, you've done everything right so far.

Now you can tighten — real good — the lock nut against the yoke. This keeps the retaining screw from coming loose during firing.

There were numerous problems keeping the early type of muzzle stabilizers tight. This excerpt from a PS manual was issued to instruct soldiers in the field on how to correctly tighten the device. A Modification Work Order was eventually issued to modify the clamping mechanism. (John M. Miller)

Latch mechanism of the M14A1 folding foregrip. (U.S. Marine Corps Museum, Quantico, VA)

was formally adopted as the U.S. Rifle M14A1, 7.62mm (the A1 suffix stood for Alteration 1). The M14A1 weighed 14.5 pounds, ready to fight.

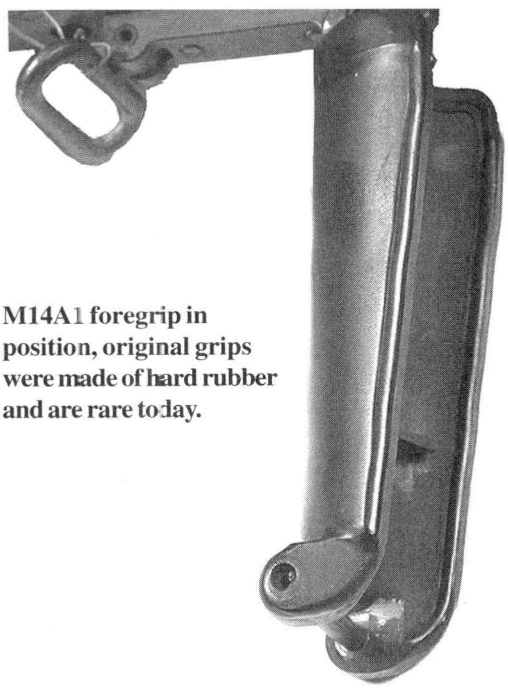

M14A1 foregrip in position, original grips were made of hard rubber and are rare today.

Historical Note: Durward Dean Gosney, who had original inspiration of what was to become the M14A1 rifle, never had the opportunity to see his idea for the inline M14 stock adopted by the Army. Captain Gosney was serving as a MACV Senior advisor assigned to the 5th ARVN Division, when he was killed after the UH-1 helicopter in which he was a passenger, was shot down in the Republic of Viet-

Maybe a few words on the "why" for where and how you should position the handgrip assembly on your M14A1 rifle will help prevent some of the busted stocks and hinges that've been haunting some outfits . . .

F'rinstance, when you do it the right way—with the second hook on the sling through the rung on the bottom of the handgrip . . . and the handgrip slightly forward in the unlocked position . . . and the sling dead tight from the front (bipod) swivel end to hold the handgrip this way—you've got it made.

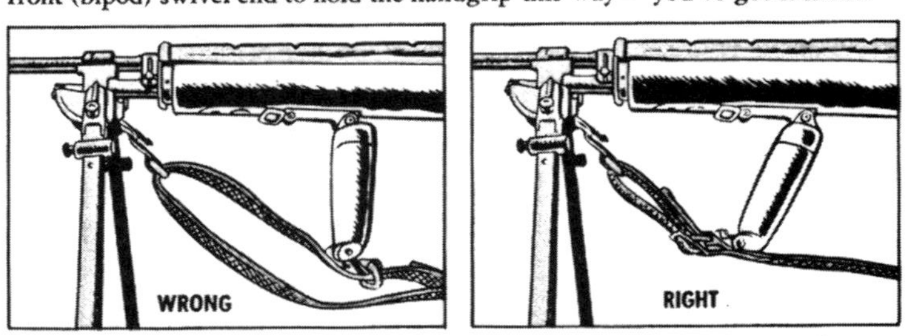

This way any blow at the handgrip will be deflected by the sling or the sling will absorb the shock through its webbing. Also this set-up will keep you from yanking back too hard and busting the hinge when firing. Besides, it'll give you better muzzle control during automatic firing.

On t'other hand, if you have the handgrip straight down and locked and leave the sling loose or allow metal-to-metal contact between the handgrip and the base, any blow to the handgrip will likely bust the stock above the handgrip.

So-oo-o!

nam on 7 October 1964. Captain Gosney was posthumously promoted to the rank of Major.

Official Description

The M14A1 rifle consists of a standard M14 receiver, barrel and trigger group modified as to meet accurate fire requirements in the role of the Infantry Squad Automatic Rifle. Enhancements included a straight-line design wooden buttstock with a rear pistol grip, and a folding forward handgrip that can be adjusted longitudinally to accommodate each individual rifleman. The rear of the stock is fitted with a hinged buttplate and rubber recoil pad. These modifications are intended to control vertical movement of the weapon. A clamp-on muzzle stabilizer, adjustable bipod and a BAR sling are also part of the M14A1 configuration.

U.S. troops take up a position with their M14 rifles. The weapon at the top of the photograph is in the standard configuration. The soldier at the bottom of the picture has the M14A1 version. (U.S. Army)

U.S. Rifle 7.62 mm M14

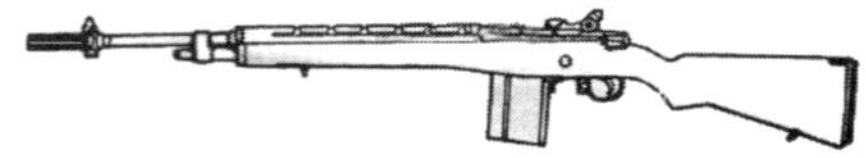

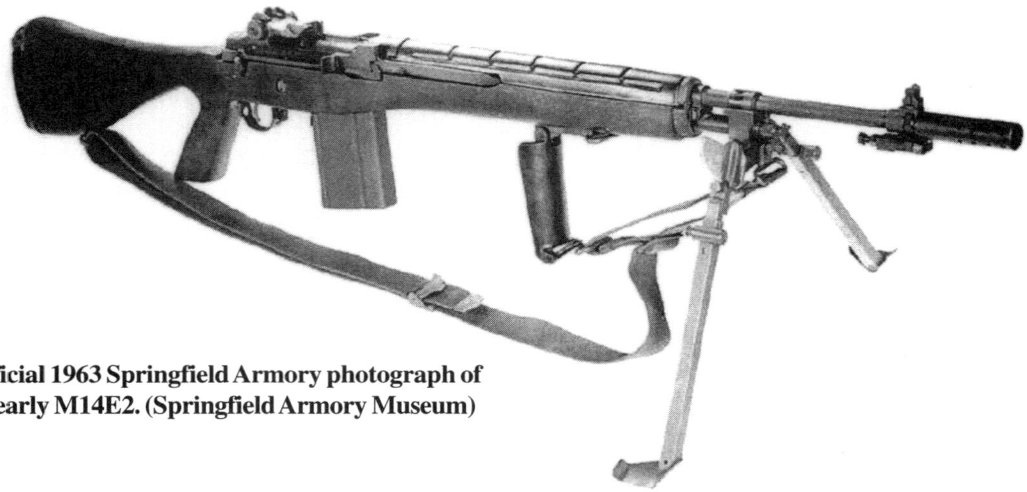

Official 1963 Springfield Armory photograph of an early M14E2. (Springfield Armory Museum)

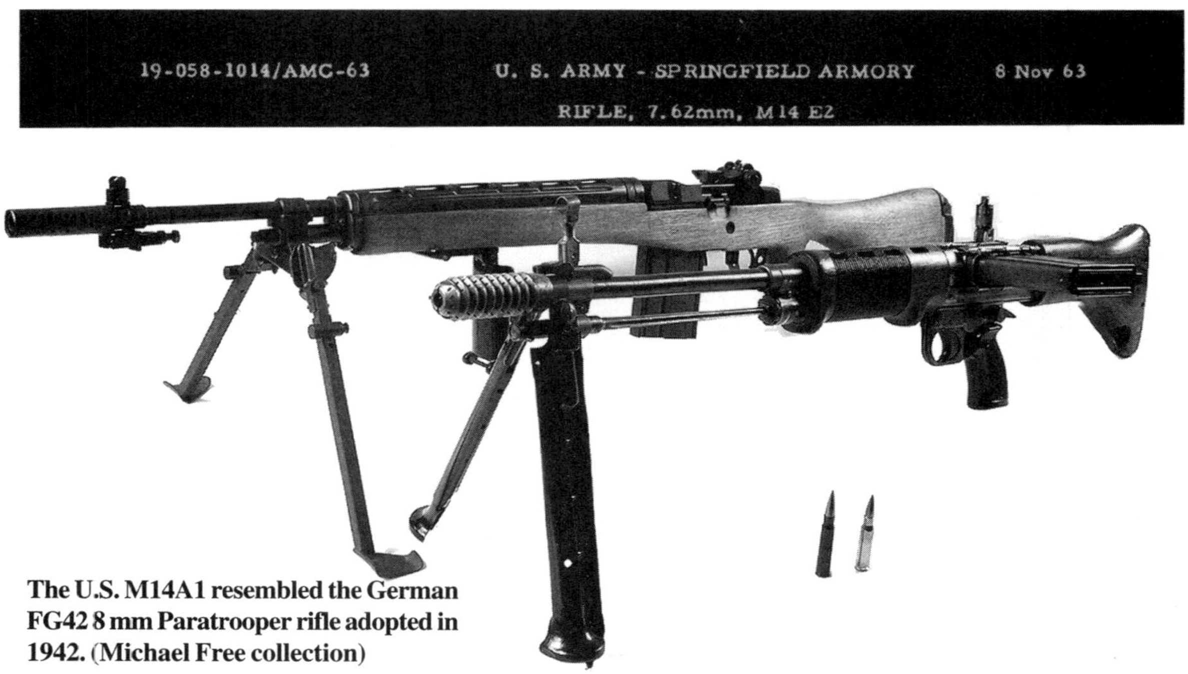

The U.S. M14A1 resembled the German FG42 8 mm Paratrooper rifle adopted in 1942. (Michael Free collection)

Chapter 9

The National Match M14 Rifle

A part of the adoption process of the M14 rifle was based on its ability to accurately engage targets at long ranges, thus enabling the rifle to uphold the rifle the marksman tradition of the U.S. military. In 1959 a requirement to develop a National Match version of the M14 rifle was established.

The National Match version of the M14 rifle differed from the standard service rifle. One of the first changes was to render the M14 to a semiautomatic-only configuration by welding the selector shaft and lock. Initially the roll-pin securing the lock would be welded over on both ends, and the lock-knob itself permanently welded onto the shaft. The barrel, one of the most important components, was machined to 1/2 the tolerances allowed in the standard issue barrel. The bores of match rifles were not chromium plated. The stock was glass bedded and individual components carefully chosen and fitted to the rifle. Specially calibrated rear sights were developed that had 1/2 minute of angle adjustments. The receivers of match rifles were marked NM to designate their match-grade construction. Minimum accuracy requirements were five-ten shot groups with 3.5-inch average extreme spread at a range of 100 yards.

A U.S. Marine Corps National Match M14 Rifle. (Courtesy of the U.S. Marine Corps Museum, Quantico)

Prior to the adoption of the U.S. M14 rifle, the M1 Garand rifle was the standard weapon for competing in National Match contests. At the 1962 National Matches held at Camp Perry, Ohio it was announced that the new M14 rifle would be introduced the

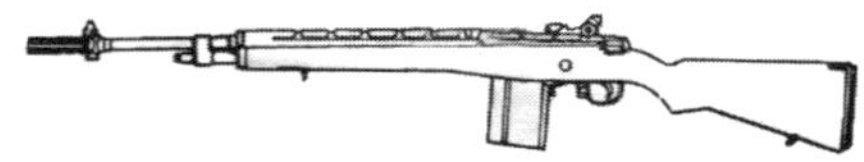

A Springfield Armory National Match receiver. Note the NM mark added after the M14 designation. (Courtesy of the U.S. Marine Corps Museum, Quantico)

After a NM M14 rifle had successfully passed all the acceptance requirements it was stamped with the Ordnance eagle acceptance mark (with white paint) on the back of the receiver heel.

following year at the 1963 Camp Perry Match for orientation purposes only. The following year the M14 was authorized to compete in the matches.

During the 1963 Matches it was announced that the new semiautomatic M14 M rifle would be utilized at the 1963 Small Arms Firing School to familiarize students with the basic rudiments of the M14 rifle. The M14 M rifle was modified by welding to disable the automatic firing capability of the weapon. The M14 M rifle was to be an item for club issue and eventually available through the Director of Civilian Marksmanship (DCM) for the civilian sales program, when produced in sufficient quantities. The National Match version would be accredited to state associations and through DCM sales at a later date.

The Springfield Armory received an order during Fiscal year 1963 to assemble 1,000 special match M14(M) rifles. The order specified that the rifles would have their selector shafts permanently welded in the semi-automatic position. Only standard grade barrels would be fitted and the stocks would not be glass bedded. Reportedly 1,009 special M14 M rifles were built.

No M14 (M) rifles were offered for civilian sales in 1964, but a National Match M1 Garand could be purchased for $148.00 each plus $4.50 packing and shipping. A standard grade M1 rifle could be purchased for $94.30 plus the $4.50 packing and shipping charge. The rifles were to be shipped from The Erie Army Depot, Port Clinton, Ohio. Only one rifle per NRA adult member was authorized.

No M14 (M) rifles were offered for civilian sales in 1964, but a National Match M1 Garand could be purchased for $148.00 each plus $4.50 packing and shipping. A standard grade M1 rifle could be purchased for $94.30 plus the $4.50 packing and shipping charge. (John M. Miller)

The Springfield Armory National Match M14 Rifle

The engineering program was initiated at the Springfield Armory in 1959 to develop and fabricate prototype M14 National Match rifles. The first National Match M14 rifles were tested during the early months of 1961. Upon receipt of the satisfactory test results, approximately 3,000 National Match rifles were fabricated at the Springfield Armory during 1962 based on the requirements developed and established during the Armory's National Match Engineering Program. During 1963 the Armory delivered 3,641 brand new NM M14 rifles. The M14 National Match Rifles were made available at the 1963 National Matches for shooter orientation and familiarization. In 1964 a total of 4,874 National

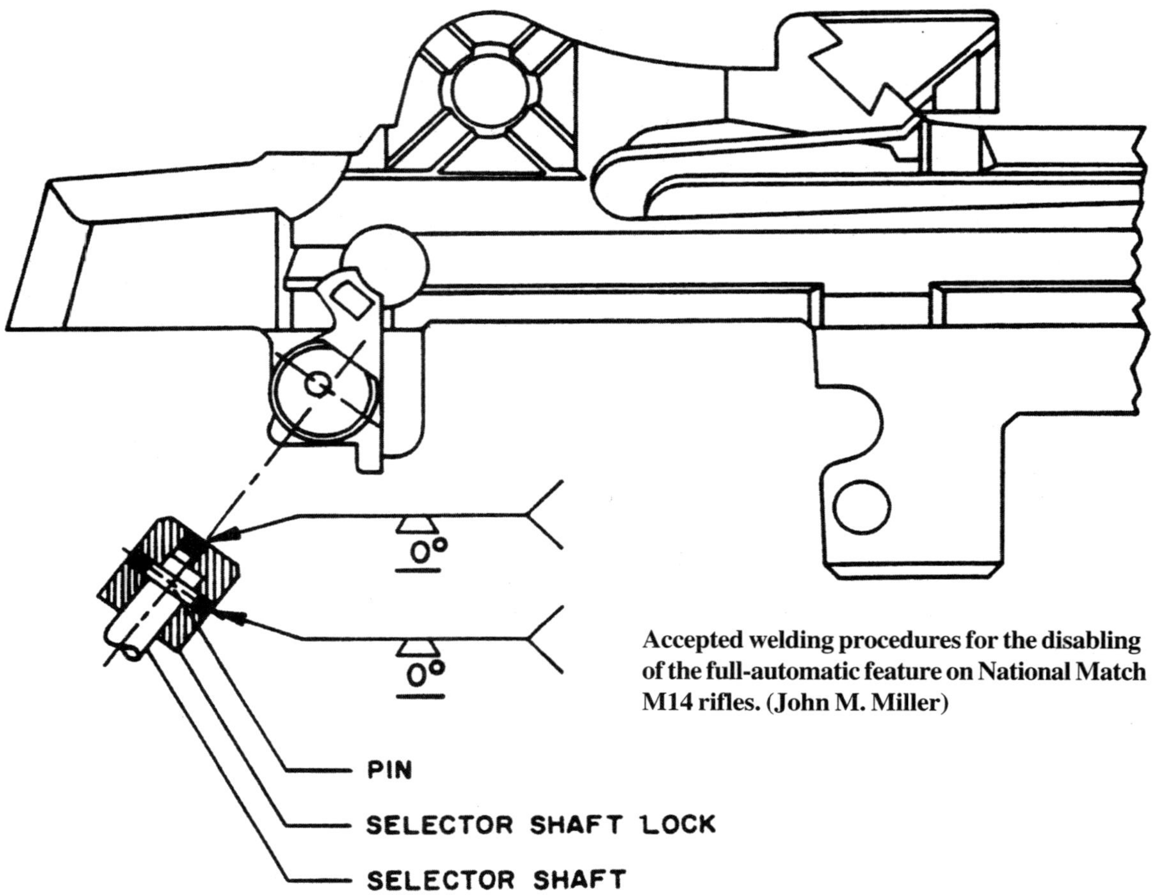

Accepted welding procedures for the disabling of the full-automatic feature on National Match M14 rifles. (John M. Miller)

Match Rifles were produced by Thompson Ramo Wooldridge, a commercial M14 rifle contractor, for the 1964 matches. This marked the first time that a commercial manufacturer produced match-grade U.S. military rifles. All of the first Springfield Armory and TRW National Match rifles were produced using brand new rifles. After production of the M14 ceased, the NM rifles were assembled from M14s that were returned to the Springfield Armory to undergo arsenal rebuilding.

The passing of the infamous Gun Control Act, implemented on 16 December 1968, determined that any firearm designed as a machine gun was in fact a machine gun irregardless of any attempts to render it to function in a semi-

TRW manufactured National Match M14 rifle serial number 1453711. During 1964 a total of 4,874 National Match Rifles were produced by Thompson Ramo Wooldridge, a commercial M14 rifle contractor, for the 1964 matches. This marked the first time that a commercial manufacturer produced match-grade U.S. military rifles. (Courtesy of Walt Kuleck and Scott Duff)

automatic-only configuration. This of course applied to the M14 rifle and thus the Gun Control Act ended any hope of future government sales of M14 rifles to civilian shooters.

There was one TRW M14 rifle, serial number 1453711 that was purchased by the late Mr. Robert Sauerman during the mid 1970s. The rifle was not registered with the BATF as a NFA firearm (machine gun). Mr. Sauerman advised the BATF that he was in possession of the M14 rifle and it was quickly seized as contraband by the authorities. After taking the matter to court it was determined that the TRW M14 rifle in question was not a machine gun, and the court ordered that the rifle be returned to Mr. Sauerman. The rifle is currently in the possession of a private collector. The court's decision unfortunately only applied to this particular M14 rifle.

National Match Barrels

Match barrels were produced in several configurations, to include standard and heavy-weight barrels, to enhance the accuracy of the M14 rifle. Close tolerance GI National Match barrels (standard weight configuration) for the M14 rifle were identified by their Ordnance drawing number 7791362, stamped on the right side of the breech end of the barrel. When the original 3,000 match rifles were assembled at the Springfield Armory a chromium-lined match barrel drawing number 7791173 was fitted, the tolerances and dimensions were that same as the unlined match barrel. The chro-

U.S. Marine Corps armorers assemble match-grade M14 rifles. (USMC)

mium-lined barrel was discontinued and replaced by the unlined version because of the added expense to chromium plate them, which contributed nothing to the accuracy of the barrel.

In addition to a drawing number present on the NM barrels was the manufacturer; date of manufacture and heat lot. To view the markings the operating rod was retracted and locked in a rearmost position. One exception were the barrels manufactured by Canadian Arsenals, these barrels were marked CA on top of the barrel under the hand guard. Government match-grade barrels were generally marked NM on the top between the gas cyl-

The NM markings on this barrel identify it as being manufactured to National Match specifications. Unfortunately such markings are occasionally added to standard-grade M14 barrels to pass them off as National Match grade. A genuine M14 NM barrel can be verified by the drawing number located on the side of the barrel. (U.S. Marine Corps Museum, Quantico)

inder and the flash suppressor. The original 1963 specifications stated that the bore diameter was specified at .300-inch + .001-inch, groove diameter .3075-inch + .001. Bore and groove dimensions were checked with an air gage that was able to pick up variations as little as .0001-inch. The maximum deviation was not allowed to exceed 2 minutes 23 seconds of angle in the total length of the barrel. Headspace for National Match barrels was

GI National Match barrels in the standard weight configuration for the M14 rifle can be identified by their Ordnance drawing number 7791362, stamped on the right side of the breech end of the barrel. This particular barrel was manufactured by Saco-Lowell (SAK). (Courtesy of Gene Barnett)

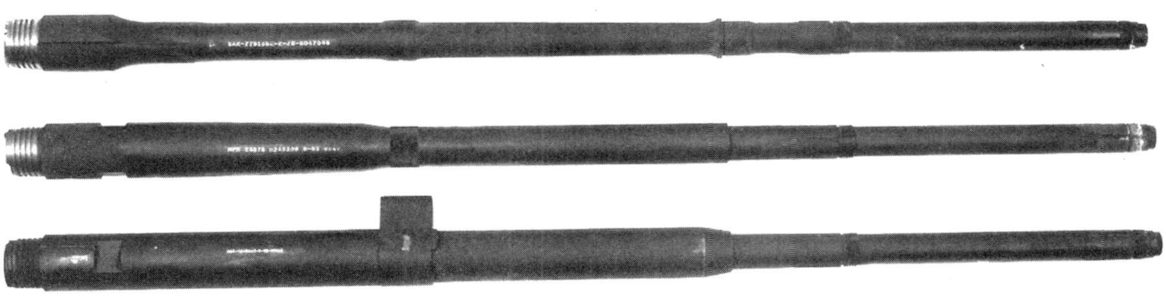

National Match M14 barrels. From the top; a standard weight NM barrel, a medium weight NM barrel and a rare heavy NM barrel. (Courtesy of Gene Barnett)

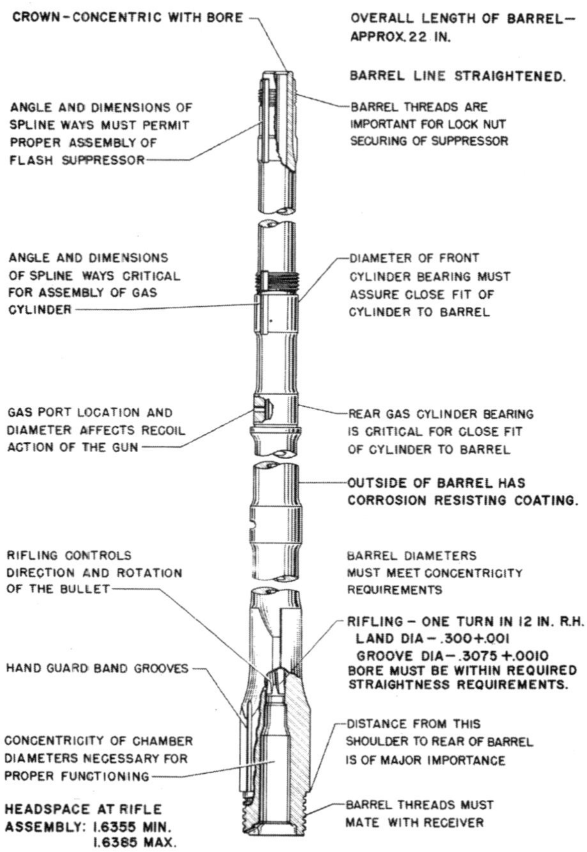

During 1975 the National Board for the Promotion of Rifle Practice authorized the use of a medium weight National Match barrel. Government contract "medium" weight configuration match barrels were marked 9345206. The manufacturer's CAGE number 26978 indicates that this barrel, dated 8-83, was made by Saco Defense. (Courtesy of Gene Barnett)

to fall within a range of 1.6355-inch to 1.6385-inch.

Companies that have manufactured National Match M14 barrels over the years have been; Canadian Arsenals; Nomura Machine; Saco-Lowell; SGW; TRW; Gene Barnett; The Springfield Armory; Hart and Kreiger. There were some problems with improper heat treat-

A very small number of "heavy" configuration National Match barrels were procured by the government during the 1990s. Reportedly less than 1,000 were made due in part to their high cost. The barrels were manufactured by the New Arc Welding and Steel Company (NOA) and can be identified by the drawing number 9349847. (Courtesy of Gene Barnett)

ACTION LOV
COORDINATE SM
RECEIVED
INFO

****
* U N C L A S S I F I E D *
****

'AGE 0 RUCIAFB UNCLAS DEC 15 9 31 AM '83
3T
JNCLAS
SUBJ M14 NATIONAL MATCH AND M21 SNIPER 7.62MM RIFLE BARRELS,
ISN 1005-00-018-3255
INVESTIGATION OF SUBJECT BARRELS MANUFACTURED BY SGW, INC.
NDICATE THE MAJORITY WERE NOT HEAT TREATED (HARDENED) AS REQUIRED

BY THE DRAWING. FAILURE TO HEAT TREAT REDUCES THE BARREL STRENGTHH
TO THE DEGREE THAT THE BARREL MAY RUPTURE DURING NORMAL FIRING AND
CREATE A POSSIBLE SAFETY HAZARD TO THE USER.
2. REQUEST URGENT ACTION BE TAKEN TO SCREEN ALL STOCKS OF REPLACE-
MENT BARRELS AND M14 NATIONAL MATCH AND M21 SNIPER RIFLES FOR THE
SGW PRODUCED BARRELS AND RESTRICT THEM FOR ISSUE OR USE IN RIFLES
THE SUSPECTED BARRELS ARE STAMPED WITH THE MANUFACTURER'S
IDENTIFICATION "SGW" LOCATED FORWARD OF THE RECEIVER AREA
APPROXIMATELY 3-1/2 INCHES. ON A SAMPLE OF RIFLES OBSERVED, THE
SGW MARKING ON THE BARREL COULD BE SEEN THROUGH THE STOCK
OPERATING ROD SLOT WITH THE OPERATING ROD IN THE REAR POSITION
IF NOT, THE MARKING CAN BE SEEN AFTER REMOVAL OF THE HANDGUARD
AND STOCK.
3. THESE BARRELS WERE DISTRIBUTED THROUGH NORMAL SUPPLY CHANNELS

PAGE 03 RUCIAFB6628 UNCLAS
AS FOLLOWS:
A. AS A BARREL INSTALLED ON A M14 NATIONAL MATCH RIFLE, NSN
1005-00-678-9828.
B. AS A BARREL INSTALLED ON A M21 SNIPER RIFLE, NSN 1005-00-
179-0300.
C. AS A REPLACEMENT BARREL FOR EITHER THE M14 NATIONAL MATCH OR
THE M21 SNIPER RIFLES, NSN 1005-00-018-3255.
4 RIFLES WHICH WERE REQUISITIONED AS END ITEM M14 NATIONAL MATCH
AND M21 SNIPER RIFLES SHOULD BE RETURNED AS ASSEMBLED RIFLES.
5 REQUEST END ITEM RIFLES WITH SGW BARRELS AND SGW REPLACEMENT
BARRELS BE RETURNED TO COMMANDER, ANNISTON ARMY DEPOT, ATTN:
SMCAN-DSP-WD, DARROL O'BRIEN, ANNISTON, AL 36201. ANNISTON ARMY
DEPOT IS TO BE PROVIDED BY MESSAGE PERTINENT SHIPPING INFORMATION
PRIOR TO SHIPMENT, WITH AN INFORMATION COPY TO THIS COMMAND.
REQUEST WIDEST DISSEMINATION OF THE ABOVE INFORMATION AS SOON
AS POSSIBLE TO INCLUDE LAO'S TO ASSURE ALL AMCCOM LAR AND SSTR'S
RECEIVE A COPY OF THIS MESSAGE.
BT
#6628

U.S. Army memo regarding problems encountered with SGW manufactured barrels. (Courtesy of John M. Miller)

A civilian Armscorp National Match M14 receiver fitted with a NM rear sight. (Courtesy of Drew Paskovitch)

ing of SGW barrels manufactured during 1982.

After experiencing several barrel failures the Army put out an urgent bulletin dated 15 December 1983, that directed all facilities to perform an inventory check for the SGW match-grade barrels. An inspection of all M14 National Match rifles and M21 sniper rifles for the presence of SGW manufactured barrels was also ordered. All M14 and M21 rifles with SGW barrels and individual SGW barrels from depot stores were to be returned to the Anniston Army Depot attention of the Commander. The barrels were eventually returned to the contractor SGW, who tested the barrels. Those found to be acceptable were marked OLY.

During 1975 the National Board for the Pro-

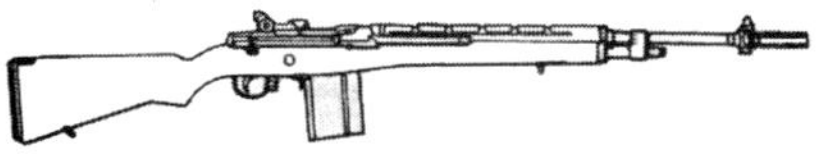

motion of Rifle Practice authorized the use of a medium weight National Match barrel. Through use of the heavier barrel, improvements in accuracy up to 30-percent were possible. The new barrel weighted 1.5 pounds more than a standard issue barrel. The medium-weight barrel was approved for use in all NBPRP regional and National matches. The barrel was manufactured much like a standard

A very small number of "heavy" configuration National Match barrels were procured by the government during the 1990s. Reportedly less than 1,000 were made due in part to their high cost. The barrels were manufactured by the New Arc Welding and Steel Company and can be identified by the drawing number 9349847.

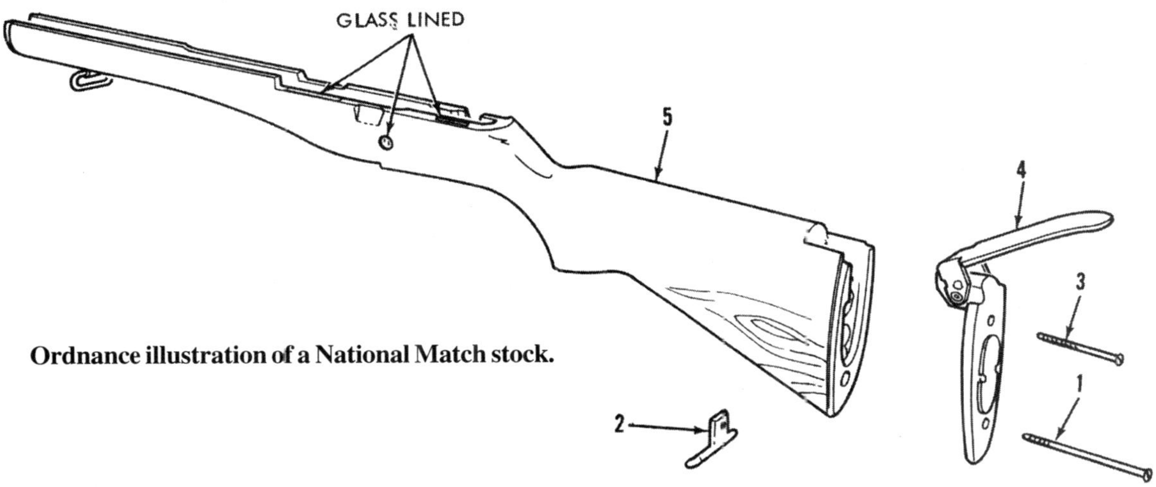

Ordnance illustration of a National Match stock.

match barrel, except there was less exterior machining. The heavy barrel provided increased stability. The barrel retained the use of the standard M14 gas system and parts. The U.S. Government generally only contracted standard or medium weight match-grade M14 barrels. Government contract "medium" weight configuration match barrels were marked 9345206.

Over the years the demand for match grade M14 barrels has been increasing. Many barrels have been made for the civilian market by a number of individuals and commercial companies. The demand continues and M14 barrels are still being produced today in substantial numbers.

A U.S. Marine Corps built National Match M14 rifle and scoring record book. (U.S. Marine Corps Museum)

1963 National Match M14 Rifle Specifications

The original 1963 Nation Match requirements specified that the gas cylinders were selectively hand fitted to the barrel to insure that no rotation was possible, the gas cylinder lock was also hand selected and fitted to insure a secure fit.

The trigger pull of NM M14 rifles was specified to fall between 4.5 and 6 pounds, but as close to 5 pounds as possible. No trigger "creep" was allowed. The trigger pull was adjusted by skilled armorers who would carefully hone the trigger hooks. After tuning the trigger group the housing was engraved with the last five digits of the host rifle's receiver serial number.

Fitting of the operating rod was found to be less critical in the M14 versus the M1 rifle. The rod was to be free of the barrel through its entire length. The freedom of movement between the bolt and operating rod were checked by first removing the operating spring and guide. The muzzle of the rifle would be raised to a 60-degree angle. At this position the bolt should open under its own weight. Turning the rifle's muzzle down 60-degrees should result in the bolt falling and locking under its own weight. The operating rod was not allowed to touch the stock at any point.

The National Match front sight post was thinner than that found on the standard M14 service rifle with the post being 0.065 - .005 inch wide. The thinner sight post allowed a better sight picture and covered less area of the target. The top of the sight blade was machined at a 5-degree angle tapering to the front. The purpose of this configuration was to reduce reflection from the sun. The letters NM and .062 were marked on the right side of the sight base.

The NM rear sight consisted of a special National Match base, windage knob and aperture assembly. The National Match base and windage knob had 5/16x64 threads in place of the standard 5/16x32 threads found on the standard sight. The finer threads reduced the lateral movement from 1 minute of angle to 1/2 minute of angle for each click. The slot dimension in the base was also reduced to provide a tighter fit at the 600 meter setting. The aperture assembly consisted of a rack adjustable in 1/2 minute steps,compared to the 1-minute steps of the standard service rifle. The aperture hole in the sight was drilled .002 inch off of center, providing .004 inch vertical movement of the hole when the hooded eyepiece is rotated 180 degrees. Two aperture diameters were available. Standard was .0595 inch while the optional one was .0520 inch.

National Match stocks were carefully chosen from selected hardwood blanks. The grain was required to be straight and parallel to the longitudinal axis of the stock. Prior to the bedding of the stock the serial number of the M14's receiver was marked on the stock. The receiver contact points of the stock were then glass bedded.

The receiver was bedded squarely in the stock and the barrel was maintained centered to .006 inch to the forearm portion of the stock. To achieve optimum accuracy potential the barreled receiver must be so fitted in the stock with no perceptible lateral or longitudinal movement was possible. During the curing process of the fiberglass, the trigger guard was not locked. This insured a tight lockup of the trigger group after curing.

After assembly the M14 National Match rifle was tested using a high-pressure proof round. After proof-firing, the headspace was rechecked to insure it was still within the acceptable standards. The rifle would then be function tested by firing twenty rounds. The rifle was then placed in a machine rest for accuracy acceptance testing. Two rounds were fired for seating and were followed by three 10-shot strings at a range of 100 yards. During targeting the rifle remained in the rest and was loaded through the top of the receiver via the cartridge clip guide.

After the NM M14 rifle had successfully passed all the acceptance requirements it was stamped with the Ordnance eagle acceptance mark (white paint) on the back of the receiver heel. If the rifle failed any of the tests it would be returned to the shop to be rechecked by the armorers who would make the necessary adjustments. If the rifle failed a second time it would be taken apart and rebuilt.

Since the first National Match rifles were built in 1963, much progress has been made through trial and error, and the sophistication of modern equipment that reduces tolerances to specifications that were unimaginable in 1963. Ever increasing interest in match shooting has introduced new shooters, and companies with diversified talents and ideas, further enabling progress in the accuracy of match-grade rifles

National Match Ammunition

The M118 National Match cartridge was originally developed at the Frankford Arsenal for use in the NM M14 rifle. The ball projectiles used were a 173 grain, 7.62 mm boat-tail design using an uncannelured gilded metal jacket. This was exactly the same bullet used for the .30 caliber M72 National Match cartridges for the M1 Garand rifle. During manufacture of the projectiles rigid requirements in dimensional tolerances of both tooling and components were necessary to insure consistently accurate ammunition.

Early 1964 testing of the 7.62 mm propellants indicated that WC 846 was best suited for the cartridge. Subsequent testing indicated that by close control over the internal volume of the cartridge case, IMR 4895 was suitable to produce the required velocity level of 2,550 feet per second with an average chamber pressure of 50,000 psi. Accuracy of the M118 NM ammunition was 3.5 inch mean radius at a range of 600 yards.

Chapter 10
Magazines

During a break at the firing range a group of GIs reload a table-full of M14 magazines. (U.S. Army)

During the study of military weapons, the detachable box magazine's importance as a vital component is often overlooked. Each individual magazine produced must fit and function perfectly in any one of hundreds of thousands, and in many cases millions, of weapons (of the same design). Additionally each individual part of a magazine must fully interchange with other like magazines often manufactured by many different contractors. The inside and outside dimensions of each magazine as well as the follower must be held to close manufacturing tolerances. Perhaps the most important element of the magazine is the follower spring. The spring tension must be precisely calibrated to provide the exact amount of lift to feed the weapon on both semi-automatic and full-automatic fire, and must be durable enough to retain its strength through countless loading and cycling through a weapon. A magazine must be capable of all the aforementioned after being subjected to the worst conditions imaginable, including heat, humidity, extreme cold, dampness, mud, sand and even exposure to corrosive sea water.

One of the most noticeable changes between the M1 Garand Rifle and its direct descendant the M14, is the latter's twenty-round box type magazine that replaced the Garand's eight-round en-bloc clip. There were several disadvantages to the Garand's system. First

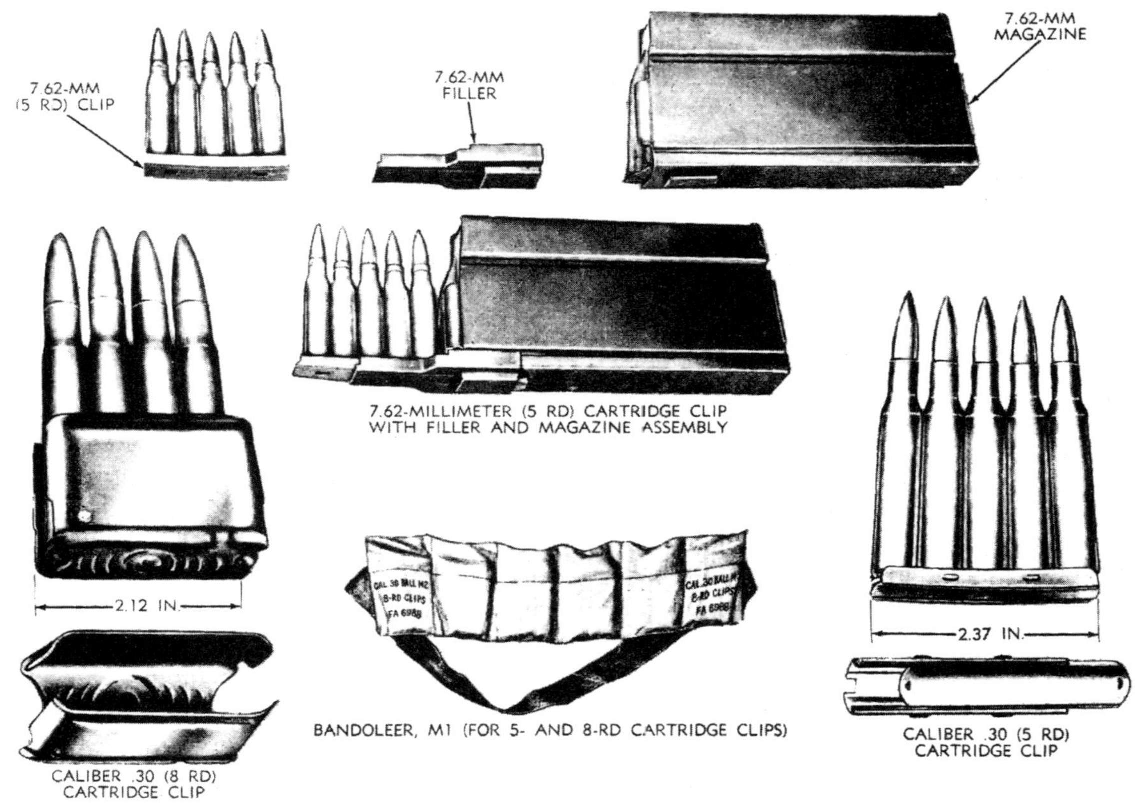

This Ordnance Corps drawing illustrates the primary differences between the M14's 20-round magazine system and the 8-round clip of its predecessor; the M1 rifle. (Donald G. Thomas)

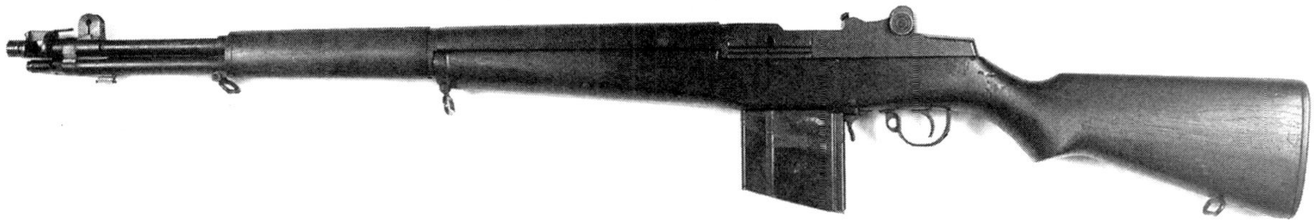

The early T20 series rifles used the 20-round box magazine from the BAR. Many problems were encountered requiring a completely new magazine to be designed. (U.S. Marine Corps Museum)

and foremost the clip was limited to holding only eight cartridges. Secondly the clip could not be topped off after firing a few rounds. The clip had to be emptied upon which it automatically ejected from the weapon, making a distinct sound, the third disadvantage.

When the prototype select-fire M1 Garand, the T20, was being developed, the obvious choice for a magazine was the existing twenty-round box magazine used for the U.S. Browning Automatic Rifle. However, it was soon discovered that the BAR magazine would not function as hoped. One problem was found to be that the increased velocity of the T20 rifle's bolt was deforming the magazine's feed lips. A second problem was that the follower spring of the BAR magazine lacked sufficient tension for the lighter, fast cycling bolt of the T20. The bolt would fail to strip fresh rounds from the magazine resulting in jams. This presented the problem of developing a new magazine. Having one magazine interchange with two weapons would be preferred, but it was not possible. The BAR magazine could be easily altered to function in the T20, but then it would not work in the BAR and the original BAR magazine would not work in the T20, even though either magazine would fit in both weapons. Having two identical magazines that were not interchangeable could prove disastrous on the battlefield. A new magazine designed specifically for the new select-fire rifle was needed. An entirely new magazine was designed for the next prototype rifle, the T-20E1. The new magazine would not fit into a BAR.

Many of the T20 and T44 series of experimental rifles used magazines that differed slightly from those eventually used in the M14 rifle. During the search for a suitable lightweight rifle, famous Springfield Armory designer John Garand had conceived the .30 caliber T31 rifle. The T31 was an odd bullpup design that had remained uncompleted at the time of Mr. Garand's retirement in 1953. Due in part to lack of funding the T31 rifle project was abandoned. Because of the low stripping force of the T31 rifle's bolt, a new magazine had to be

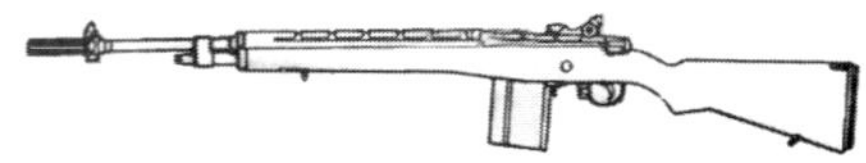

The magazine designed for T31 had two separate sets of lips formed on each side of the magazine. The rearmost lip was designed to retain the rounds in the magazine, while the forward lips were used to guide the round into the rifle's chamber. The front lips proved to cause an excess amount of friction.

The original floor plate design of the T44 rifle magazine had retaining lips that protruded from the side. (John M. Miller)

designed. This new magazine was also conceived by John Garand. After new Ordnance Department drawings were completed, two-hundred magazine tubes, bases and followers were fabricated by the Bay State Tool and Machine Company in late 1951 for the T31. The magazine designed by Garand was very reliable and later adopted, in a modified form, for use with the early T44 rifle.

After retiring from the Springfield Armory John Garand worked as a consultant to the Mathewson Tool Company, who fabricated the early T44 rifles. During his association with Mathewson he was able to further refine his original T31 magazine design. Originally the magazine designed for T31 had two separate sets of lips formed on each side of the magazine. The rearmost lip was designed to retain the rounds in the magazine, while the forward lips were used to guide the round into the rifle's chamber. The front lips proved to cause an excess amount of friction, requiring a lot of force from the forward moving bolt, to strip a round and chamber it into the rifle's breech. Garand modified the magazine into a single-lip design (the lip was lengthened by 1/32 of an inch) that required much less force to strip cartridges from the magazine, relying on the receiver to guide the cartridges into the chamber. A new redesigned spring with a larger .060 inch diameter wire was used. New magazine tubes were procured from the Bay State Tool and Machine Company that were lengthened by 3/8 of an inch. It was this design that was eventually chosen for the M14 rifle.

Aluminum Magazines

During most of the design phase of the M14 rifle, engineers were always searching for ways to lighten the weapon without compromising functioning or reliably. The magazine was no exception, and a number of magazines made from aluminum were designed and tested. An early 1959 dated M14 Field Manual even listed the M14 rifle's weight with an aluminum magazine.

Because of the pending, imminent standardization of an aluminum magazine for the M14, the Springfield Armory began contract negotiations with the Mathewson Tool Company

A prototype aluminum M14 magazine. One of the major problems that arose with the use of aluminum for fabricating magazines was the distortion of the aluminum that occurred during the heat-treating process. The problem was resolved by implementing a sequence of operations in which the forming of the aluminum magazine body was formed immediately after heat-treating and before age hardening of the metal. (U.S. Marine Corps Museum)

Manufacturing problems continued to be encountered with the aluminum magazine (left) concept, resulting in all production M14 magazines being fabricated from steel. The steel magazines (right) were finished with a gray-black phosphate finish, and most were stamped with the contractor's code or logo on the rear of the magazine body, approximately .75 of an inch from the base of the magazine.

in 1960. The contract was for research and development studies for large-scale production of aluminum twenty-round box magazine for future incorporation with the M14 rifle.

One of the major problems that arose with the use of aluminum for fabricating magazines was the distortion of the aluminum that occurred during the heat-treating process. The problem was resolved by implementing a sequence of operations in which the forming of the aluminum magazine body was formed immediately after heat-treating and before age hardening of the metal.

The dies being used to form the aluminum magazines were altered during July of 1960 to conform to the most recent design changes and drawings. Two hundred magazines were then fabricated under contract DA-19-020-504-ORD-5115, and shipped to the Armory for firing test and a metallurgical evaluation. After the magazines were received it was discovered that minute cracks had developed along the two longitudinal bends at the rear of the magazine. Further investigation revealed that the cracks were being caused by the sharpness of the bends. The contract with Mathewson Tool was extended to allow for further tooling alterations required to eliminate the stress-cracking problem. Two hundred additional magazines were ordered and scheduled to be forwarded to the Armory upon completion. During March of 1962 the final shipment of 200 magazines fabricated from aluminum were received at the Armory. A detailed report from the contractor with the recommended fabrication procedures was included.

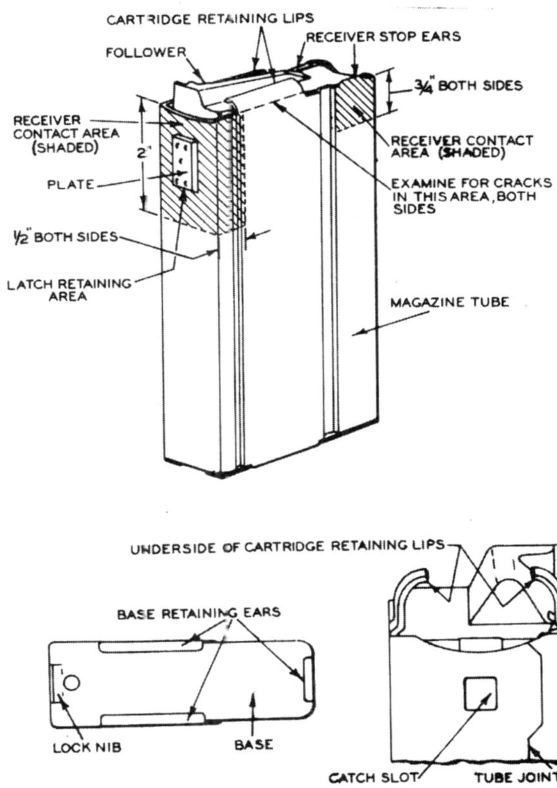

Component parts of the 20-round M14 steel box magazine. (U.S. Army)

Some of the magazines underwent extensive metallurgical evaluation, and a large portion of the magazines were gaged for their conformity to the latest Ordnance drawings. A number of the aluminum magazines were scheduled for firing tests and further evaluation. The final results from the ongoing testing were expected to be completed by the end of July 1962. A dimensional check of 50 magazines revealed that discrepancies existed. In most cases it was determined that the discrepancies could be eliminated when more permanent forming dies were developed. In the firing test it was found that the aluminum magazines functioned as well as those fabricated from steel. The acceptance ratio of the aluminum magazines compared favorably with that of the steel magazines. The aluminum M14 magazine was 4.3 ounces lighter than the standard steel version. The only disadvantages to the aluminum magazines were that the lips were easily deformed during hard field use,

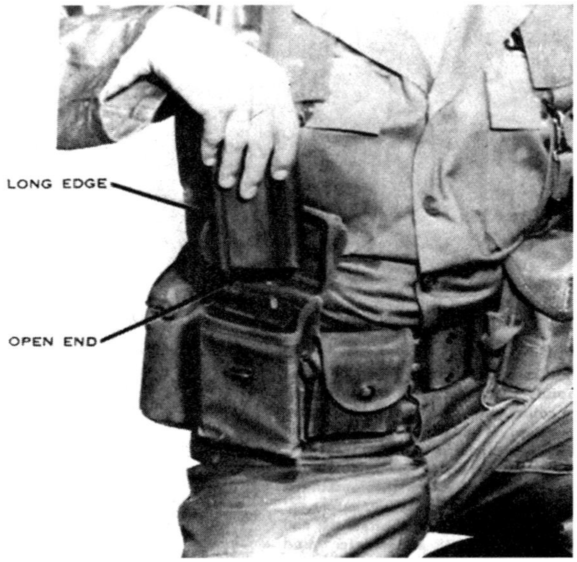

Troops were trained to place magazines into their carrying pouches with the cartridge end down to avoid contamination from dirt and debris.

Early 1960 dated magazine wrapper. The magazine contained inside this particular packaging was made by Olin-Mathieson the parent company of Winchester, a primary contractor of the M14 rifle. (Courtesy of Bill Ricca)

and the magazine body was easily dented, often jamming the follower in the magazine tube. Aluminum magazines observed during the research of this book had no manufacturer's makings on them. The magazine's had a bronze anodized finish. Other than the color and weight they appeared identical to a production M14 steel magazine.

There were ongoing problems with the manufacture of the magazine tube (body). The main problem was keeping the magazine tubes within the specified manufacturing tolerances. During 1962 the Springfield Armory entered into a Bailment Agreement with the Remington Arms Company Inc, to experiment on the feasibility of explosive forming of the M14 magazine tube. Blanks and dies were shipped to Remington to assist them in their development efforts. The agreement was eventually terminated after it was determined that Remington's background experience could not be applied to the M14 magazine. All funds allocated for the study were exhausted. The Springfield Armory working with the Research and Development of the Frankford Arsenal implemented a program to further study explosive forming procedures on steel and aluminum magazines and rifle barrels. The Frankford Arsenal was assisted by the Westinghouse Corporation.

Manufacturing problems continued to be encountered with the aluminum magazine concept, resulting in all production M14 magazines being fabricated from steel. The M14 magazines were the proven reliable double-stack, double-feed design with a capacity of twenty-rounds. The magazines were finished with a dark gray-black phosphate finish, and most were stamped with the contractor's code or logo on the rear of the magazine body, approximately .75 of an inch from the base of the magazine.

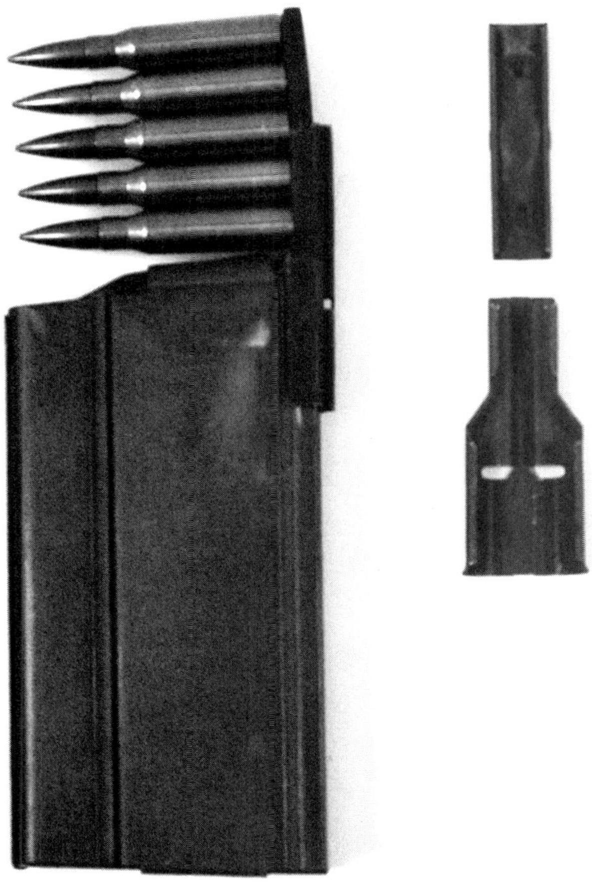

M14 magazine shown with its 5-round stripper-clip and loading guide. The 5-round clips were used for NATO compliance.

Shortly after the M14 was put into service another magazine related problem surfaced. The problem was that the bolt would fail to pick up a round from the magazine and close on an empty chamber, or would pick up a round and jam. The problem was studied by firing a rifle while using a high-speed camera to record the action. By studying the film footage it was determined that the cartridge in the magazine was not being raised upward quickly enough. The bolt was closing before the magazine could raise a round up for the forward-moving bolt to strip and chamber. The problem was addressed by reshaping the magazine spring to place more pressure on the back end of the follower. Although the alteration to the magazine spring helped somewhat, it failed to completely solve the problem. Further study of the malfunction determined that the recoiling bolt could have been bouncing off of the back of the receiver (then moving forward) and then striking the operating rod that was still moving rearward slowing down the action. The problem was eventually solved by slightly reducing the length of the operating-rod. It is believed that the problem was directly related to the use of ball powder as an alternate propellant to load cartridges, thus increasing the bolt's speed.

Virtually all of the early M14 steel magazines initially obtained from outside contractors were rejected. A number of the magazines had to be reworked at the Armory to function in the test rifles. Some of the inspection and repair work on the magazines was subcontracted out to the Rochester Manufacturing Company.

Prime contractors of M14 twenty-round magazines were:

AV (stylized logo) Atwood Vacuum Machine Company, Rockford, Illinois

Typical manufacturer's logo and ID markings located on the back of U.S. GI M14 contract magazines. From left; Stylized AV of Atwood Vacuum Machine Co.; B-R-W B2, Borg Warner; BRW S-1, Borg Warner Spring Division.

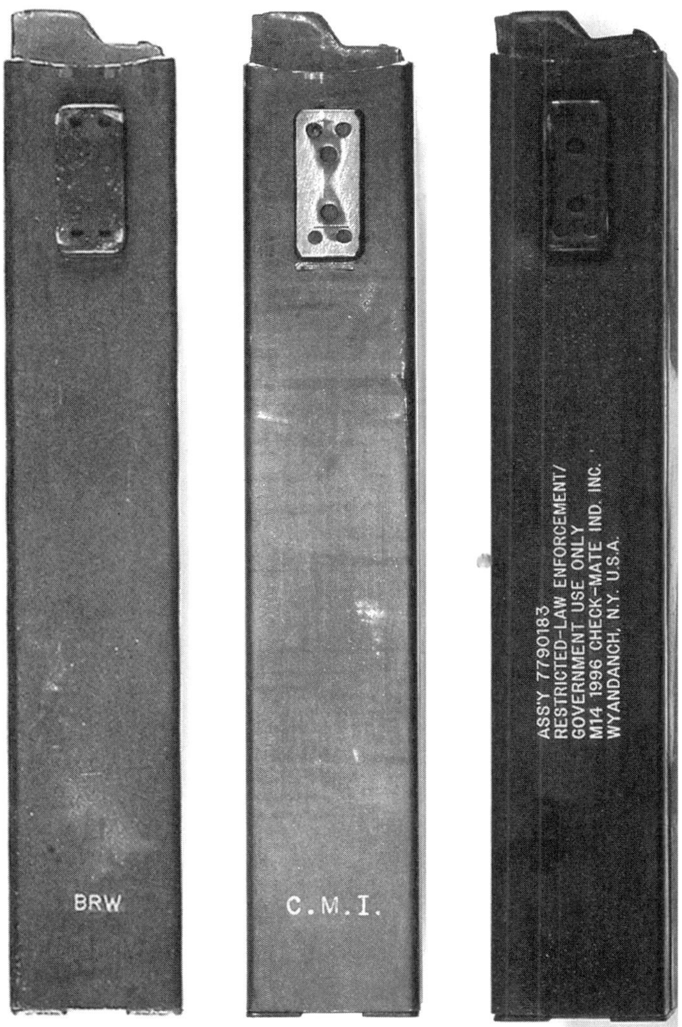

U.S. GI magazines from left; BRW, Borg Warner; C.M.I. Check-Mate Industries; Check-Mate Industries, restricted magazine marked for Law Enforcement or Government use only. (Made after the 1994 magazine ban)

BRW Borg Warner

BRW S-1 Borg Warner's Spring Division

BRW B2 Borg Warner (stylized logo)

C.M.I. Check Mate Industries, LE only marked and dated 1996

HR Harrington and Richardson

H-R Harrington and Richardson (unknown subcontractor)

HR-R Harrington and Richardson (unknown subcontractor)

KMT (stylized logo) Kileen Machine and Tool Company

OM Olin-Mathieson

R Rochester Manufacturing Company, Rochester, New York

SA The Government Springfield Armory

TRW Thompson Ramo Woolridge

UHC (stylized logo) Union Hardware Company

W Winchester Arms

WRA Winchester Repeating arms

Unmarked Borg Warner, Lot BRW 442 1/63; Lot BR4 (BRW) 4/80

Note: only the prime contractor is listed as a magazine manufacturer, although a subcontractor may have been engaged to manufacture the magazines for the prime contractor, as were many component parts for the M14 rifle. Harrington and Richardson marked magazines for example have three different markings. At least two or perhaps all three were believed to be made by subcontractors.

All magazines manufactured under contract for the U.S. government were generally marked with alpha or alpha-numerical characters or logo to identify the contractor. The marking were needed to identify the manufacturer in the event that a problem would occur with any particular contractor's magazines. If a problem did arise then only magazines produced by the one manufacturer in question would be condemned, rather than the entire supply. Whether or not original GI magazines were all marked or not has generated some controversy among collectors. However there have been some Borg Warner manufactured magazines

From left; Variations of Harrington & Richardson contract markings, HR and H-R there was also a third variation (not shown) marked HR-R; Stylized KMT, Kileen Machine and Tool Company.

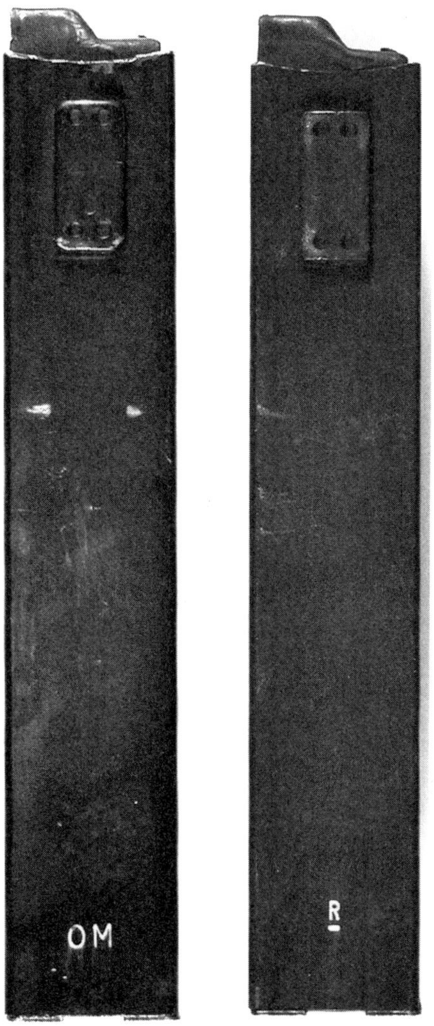

From left; OM for Olin-Mathieson (Winchester); R, Rochester Manufacturing Company.

that were removed from the sealed original packages, upon close examination the magazine bodies had no obvious manufacturer identification markings on them. The magazines were closely examined for the presence of markings using a commercial jeweler's loupe.

Often the contractor marks were lightly stamped and difficult to see. Rubbing oil on the area were the markings should be located will usually make the marks visible. Remanufactured magazines that have undergone Ordnance rebuild programs were usually sandblasted and refinished. This procedure often obliterated the contractor ID marks.

Magazine Packaging

There were several different configurations of packing used for M14 magazines. One of the most

SA, The government Springfield Armory.

common variations was four magazines packed together. Also seen are two magazines packed together and individually wrapped magazines. The outside wrappings included the nomenclature of the magazine, quantity, the part number, date wrapped and the wrapper manufacturer's name. (Often incorrectly thought to be the manufacturer of the magazine itself). The inside of the wrapper was a silver colored vapor barrier. Most magazines were packed dry relying on the vapor corrosion inhibitor (VCI) wrapper to protect them from the elements.

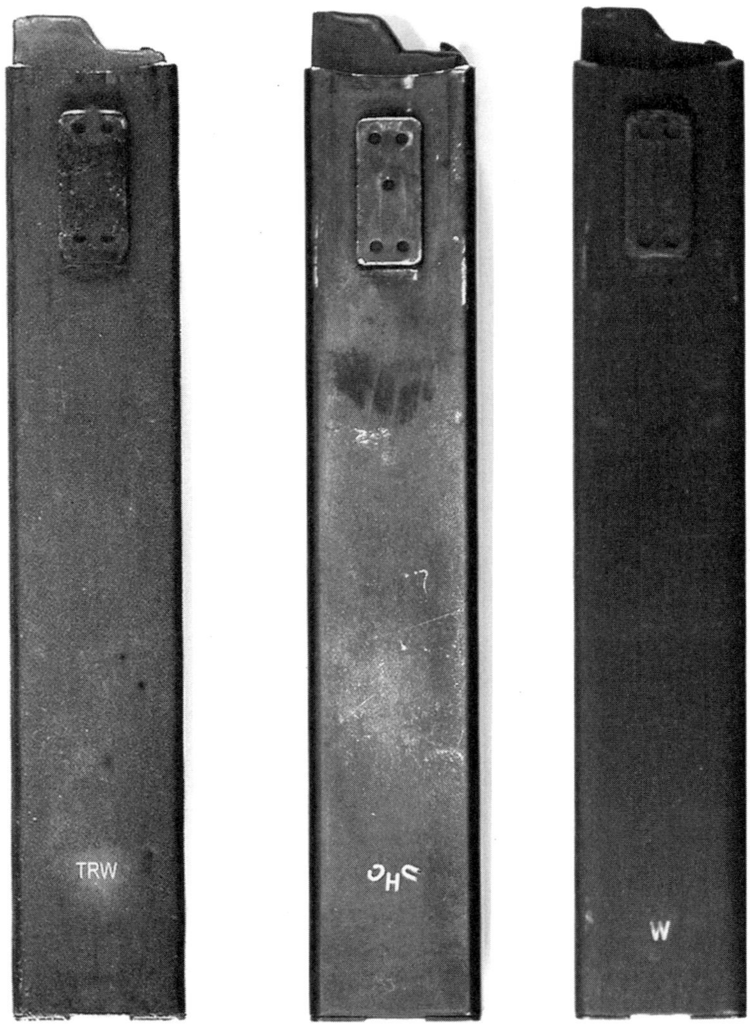

From left; TRW, Thompson-Ramo-Wooldridge; Stylized UHC, Union Hardware Company; W, Winchester.

VCI technology is based upon the dissemination of a corrosion blocking material through the air that is deposited upon exposed metallic surfaces. The deposited material, perhaps as thin as a single molecule, forms a mono molecular film that disrupts the flow of electrons needed to form a corrosion cell at the point of corrosion. Using the VCI wrapping material allows the part wrapped in the material to be used immediately. There was no messy grease or Cosmoline to remove. A few known VCI wrapper manufacturers used to package M14 magazine were RAP industries, and the Plastic Film Company.

Disposable Magazines

During the early days of the M14 rifle there was a program instituted for the development of disposable magazines. This desire originated from the M1 rifle clips that were considered expendable in the field. The Ordnance Corps felt that the sustained rate of fire of the M14 would be hampered by attempts to salvage empty magazines. The Remington Arms Company Inc. was enlisted to perform a feasibility study to develop a 20-round magazine for the M14 rifle at a cost sufficiently low enough that the magazine could be considered expendable. Details of the program are listed in the experimental chapter of this publication.

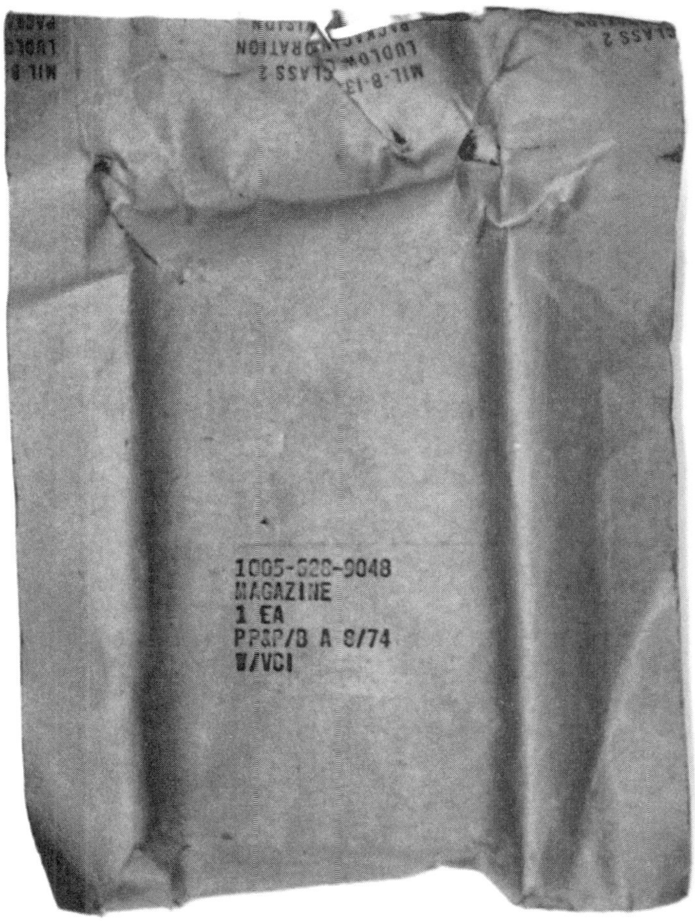

Typical U.S. GI wrapped M14 magazine. Original appearing packaging does not always guarantee authentic original magazines. Unscrupulous dealers have been known to wrap aftermarket and foreign-made M14 magazines in GI-style paper and pass them off as original. (Courtesy of Bill Ricca)

Aftermarket and Foreign Made Magazines

The most important aspect of being able to properly identify aftermarket or foreign made M14 magazines is

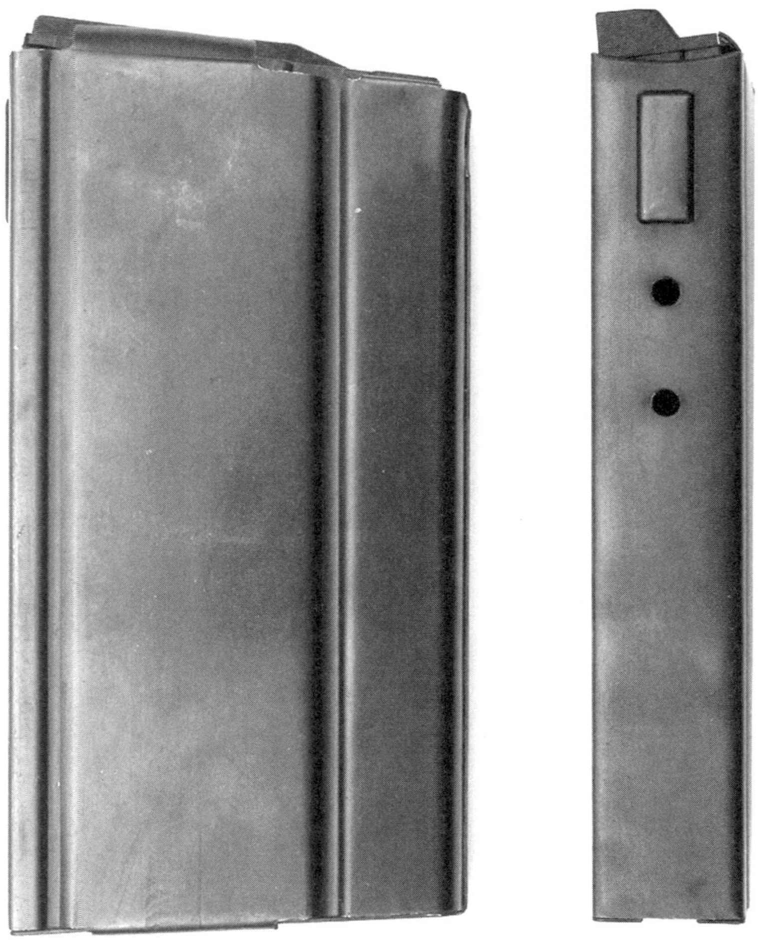

Blued aftermarket M14 magazine. Aftermarket M14 magazines are generally unreliable. Note that the boss for the magazine latch is pressed into the sheet-metal rather than a steel block attached by spot-welding.

that you don't pay an excessive amount of money for them. They should always be much less expensive than original GI contract magazines. Foreign and aftermarket magazines are often mistakenly (or not) advertised as genuine US GI magazines. Beware of such advertising verbiage as GI type or just GI. GI represents Government Issue, the question is; which government? Often it is not the United States government.

Aftermarket magazines are generally produced when there is a shortage of original magazines, originals are expensive or original magazines are not available at all. Although original M14 magazines certainly could not be considered rare, there have been quite a few aftermarket copies made. The aftermarket magazines are generally of poor quality, out of spec and unreliable. There have been both twenty and thirty-round aftermarket magazines made. Aftermarket magazines can generally be identified by comparing them to a known US GI original, the difference in construction and quality will usually be immediately obvious. Additionally, aftermarket magazine will not be marked with a contractors ID letter or logo. When in doubt

An M14 magazine made in China (left) is compared to an original HR-R marked Harrington & Richardson contract magazine. Chinese magazines are of very good quality and can be difficult to tell from an original GI magazine. When new, the Chinese magazines are usually heavily coated with grease inside and out. The finish is a light gray Parkerizing, there are no manufacturer markings.

An ad from 1988 offering new-in-the-wrap GI M14 magazines for $6.00 each. (Donald G. Thomas)

get a second opinion from someone who is familiar with the magazines or pass on the offer. Regardless of how cheap they can be purchased, they are not worth the trouble and frustration that you will most likely encounter.

Surplus Foreign Government Magazines

Also found on the surplus market are magazines that have been imported from foreign countries. In some cases the magazines were made under a government contract to military specifications. A few unscrupulous surplus dealers have attempted to market the foreign magazines as genuine U.S. GI, by packaging them in the same type of wrappings used by government contractors.

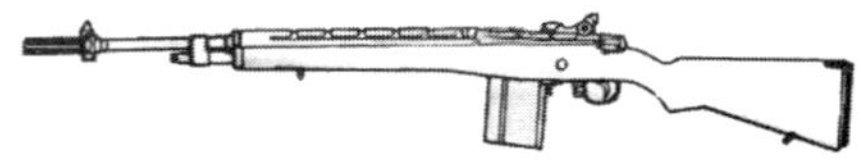

Thermold Magazines

These are twenty-round capacity M14 magazines on the market that were made from special plastic, by the Thermold Company. Most Thermold magazines do function reliably, although there have been problems reported of the magazines not fitting into some rifles without some filing and fitting required. The Thermold Company located in North Carolina, manufactured a number of C1 (M16) plastic magazines for the Canadian Government

Taiwan M14 Magazines

Between 1980 and 1990 some high-quality military surplus Type 57, M14 magazines were imported from Taiwan, Republic of China. The magazines as well as the Type 57 (M14) rifles were manufactured in Taiwan at the National Arsenal using surplus U.S. M14 tooling from the Harrington and Richardson factory. The magazines are very close copies of U.S. made magazines, except for their lack of markings and dull blue finish. The Taiwan magazines are generally reliable, but there have been problems reported with the magazine followers not

A rare brand new M14 magazine in a Springfield Armory marked wrapper. Many of the magazines in the SA wrappers were manufactured by the Atwood Vacuum Machine company (marked AV) who was subcontracted by the Armory to manufacture magazines. (Courtesy of Bill Ricca)

Vietnam 1966, this U.S. soldier has two M14 magazines taped together for additional firepower. After one magazine was emptied the taped unit could be quickly removed from the weapon, rotated 180 degrees and the fresh magazine inserted. There were two disadvantages to this system. One was weight, the other was that the rounds in the second magazine were exposed to dirt and could possibly jam the weapon. (NARA)

holding the bolt rearward after the magazine is emptied.

Chinese M14 Magazines

These magazines are of very good quality and are difficult to tell from original GI magazine except for two things. They are finished in a light gray Parkerizing and they lack any manufactures' markings that are located on the back spine of the magazine's body on U.S. GI magazines. When new, the Chinese magazines are usually heavily coated with grease inside and out. The magazines are quite reliable and have few, if any, functioning problems. The Chinese magazines are generally one-half the price of U.S. GI magazines. Although they offer excellent value as shooter magazines, many collectors shun them because they have been reportedly manufactured in Communist China

Plastic M14 magazine made by the U.S. Thermold Company. Most Thermold magazines do function reliably, although there have been problems reported of the magazines not fitting into some rifles without some minor filing and fitting required.

Manufacturer markings on a Thermold M14 magazine.

labor camps. 6000 of these were imported by LMO in 1995, and sold on the surplus market. The majority were converted to use for the Armalite AR10 rifle.

Conversions of M14 Magazines

In days now long past, original US GI M14 magazines could be purchased new in the wrap for as little as $5.00 each. This low price and the (then) seemingly unlimited supply, unfortunately made the magazines ideal for adaptation to other 7.62 mm caliber rifles. This usually occurred when the original magazines were difficult or expensive to obtain. Magazines for the first Israeli Galil rifles imported into the United States used modified M14 magazines. The original locking plate was removed and new lugs for the Galil rifle attached by welding.

The modern Armalite Company introduced a new version of the 1950s AR10 in a semiautomatic-only civilian model during 1996. Like

Variations of M14 magazine latch plates and spot-welding techniques. From left; Aftermarket (embossed plate); Taiwan military; Mainland China; Unknown aftermarket with plastic follower; Atwood Vacuum Machine Co.; Kileen Machine and Tool; Olin-Mathieson; Rochester Manufacturing.

the original, the modern version of the AR10 is chambered for 7.62 NATO cartridge. Due in part to the over ten-round magazine ban (since expired) original US GI M14 magazines were Armalite factory converted to fit and function in their AR10 rifle.

Although they may have once been US GI M14 magazines, after being converted or altered to fit other rifles, the magazines usually no longer fit or function in the M14 rifle.

Aftermarket 30-round M14 magazine in plastic wrapper. No U.S. GI 30-round M14 magazines were ever manufactured for issue.

M14 magazines from left; Standard 20-round; Sniper 7-round; and 5-round for hunting purposes in some areas. (Courtesy of Scott Duff)

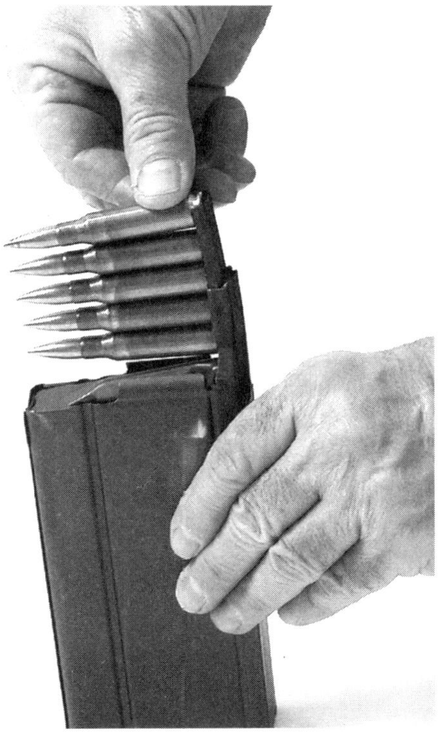

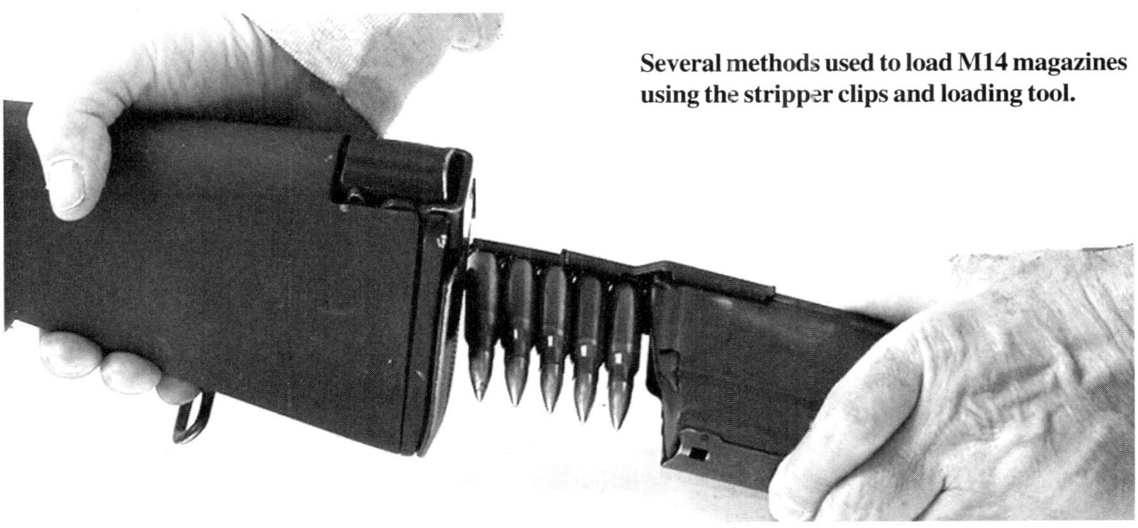

Several methods used to load M14 magazines using the stripper clips and loading tool.

Procedure for installing a magazine in the M14 rifle.

Chapter 11
Parts and Accessories

One of the accessories for the M14 was the magazine carrying pouch. The M14 rifleman generally carried five loaded 20-round magazines. Spare ammunition was carried on 5-round stripper clips in cotton bandoleers.

Pouches, Pockets and Cases

There were several nomenclatures used for "magazine pouches". Many magazine carrying devices were described as a "pocket". A pocket was a soft-body item used for carrying a specific magazine for a specific weapon. A case was an item also used to carry a specific item, but was reinforced, usually with semi-rigid nylon or plastic plates, sewn inside the material to give it a specific shape.

Early web gear issued to troops in Vietnam was usually the M1956 web gear. Many of the components of this gear were made from cotton-duck material. While the material was

very rugged it proved to mildew very quickly in the hot humid tropical climate encountered in Vietnam. To solve the problem of the fast disintegration of the M1956 web gear the M1967 gear was introduced. The M1967 gear was made of nylon, which is very mildew resistant. The metal hardware used on the M1956 gear was replaced by components made of black plastic. It was not uncommon for a mixture of M1956 and M1967 gear to be issued and worn. All of the components were completely interchangeable.

The cotton duck M1956 universal pouch used for carrying two M14 magazines.

M1956 Lightweight Load Carrying Equipment

First issued during 1956 was a new style web gear to replace the old World War II era individual equipment. Made of a heavy-duty cotton-duck material the equipment included: Pistol belt, load bearing suspenders, field pack, entrenching tool carrier, canteen cover, first aid and compass case, sleeping bag carrier and an ammunition carrying pouch. A new feature of the M1956 equipment was the slide clip fasteners that replaced the older M1910 wire hooks.

M1956 Universal Pouch

The early magazine pouch issued with the M14 rifle was the M1956 Universal Pouch. This pouch was made of a cotton duck material and

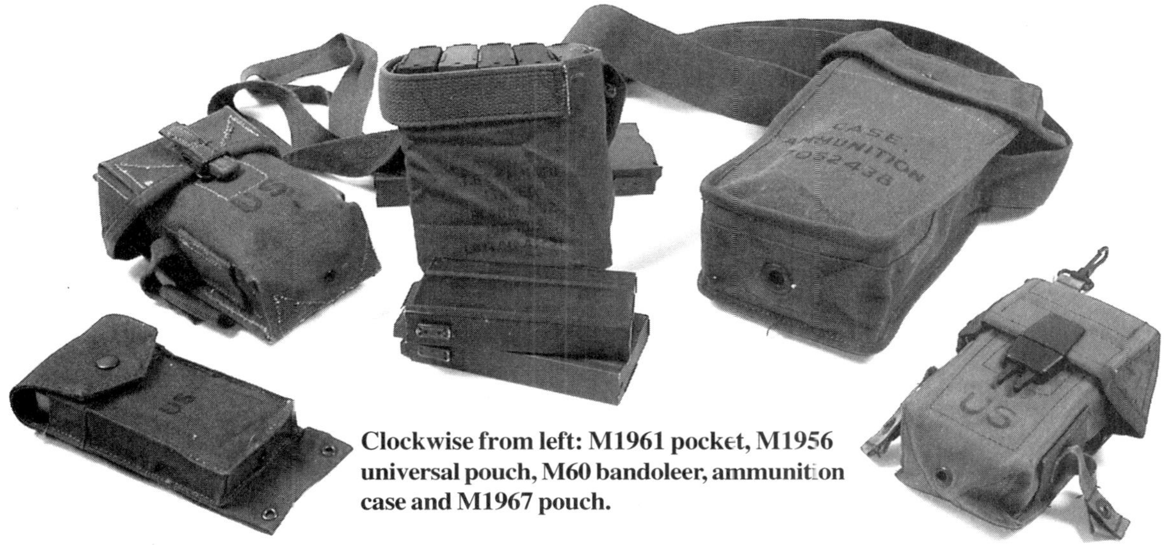

Clockwise from left: M1961 pocket, M1956 universal pouch, M60 bandoleer, ammunition case and M1967 pouch.

was able to carry two M14 20-round magazines. The official nomenclature was Pouch, Small Arms, Ammunition, Universal FSN 8465-647-0852.

This pouch was introduced in 1956 and commonly called the M14 pouch, although it was used to carry magazines for other weapons including the BAR, M1 Carbine and later the M16. There were two variations of the M1956 pouch. One was made of canvas and had a metal plate in the front to protect the magazines from shell fragments. The second variation appeared in 1962, which was similar, but the metal plate was no longer used.

M1961 Pocket, Ammunition, M14

The M1961 "pocket" pouch was designed to carry one 20-round M14 magazine. The pouch had a single flap with a metal snap to secure the magazine inside, and had two separate belt loops on the rear. The U.S. Marine Corps primarily issued the single M1961 magazine pouch. The M1961 pouch

The M961 magazine "pocket".

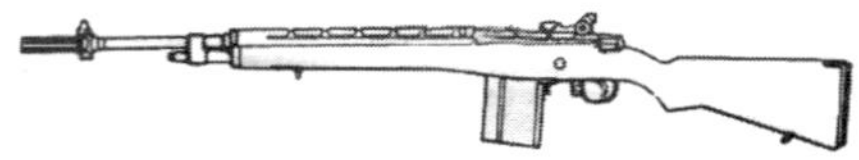

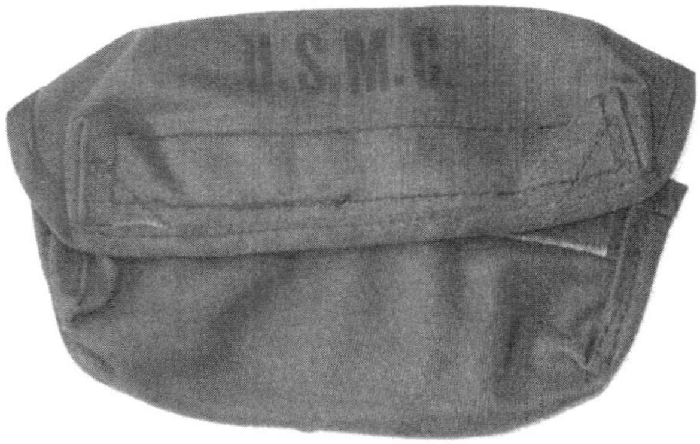

Marine Corps Horizontal style experimental nylon pouch (Courtesy of Craig Pickrall)

Marine Corps Vertical style experimental magazine pouch (Courtesy of Bill Ricca). Note: pouches shown are not to scale.

could be placed on the GI pistol belt by slipping the belt through either loop depending on how high or low the Marine wanted to carry the pouches. The loops each had a snap to secure the pouches from sliding around the belt. The Marine issue M1961 pistol belt had a row of female snaps to accommodate the male snaps on the pouch loops. The snaps were not normally seen on the M1956, or M1967 pistol belts.

USMC Horizontal and Vertical Magazine Pouches

There were a few prototype magazine pouches made for the U.S. Marine Corps as a part of the M64 experimental pack. One pouch was marked as the M14V, V designating vertical for the position of the magazines in the pouch. The nylon pouch was designed to carry two magazines and was designed with a divider separating the magazines. The other pouch was designated as the M14H for horizontal carry. The USMC marked magazines pouches described above are very rare.

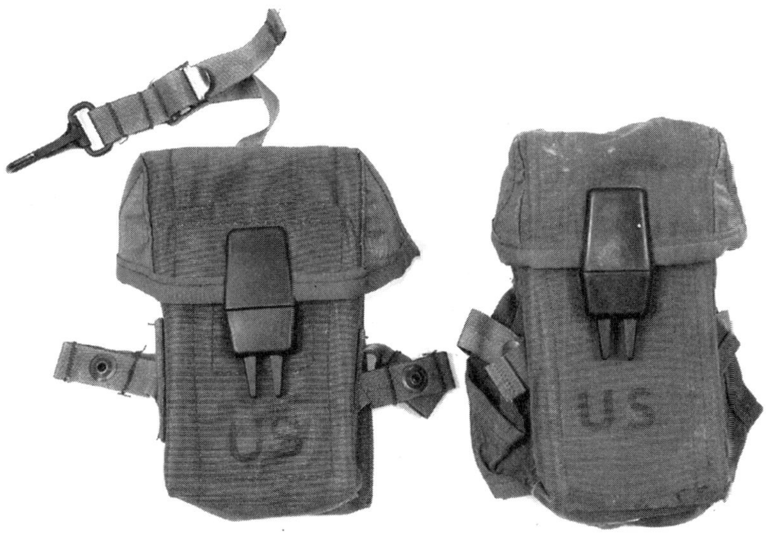

A nylon M14 magazine carry pouch (left) is displayed with the slightly taller and much more common M16 pouch (right).

M1967 Individual Load Carrying Equipment

During the fighting in the hot and humid climate of Vietnam, the cotton material and metal fittings of the M1956 gear proved subject to mildew and failed to hold up for very long in the field. To address the problem the modernized M1967 gear was designed and introduced. The primary advantage of the M1967 was its construction of mildew resistant nylon. Metal fittings and snaps were replaced by plastic and Velcro. The pistol belt and ammunition pouches were both fitted with the "Davis" quick-release plastic latches. All of the older M1956 equipment was not immediately replaced, but all of the gear was interchangeable. While the M14 was on its way to being obsolete by the time the M1967 gear was issued, there was a nylon ammunition case issued.

The nylon variation of the M1956 pouch that was part of the M1967 web gear. The case was designed to carry two M14 magazines. The pouch was fitted with two loops for carrying gre-

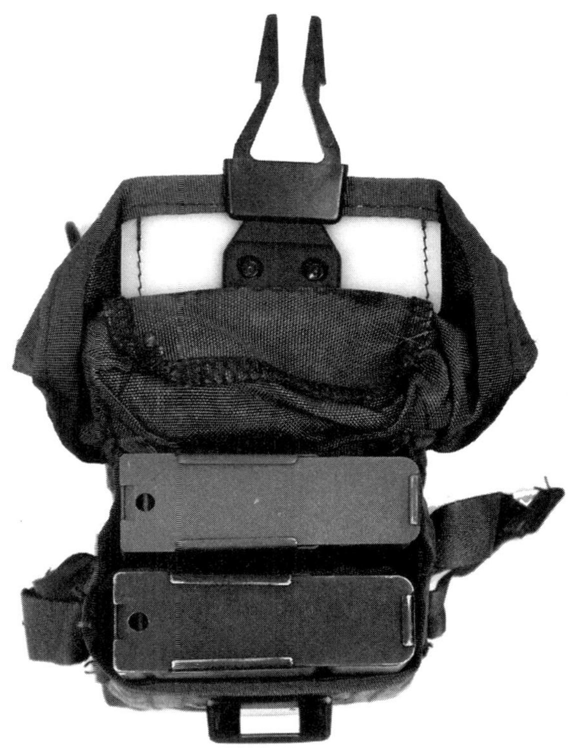

This magazine pouch made for the M14 lacked the grenade carry loops positioned on the sides of the pouch. (Bill Ricca)

The M60 machine gun ammunition bandoleer was often utilized as an expedient magazine carry pouch for the M14 rifleman. This illustration is from an U.S. Army manual.

nades and an LC2 strap. The inside of the cover was usually stamped M14 RIFLE DSA 100 4597 8465 382 6838. This pouch was not widely issued and is often confused with the M16 magazine pouches. The FSN was 84565-00-141-0920 and the contract for manufacture was cancelled in 1975. There was also a similar magazine pouch made for the M14, but it lacked the grenade loops positioned on the sides of the pouch.

The M60 machine gun ammunition bandoleer was often utilized as an expedient magazine carry pouch for the M14 rifleman, and could be used to carry up to four loaded magazines. The pouch was made of cotton and fitted with a shoulder strap.

Sixty rounds of spare 7.62 mm ammunition pre-loaded on five-round steel stripper clips, with protective cardboard sleeves, were carried by the infantryman in six-pocket cotton bandoleers. The stripper clips were used to charge empty magazine through the use of the magazine charger issued with each bandoleer. The cotton bandoleers were only issued for combat use. The stripper clips could also be used to load the magazine while still in the rifle through the receiver mounted guide. The combination tool could be used to apply pressure to the fresh rounds making them easier to load from the stripper clips into the magazine.

Originally ten-round stripper clips were specified for loading the M14 rifle through the receiver guide or loading magazines using an adapter. In the interest of NATO standardization it was decided to use a common five-round stripper clip for compatibility with other NATO members.

Troops were trained to carry their magazines with the cartridge end facing down in the pouch, to keep dirt, mud and debris from col-

The M60 machine gun ammunition bandoleer as an expedient magazine carry pouch for the M14 rifle could be used to carry up to five loaded magazines. The pouch was made of thin cotton and fitted with a shoulder carry strap.

lecting on the rounds, and causing a weapon malfunction. The long edge or rear of the magazine was to be stored facing the rear of the pouch. This provided a systematic method of removing magazines from their pouches and inserting them into the weapon. Right-handed shooters were trained to use the magazines carried on the right side first, using the right hand to remove the empty magazine and place the fresh magazine into the weapon, the left hand was never to be taken off of the weapon. Left-handed shooters were taught the opposite.

TOTE THAT AMMO

You say you've got an ammo carrying problem that's got you down?

And you're tired of luggin' extra magazines in your pocket?

And you're fed up with being speared everytime you sit down?

Well . . . take five, friend, 'cause here's the answer.

It's the ammunition magazine case, FSN 8465-705-2438, a QM item listed in Federal Supply Catalog C2-15A, dated March 1960.

The case is made of cotton duck and webbing and comes with a shoulder strap to make things easier.

It'll carry a pretty good load of extra ammo for all of your small arms—like .45 caliber for your pistol and submachine gun; .30 caliber for the carbine and M1 and extra clips for your 7.62-mm M14 rifle.

The cases are mighty handy to have around—but they're not up for grabs.

If you have a real need for some but they're in excess of your TOE and TA, tell your supply man to use the procedure outlined on page 12, paragraph 15h in AR 725-5 for requisitioning them.

Also issued for carrying spare ammunition and magazines was the "Ammunition Magazine Case". The case was made of cotton-duck webbing or nylon and was fitted with a shoulder strap for easy carry. The case was quite large and could carry a sizable amount of spare ammunition.

Spent magazines were to be kept for reloading. A field expedient method of transporting used magazines was in an empty sandbag attached to the load-bearing equipment.

Also issued for carrying spare ammunition and magazines was the "Ammunition Magazine

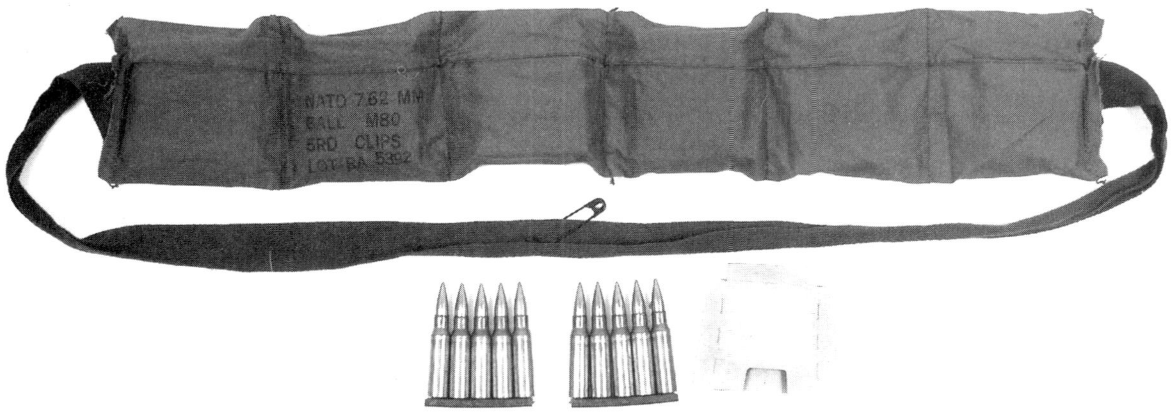

The M14 bandoleer was a convenient way for troops to carry spare ammunition. The 6-pocket bandoleer for the M14 rifle came with 60 rounds of 7.62 ammunition on 5-round stripper clips.

Case". The case was made of cotton-duck webbing and was fitted with a shoulder strap for easy carry. The general-purpose pouch was used to carry magazines and ammunition for the carbine, M1 Garand, submachine gun and .45 caliber pistol. A similar pouch was later issued made of mildew resistant nylon. The case was rather large in size and could carry a sizable amount of spare ammunition.

Carry Slings

Like other early web gear issued with the M14 rifle the Sling, Rifle, M1 was made of cotton web, dyed varying shades of an olive green color. The M1 cotton sling added .31 pounds to the weight of the rifle. The M1 cotton sling was the same one that was

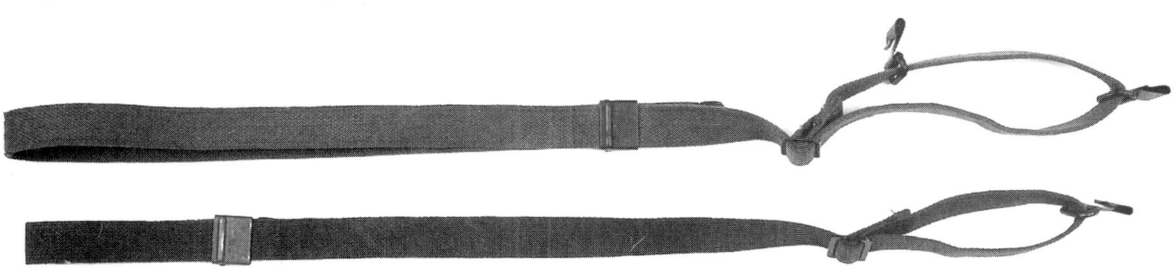

The cotton sling issued with the M14A1 rifle variation (top of photo) was very similar in deign to the M1 sling, but was longer at 64.5-inches and had 2-spring steel catch clips. One clip was for attachment to the sling swivel on the bipod, the other to a steel pin located at the base of the foregrip.

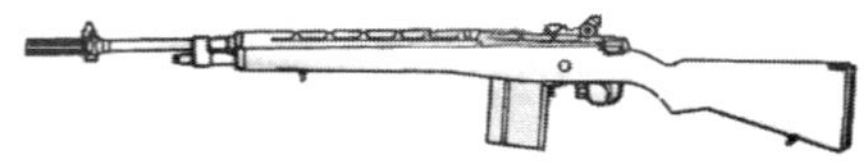

The tropical climate of Vietnam would cause the cotton M1 sling (top) to quickly mildew and rot, renedering it unserviceable. To solve the problem a mildew resistant sling (bottom) made of nylon was designed to replace the earlier slings. The steel hardware on the nylon slings was treated with corrosion resistant cadmium plating.

issued for the M1 rifle. Some M1 slings made in the 1950s era were fitted with brass fittings; later slings had Parkerized steel hardware. The standard M1 sling had an overall length of 46-inches. In a tropical humid climate as was encountered in Southeast Asia the cotton M1 sling would quickly rot and become unserviceable. To solve the problem a mildew resistant sling made of nylon was issued to replace the

The cotton sling issued with the M14A1 rifle variation was very similar in design to the M1 sling, but was longer at 64.5-inches and had 2-spring steel catch clips. One clip was for attachment to the sling swivel on bipod, the other to a steel pin located at the base of the foregrip. This arrangement served two purposes. One was to place forward tension on the foregrip to prevent breakage of the grip.

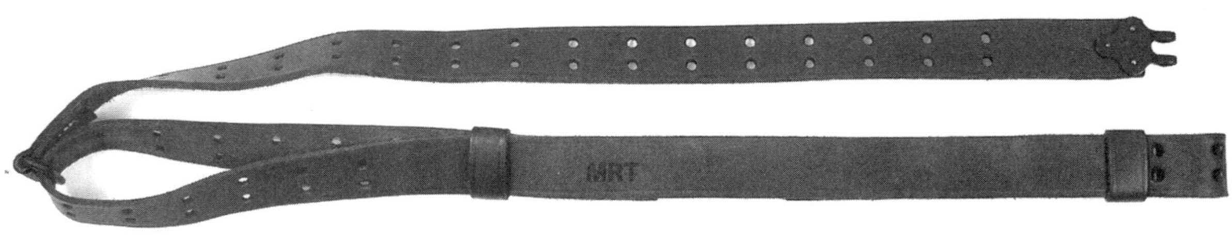

The M1907 leather sling was primarily issued for M1 and M14 match rifles and the M21 sniper weapons. The origins of the M1907 sling date back to the pre-World War I days.

earlier slings. The steel hardware on the nylon slings were treated with corrosion resistant cadmium plating. The same nylon slings were also issued for the M16 rifle.

The other feature was that while pulling rearward on the foregrip the operator would increase the down-pressure on the bipod.

The MRT markings often seen on slings and

The M6 bayonet and the accompanying M8A1 scabbard were developed and issued with the M14. (Courtesy of Ronald Maxson)

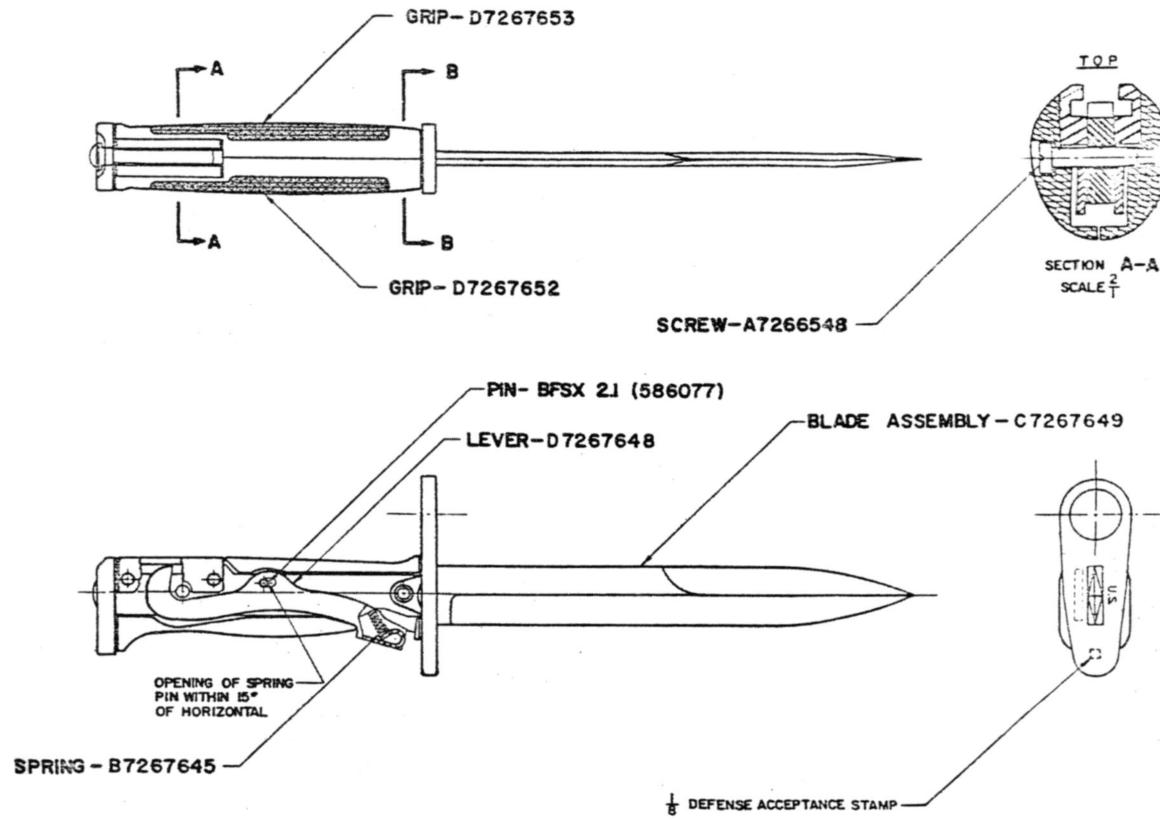

Fig. 1 - Bayonet-knife, T12

Drawing of the prototype T12 bayonet, which was adopted as the U.S. M6 bayonet for the M14 rifle. (Courtesy of Bill Ricca).

other items represent that the item has been Mildew Resistant Treated.

The M1907 leather sling was primarily issued for M1 and M14 match rifles and the M21 sniper weapons. The origins of the M1907 sling date back to the pre-World War I days of the M1903 and M1917 service rifles. The leather M1907 sling was also widely used during World War II. The leather sling consisted of two sections of leather, a long piece and a shorter piece. Two brass adjustment frogs (clamps with two hooks) were provided for changing the length of the sling, there were

M6 bayonet manufactured by Imperial. (Courtesy of Ronald Maxson)

corresponding holes in the leather for placement of the hooks. There were two leather keepers on the long portion of the sling. The shorter strap was fitted with a D-ring that was used to attach it to the longer piece. Production continued through World War II, the only difference being that the hardware on later manufacture M1907 slings was made of steel rather than brass. The issue leather slings of the era were replaced by those made of cotton web, which was less expensive and more durable for combat use. The M1907 leather sling remains popular with competition shooters today.

Bayonet

Like all U.S. infantry rifles before it, a bayonet was designed for use with the M14. The bayonet was primarily designed for close-in combat, guarding prisoners and riot control. The prototype T12 bayonet was adopted as the M6 bayonet. The M6 bayonet and the accompanying M8A1 scabbard were developed and issued with the M14. The M14 and its bayonet would experience most of its use by National Guard troops in the United States, during the many demonstrations against U.S. involvement in Vietnam War.

The M6 bayonet was made of steel with two black plastic grip halves attached with two machine screws. The metal of the bayonet was manufactured from AISI 1080 steel and was Parkerized a dark gray color. The bayonet itself had an overall length of 11.50 inches, while the blade had an overall length of 6.75 inches. A spring-loaded catch was used to install and

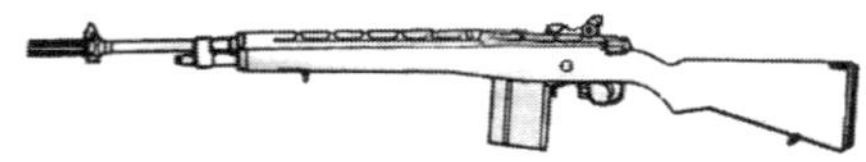

remove the bayonet from the rifle. The bayonet was installed on the M14 rifle by aligning the groove in the handle with the corresponding stud on the bayonet lug. The loop of the bayonet was designed to slide over the flash suppressor. The bayonet is then slide rearward until it latches onto the bayonet lug. To remove simply grasp the handle and depress the catch button and slide the bayonet forward. The bayonet was designed to have a loose fit when attached; this was so that the bayonet would not affect the weapon's accuracy. Known manufactures of the M6 bayonet were; Imperial Knife; Milpar and Aerial Cutlery.

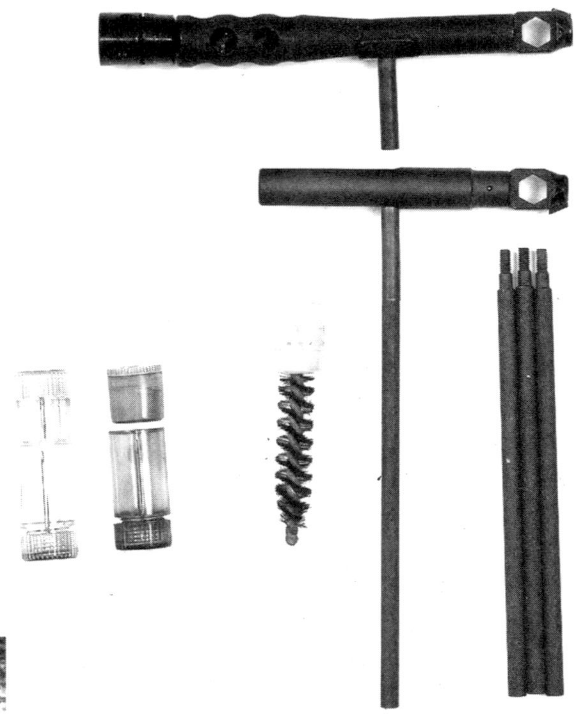

Cleaning equipment for the M14, a combination tool is displayed at the top of the photograph. (Bill Ricca)

The M8A1 scabbard was designed to carry the M6 bayonet when not in use. The scabbard was made of plastic with metal hardware. A cotton web strap was riveted to the metal part of the scabbard. Attached to the web strap was a metal clip for attaching it to the standard pistol belt. A separate strap with a metal snap was riveted to the primary strap to secure the bayonet inside the scabbard. There were several firms that manufactured the M8A1 scabbard including; Victory Plastics Company; Viz Manufacturing; TWB.

A U.S. Army soldier cleans his M14. (NARA)

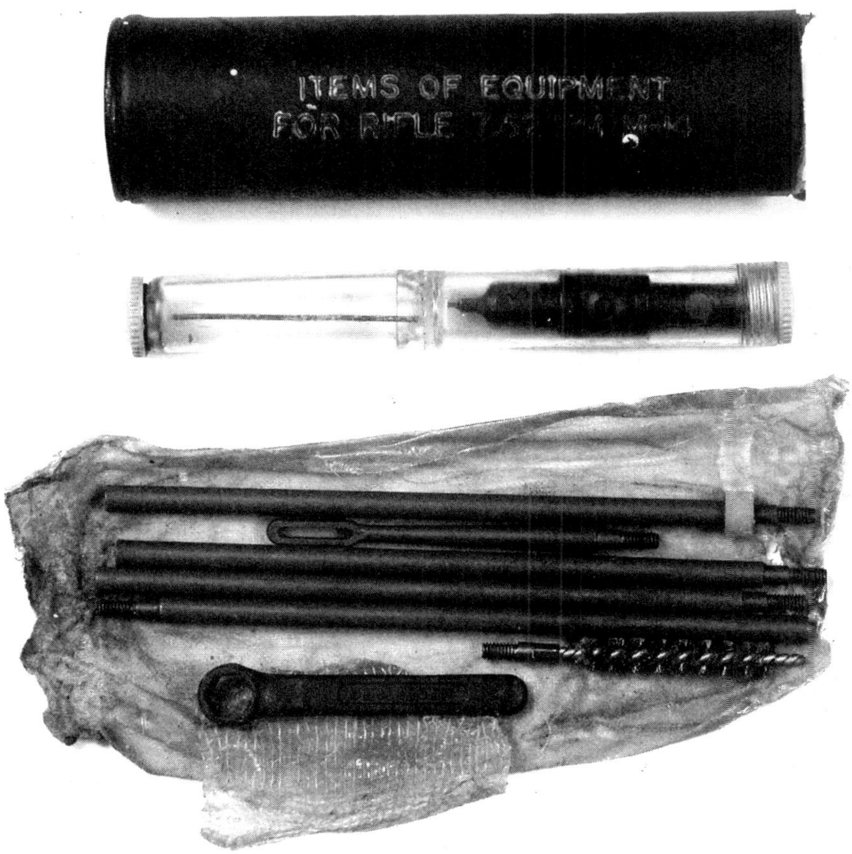

Early cleaning kit for the M14 rifle, note the separate wrench. Later kits included a combination tool that served as a wrench, screwdriver and cleaning rod handle. (Courtesy of Bill Ricca)

There were a few foreign governments that were awarded contracts for the production of the M8A1 scabbard, but these were not U.S. issue items.

Cleaning Equipment

The Springfield Armory Engineering Division designed a rotating bronze cleaning brush for cleaning the barrel's chamber on the M14. The brush was designed with a plastic end that had teeth so it could be turned or "ratcheted" in the chamber with a section of the cleaning rod. The cleaning rod screwed into place in a threaded hole provided at a 90-degree angle allowing the operator to fully rotate the brush without removing it. The bolt was then to be slowly eased forward on the brush assembly in order to seat the brush fully into the chamber, the brush assembly was then rotated using the brush's ratcheting feature. (never allow the bolt assembly to slam forward on the chamber brush). To remove the brush there were two recommended procedures. The first method was to grasp the cleaning rod with the thumb pushing the rod in the opposite direction. Then by simultaneously pulling the cleaning rod and operating rod handle to the rear, while maintaining forward thumb pressure on the cleaning rod. The second method was to lock the bolt in the rearward position and simply pull rearward on the cleaning rod while attached to the chamber brush assembly. The prototypes of the 7.62 mm chamber cleaning tool were evaluated by the U.S. Army Infantry Board. The board recommended the new chamber brush for adoption on 17 June 1959.

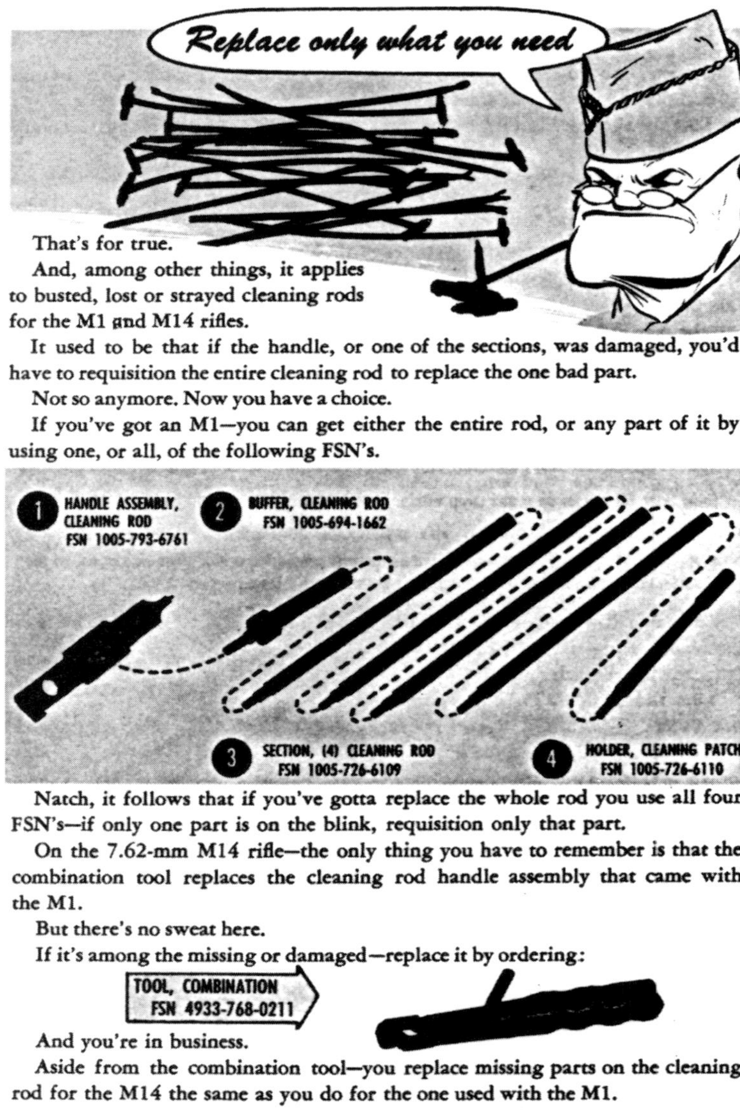

That's for true.

And, among other things, it applies to busted, lost or strayed cleaning rods for the M1 and M14 rifles.

It used to be that if the handle, or one of the sections, was damaged, you'd have to requisition the entire cleaning rod to replace the one bad part.

Not so anymore. Now you have a choice.

If you've got an M1—you can get either the entire rod, or any part of it by using one, or all, of the following FSN's.

Natch, it follows that if you've gotta replace the whole rod you use all four FSN's—if only one part is on the blink, requisition only that part.

On the 7.62-mm M14 rifle—the only thing you have to remember is that the combination tool replaces the cleaning rod handle assembly that came with the M1.

But there's no sweat here.

If it's among the missing or damaged—replace it by ordering:

And you're in business.

Aside from the combination tool—you replace missing parts on the cleaning rod for the M14 the same as you do for the one used with the M1.

The M14 brush assembly FSN 1005-6990-8441 (7990463) was similar to the earlier brush assembly issued for the M1 rifle except that its overall length was 3.25-inch as compared to the M1's brush that was 3.75-inch in its overall length. Using the M1 brush to clean the M14 allowed the tip of the chamber-sized brush to extend into the rifle's bore, possibly causing damage to the barrel.

Basic Issue Items (BII)

The M14 rifle was initially issued with the same cleaning kit as was issued with the M1 rifle during the Korean War. Including the four-section M10 cleaning rod with nylon buffer to prevent damage to the crown of the barrel during cleaning. A separate 3/8-inch box-end wrench was added for removing or tightening the gas plug of the M14.

New in-the-tube M14 combination tool. (Courtesy of Bill Ricca)

When finally developed and fielded, the M14 kit included; a combination tool; 7.62 mm chamber brush; bore brush; slotted patch tip and lubricant container. Four steel cleaning rod sections were issued along with a canvas pouch to contain the rod sections while stored in the buttstock compartment. A plastic spacer was included to keep the cleaning brush and cleaning rod patch holder lined up in side the canvas case. The spacers were usually discarded because most troops didn't know what the purpose of the spacer was. A small tab was attached to the rod pouch. Anxious GIs trying to pull the cleaning rod pouch out of the stock often tore off the tab making it quite difficult to remove.

The M14A1 stock was fitted with a rubber recoil pad and thus had no storage area for a cleaning kit. A nylon case designed for attachment to the GI web belt was issued for the M14A1 infantryman. The case contained all of the basic issue cleaning items. The cleaning equipment weighted .61 pounds.

The Combination Tool

The combination tool that was issued with the M14 rifle had a multitude of functions. The tool was designed for storage in the weapon's buttstock. The round handle of the combination tool was designed to contain and protect the chamber brush while stored in the M14 buttstock compartment. Other functions of the combination tool included; a handle for the cleaning rod; push cartridges from a stripper clip to load the magazine; turn the rear sight screws; turn the buttplate screws; rotate the spindle valve; disconnect the connector lock from the op-rod spring guide; disassemble and reassemble the breech-bolt. The wrench portion of the combination tool was used for; loosening and tightening of the gas cylinder plug; install or remove the M14A1 muzzle stabilizer; install or remove the M2 Bipod from the gas cylinder. An early version of the combination tool had a detachable wrench head. This design was not "GI proof" because the separate head could and probably would be lost rendering the tool somewhat useless. The

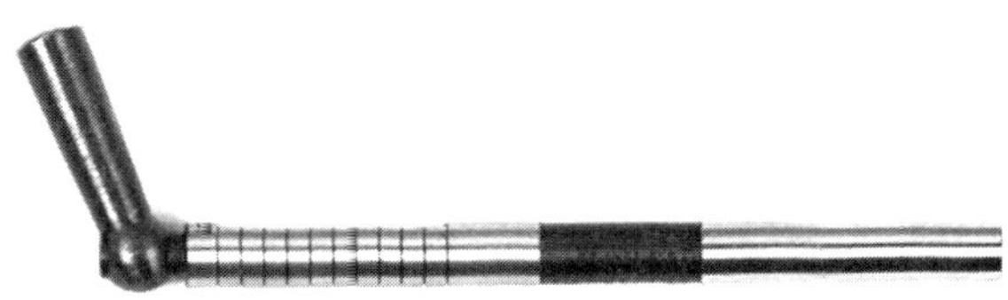

Throat erosion gage. (Scott Duff)

M14 flash suppressor alignment tool. (Courtesy of John M. Miller)

Pliers for removing and installing the flash-suppressor nut on the M14 rifle.

GI Barrel reflector for inspecting a rifle's bore. (Bill Ricca)

Gas piston gage (right) displayed with a M14 gas piston. (Courtesy of John M. Miller)

Note: Tools and gages displayed are not to scale.

Note: Tools and gages displayed are not to scale.

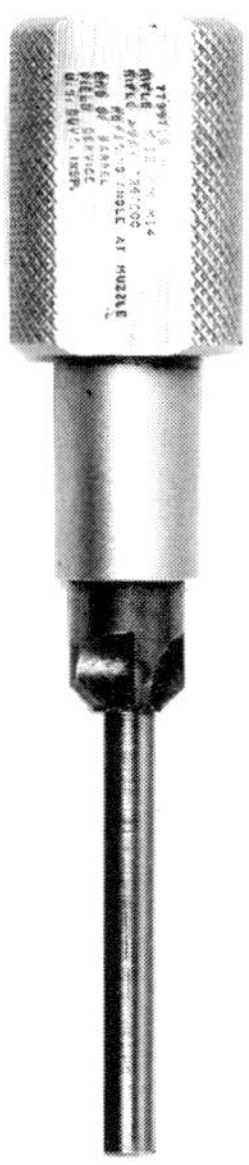

Muzzle refacing reamer. (Courtesy of John M. Miller)

Gas cylinder no-go gage. (Courtesy of John M. Miller)

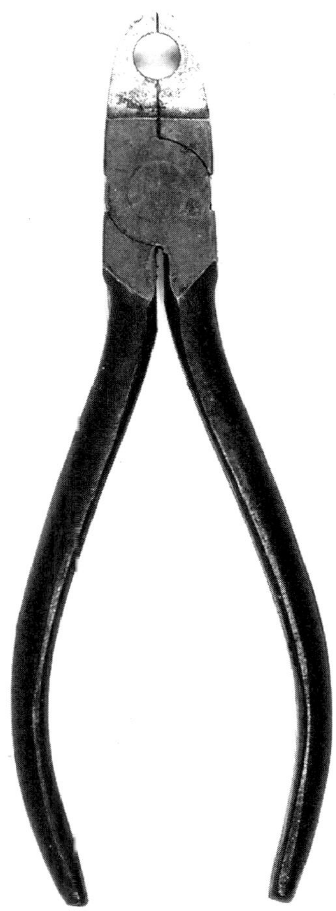

Pliers for removing the bolt-roller retaining clip. (Courtesy of John M. Miller)

GI bolt disassembly/assembly tool. (Courtesy of John M. Miller)

GI Headspace gage. (Courtesy of John M. Miller)

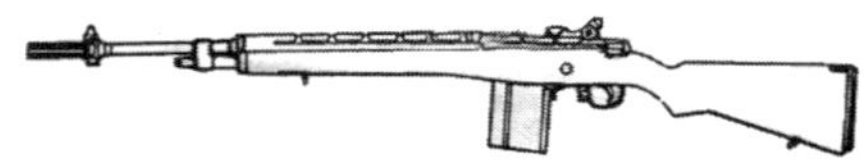

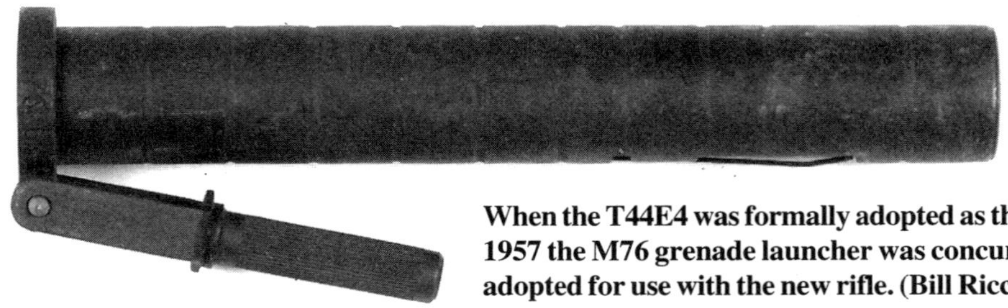

When the T44E4 was formally adopted as the M14 in 1957 the M76 grenade launcher was concurrently adopted for use with the new rifle. (Bill Ricca)

final design incorporated the head as an integral part of the combination tool, by pinning the wrench to the tool

Maintenance Tools and Equipment

Like any mechanical, device a number of tools and equipment are needed to maintain the equipment in operating condition. This is particularly important in the case of an infantryman's weapon whose life may depend on its proper function.

The most common tools seen on the surplus market today will be flash suppressor pliers, bolt roller pliers and the throat erosion (TE) gage. Not all M14 tools and accessories being offered are genuine U.S. GI. There are far more aftermarket items available than there are original government issue. While the commercial aftermarket tools and gages will suffice for civilian maintenance purposes, they are of no interest to collectors.

Tool Set, Direct and General Support Maintenance 7.62 mm Rifle

Gage, Firing Pin Protrusion. Min 0.044 Max 0.060

Pliers, Retaining Ring, Bolt Roller

Gage, Plug Piston NO-GO

Gage, Snap NO-GO Piston

Gage, Breech Bore, Field Rejection 0.310

Gage, Headspace, Field Rejection 01.6455

Bolt, Field Test

Grenade Launcher

During World War II and Korea the infantryman had the ability to launch grenades from the muzzle of his rifle. During the time period that the M14 rifle was being developed, any new infantry weapon that was being considered for adoption, the ability to launch rifle grenades was considered essential. When the T44E4 was formally adopted as the M14 in 1957, the M76 grenade launcher was concurrently adopted for use with the new rifle. The M76 Grenade Launcher evolved from the experimental T140 launcher originally developed in 1955. During 1959 six prototype M76 Grenade Launchers were fabricated, four were shipped to the Marine Corps for testing. The first M76 Grenade Launcher was produced in 1961.

The M76 launcher was designed to slide over the top of the M14 rifle's flash suppressor and was secured by a latch designed to lock on the bayonet lug. Machined on the barrel of the launcher were nine separate annular grooves. Each individual groove was marked with a number 6 to 1, 2A, 3A and 4A. These calibrations and accompanying grooves were used to launch the rifle grenade to reach different ranges. This was accomplished by sliding the grenade over the launcher's barrel until it reached the groove indicating the range estimated by the infantryman. A retainer spring was located near the muzzle of the launcher to secure the grenade in place prior to firing it from the rifle. The M14 rifle using the M76 Grenade Launcher and M64 Grenade Launching projectile could fire a grenade up to a distance of 250 meters. The M76 Grenade Launcher was capable of accepting finned rifle grenades with an inside diameter of 22-

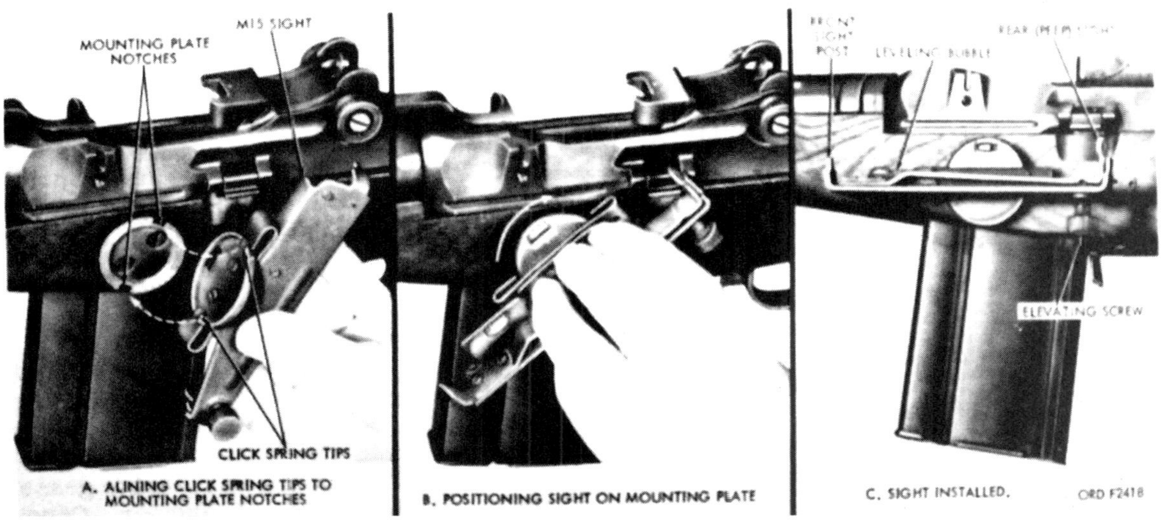

Instructions for installing the grenade launching sight on the M14 rifle.

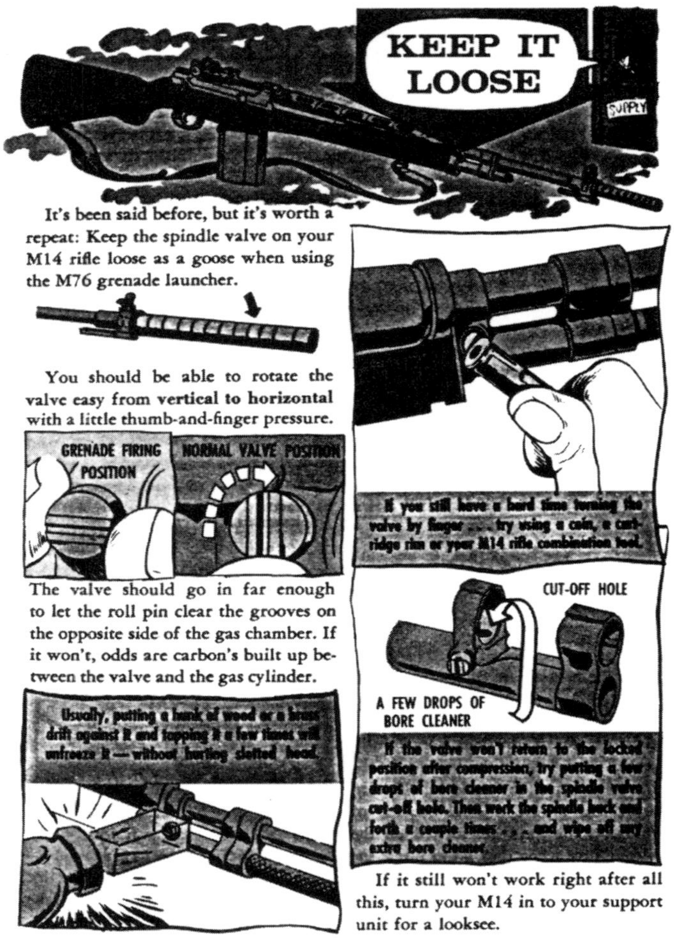

Grenade launching tips from the PS manual. (John M. Miller)

millimeters. There were several type of grenades available to the infantryman including; anti-tank; white phosphorous; smoke and inert training. The actual combat use of the M76 device was limited due to the proliferation of the M79 Grenade Launcher and LAW rockets used to fulfil the role of the rifle launched grenade. The M76 Grenade Launcher would be the last of its type adopted by the United States' military.

The Belgian firm MECAR had developed a new method of launching grenades directly from the M14 rifle's flash suppressor. This was accomplished by inserting a special steel sleeve over the flash suppressor. The outside diameter of the adapter sleeve is such that it allows the mounting of grenades with an inside diameter of 22-millimeters. The same M64 grenade-launching cartridge was used. After launching a grenade the adapter sleeve will fall to the ground a few feet in front of the muzzle and can be reused. The device was not adopted by the U.S.

M14 Spindle Valve

An adjustable spindle valve was incorporated into the design of the gas cylinder for diverting gas pressure when the M14 rifle is used to project grenades from the muzzle. The slot is placed in a vertical position when firing the rifle normally. When launching grenades the valve is rotated until the slot is parallel with the barrel. The combination tool is normally used to rotate the spindle valve into either position.

Grenade Launcher Sight

In order to accurately deliver a rifle grenade to a target, the use of an auxiliary sight was necessary. The M15 grenade sight that provided an angular measurement was used for this purpose, the sight was a carryover from the M1903 and M1 rifles. The M15 sight was attached to the left side of the rifle's stock. A separate mounting plate was carefully installed onto the stock by a unit armorer. The sight could then be attached to the mounting bracket when needed. The sight could be installed on the bracket by aligning the notches of the plate with the click-spring tips of the sight. The sight was then turned clockwise until the index line was aligned with the 0-degree index on the mounting bracket. When rotated to that position the leveling bubble should be centered. When not in use the sight was to be removed and stored in a carrying pouch provided to the infantryman.

A soldier prepares to launch a grenade from his M14 rifle. (U.S. Army)

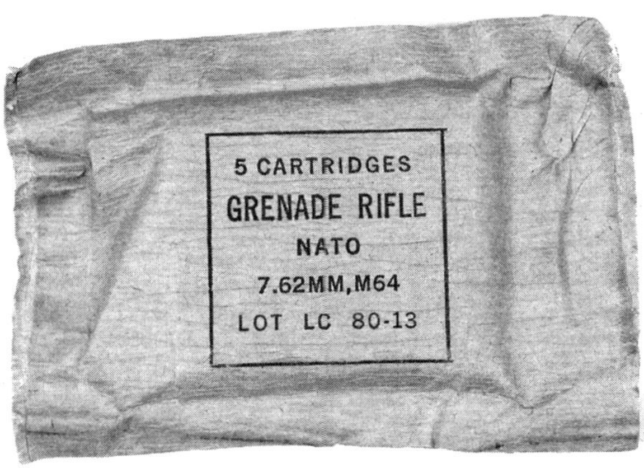

Special grenade launching cartridges. (Bill Ricca)

Grenade Launching Cartridge

A special cartridge was designed for use with the M76 grenade launcher on the M14 rifle.

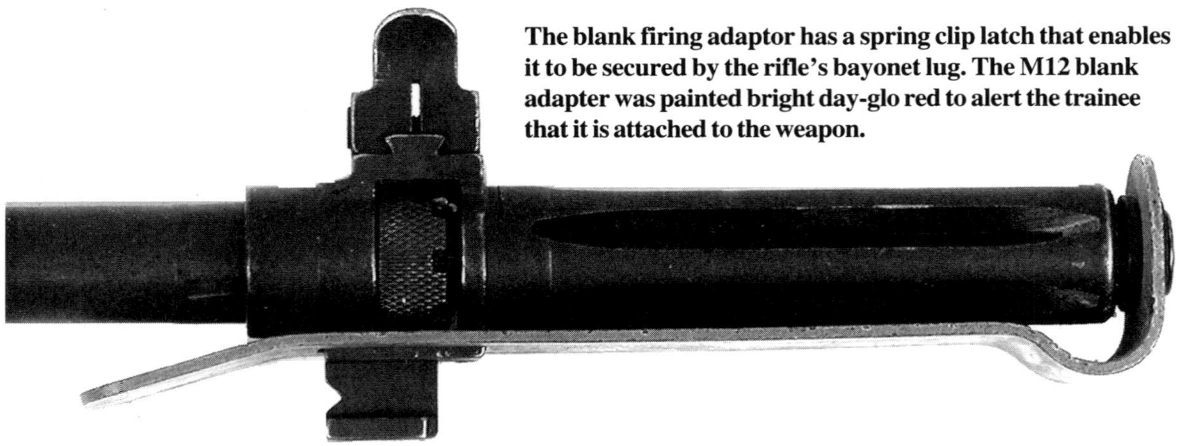

The blank firing adaptor has a spring clip latch that enables it to be secured by the rifle's bayonet lug. The M12 blank adapter was painted bright day-glo red to alert the trainee that it is attached to the weapon.

The M64 rifle grenade cartridge is identified by its 5-point rosette crimp in place of a projectile. The overall length of the M64 cartridge is 2.0 inches, it is normally loaded with 45-grains of Winchester WC 830 propellant. The cartridge is capable of propelling a rifle grenade at a velocity of 160 feet per second, measured at approximately 5.6 feet from the muzzle.

Blank Firing Adapter (BFA)

For realistic training purposes a muzzle adapter was designed for firing blank cartridges. Semiautomatic and automatic gas-operated weapons require a sufficient amount of pressure to operate their mechanisms that operate the action. While standard service cartridges provide sufficient back-pressure, blank cartridges without projectiles do not.

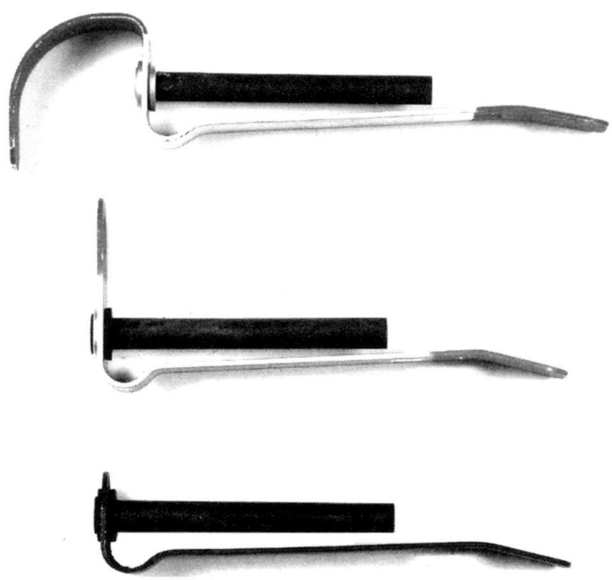

There were several variations of blank adapters designed for the M14, all were in an effort to make them safer and to reduce injuries during training. (Bill Ricca)

Therefore a muzzle-restricting device that allows back-pressure to build in order to operate the rifle is needed.

During the late 1960 and early 1961 period the XM12 blank firing device and the XM3 breech shield were developed at the Springfield Armory for use with the M14 rifle. Both units were then submitted to the Infantry Board, Fort Benning for testing. Although the initial results proved favorable, the Board voiced concern over the necessity of a breech shield. The devices were both formally adopted as Standard for the M14 rifle by the

There were two separate designs of the breech shields produced. The original shield was redesigned to feature a simplified clip guide lock for securing the shield to the rifle. (Bill Ricca)

This tired looking young soldier has a blank adapter on the muzzle of his M14 rifle (NARA)

Ordnance Technical Committee, as the Firing Attachment, Blank Ammunition: 7.62 mm Rifle M12, on 21 September 1961.

The muzzle device incorporates a calibrated orifice tube that slides into the flash suppressor and muzzle of the barrel. The tube restricts the amount of gas pressure generated from the firing of a blank round leaving the muzzle. The blank firing adapter has a spring clip latch that enables it to be secured by the rifle's bayonet lug. The M12 blank adapter was painted bright day-glo red to

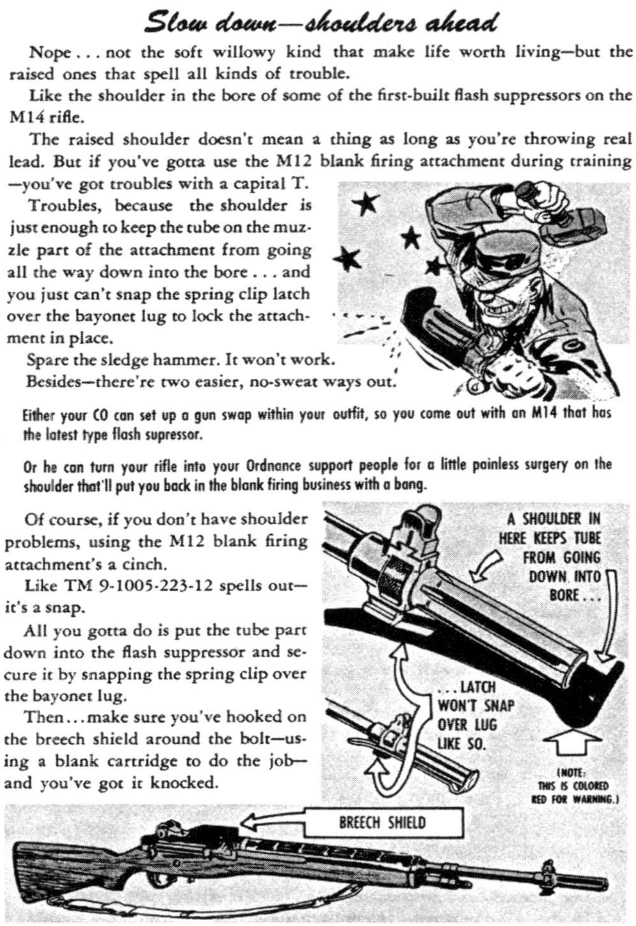

Slow down—shoulders ahead

Nope . . . not the soft willowy kind that make life worth living—but the raised ones that spell all kinds of trouble.

Like the shoulder in the bore of some of the first-built flash suppressors on the M14 rifle.

The raised shoulder doesn't mean a thing as long as you're throwing real lead. But if you've gotta use the M12 blank firing attachment during training —you've got troubles with a capital T.

Troubles, because the shoulder is just enough to keep the tube on the muzzle part of the attachment from going all the way down into the bore . . . and you just can't snap the spring clip latch over the bayonet lug to lock the attachment in place.

Spare the sledge hammer. It won't work.

Besides—there're two easier, no-sweat ways out.

Either your CO can set up a gun swap within your outfit, so you come out with an M14 that has the latest type flash supressor.

Or he can turn your rifle into your Ordnance support people for a little painless surgery on the shoulder that'll put you back in the blank firing business with a bang.

Of course, if you don't have shoulder problems, using the M12 blank firing attachment's a cinch.

Like TM 9-1005-223-12 spells out—it's a snap.

All you gotta do is put the tube part down into the flash suppressor and secure it by snapping the spring clip over the bayonet lug.

Then . . . make sure you've hooked on the breech shield around the bolt—using a blank cartridge to do the job—and you've got it knocked.

alert the trainee that it is attached to the weapon. In the event that a live 7.62 mm round was inadvertently fired with the M12 adapter in place, it was designed to break away causing minimum damage to the weapon and operator.

There were three slightly different styles of blank adapters, and two versions of the breech shields, manufactured by two separate companies. The early style blank firing adapter was quite simple. To make it safer armorers added a piece of 1/8-inch thick sheet metal strap 5/32- inch wide with a curved end that was positioned in front of the orifice, the overall length was 3-29/32nds of an inch. The modified adapter was riveted to the existing blank firing adapter. The purpose of the attachment was to dissipate burning, volatile gases discharged from the muzzle and prevent possible injury to trainees.

The second type of blank adapter muzzle device was initially fielded in 1964 and again in 1968. The design was altered to prevent injuries from the paraffin sealing wad or unburned power particles exiting the muzzle. The design featured a bright day-glo red extension that encircled the shooters view of the front sight. The thinking behind the design was to remind the shooter of the possible danger of firing blanks.

The third style of blank firing device featured a curved piece of metal that blocked the muzzle and diverted the blast, preventing injuries. This style was manufactured in 1968 through 1969.

There were two companies that manufactured blank firing adapters for the M14 rifle, one was East Moline Metal Products. Their blank firing adapters were marked EMMP.

The Springfield Armory conducted development testing for the optimum orifice size to be incorporated into the XM12 blank firing adapter. Functioning problems were being encountered with the device when used at arctic temperatures below -25 degrees Fahrenheit.

Breech Shield

To protect the trainee from excess gas pressure build-up, and sudden ignition of unburned powder collected in the breech during extended periods of firing blank ammunition, the M3 breech shield was developed. The M3 breech shield was used in conjunction with the M12 blank adapter. The metal breech shield consists of a sheet metal deflector shield and a guide lug with a spring-loaded plunger to secure the device to the receiver. The shield didn't interfere with the ejection of blank or live rounds, but was designed to prevent loading of ammunition through the stripper clip guide located on the top of the receiver of the M14 rifle. The M3 breech shield proved to especially effective when issued to left-handed shooters.

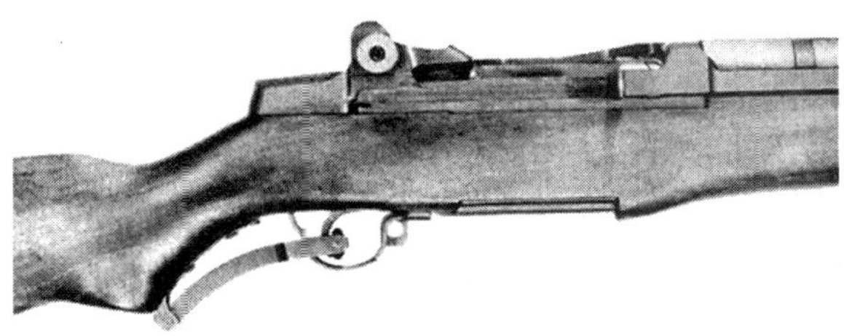

The addition of a winter trigger device allowed the rifle to be fired while wearing gloves or mittens. (Bill Ricca)

There were two separate designs of the breech shields produced. The original shield was redesigned to feature a simplified clip guide lock for securing the shield to the rifle.

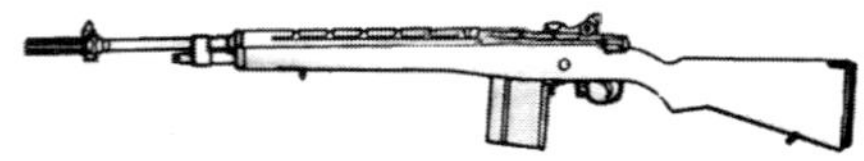

An early Type II prototype bipod. This model attached to the bayonet lug of the M14 rifle. (Courtesy of Bill Ricca)

Blank Cartridges

Special cartridges were designed for use with the M12 blank adapter. Their long narrow neck and lack of a projectile identify the M82 blank rounds.

Winter Trigger

The trigger guard of the M14 rifle was not large enough to permit efficient operation while wearing heavy gloves or mittens. The winter trigger was designed for use with the M14 rifle in arctic or extremely cold weather. The winter trigger consists of two wood-screws, the trigger and safety. Authorization for installation and use of the winter trigger was required from the theater commander.

The "Colley" design winter triggers were submitted for arctic testing, and recommended by the Arctic Test Board by their report of 21 February 1959.

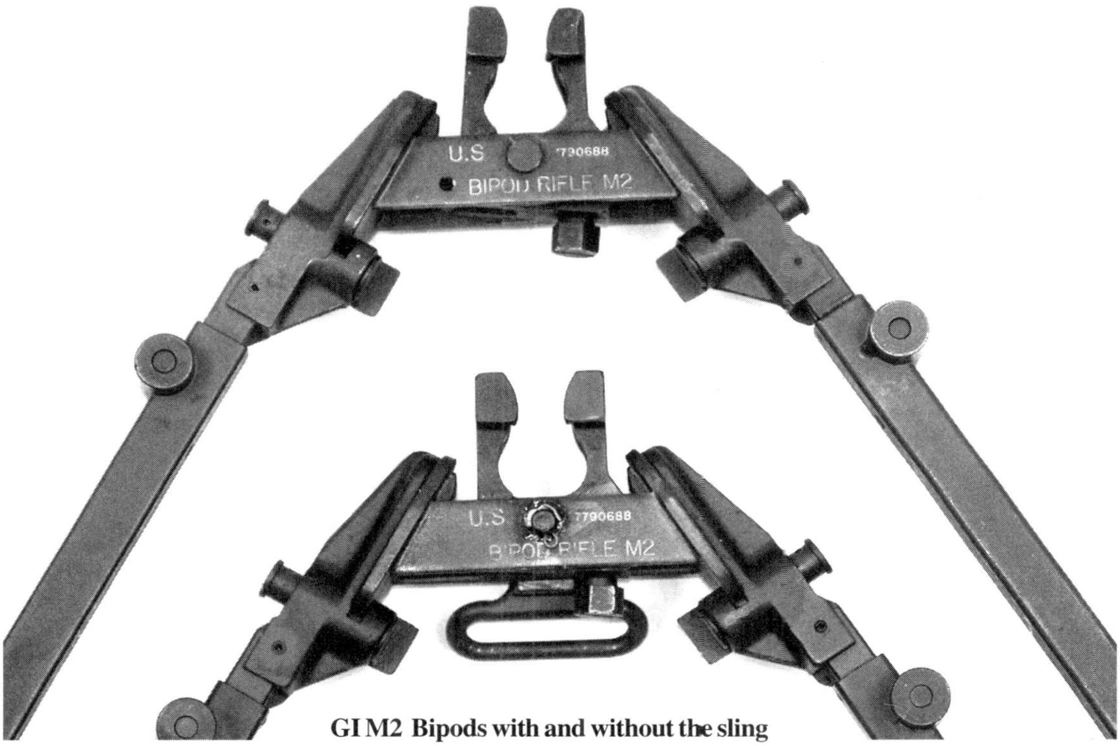

GI M2 Bipods with and without the sling swivel. (David Fassinger)

Bipod

A joint user-developer conference was held by the Infantry Board, Fort Benning, Georgia in February of 1959 to review and select a bipod design suitable for the M15 rifle modified for the Squad Automatic Weapon, replacing the BAR. There were two designs submitted by the Springfield Armory and subsequently tested by the Infantry Board from 16 March to 23 March 1959.

The two separate bipod designs were designated as the Type I and Type II. The Type I weighed .75-pound and had no adjustment for command height. The Type II bipod featured three command height adjustments, and could be stored folded rearward against the barrel.

The soldier on the left has an M2 bipod mounted on his M14, the man in the center has a blank firing adapter on his weapon. (U.S. Army)

The Type II bipod weighed 1.28 pounds. Both prototype bipods were designed to be attached to the weapon's bayonet lug. The bipods were tested with the heavy-barrel version of the M14 rifle, designated as the "M15, modified for the BAR role."

As early as 1958 the U.S. Marine Corps, while testing the T44E5 (M15), removed the weapon's bipod and placed in on a T44E4 rifle (M14). The Marines liked the lighter weapon with the bipod, which enhanced accuracy and range, and requested that a similar

M14 rifle Modified, with slotted fiberglass hand guard, flip-up buttplate, bipod and full-automatic selector switch.

bipod be authorized for use with the rifle in the standard configuration.

A third type of bipod designated as the Type III was offered to the user services by the Armory as an improvement over the Type II. The Type III bipod was readily attachable by means of a latching system that clamped over the gas cylinder and lock, as opposed to the earlier styles that utilized the bayonet lug. The attachment to the gas cylinder provided a more stable platform for the bipod. The Type III bipod was tested by the Army Arctic and Infantry Boards, and eventually Standardized as the Bipod, M2 in August 1959. During January of 1961 modifications were made to the M2 Bipod to provide a more positive locking of the legs in the extended position, by incorporating cylindrical plungers in lieu of the tapered plungers used in the original design.

Representatives from the Springfield Armory, the Ordnance Weapons Command, the St. Louis Ordnance District and the contractor Farmer Tool Supply, attended a meeting from 23 to 24 February 1961, in order to discuss the changes being incorporated into the M2 Bipod. Two engineers from the Springfield

Armory visited the contractor's plant in Denver, Colorado to assist the contractor in resolving problems being encountered in the metal finish process of the bipod. The Armory engineers worked at the Denver plant from 28 February to 3 March 1961 to resolve the problems. Initial samples of the bipods were received at the Armory from the vendor and tested from 8 March to 9 March 1961. The testing found the bipods to be acceptable. Confirmatory tests were conducted by the Infantry Board on 14 March through 17 March 1961.

Two different adapters were developed to modify the M2 Bipod for possible use with the M60 General Purpose Machine Gun. Accuracy tests were conducted with both the standard M60 bipod and the modified M14 rifle, M2 bipod.

A group of soldiers train on the firing line. All of their M14 rifles are equipped with M2 bipods. (U.S. Army)

The M15 version of the M14 rifle was adopted on 1 May 1957 and declared obsolete just seven months later on 17 December 1959. The Type III bipod, aluminum hinged buttplate, and fiberglass hand guard initially used for the M15 weapon were to be incorporated into the M14, and designated as the M14 Rifle (M) or Modified. To extend the effective range of fire of the M14 rifle out to 700-meters the detachable M2 bipod was authorized.

The M2 bipod was primarily issued with the M14A1 and M14 M (modified) that had a fire selector lever installed. There were two different variations of the M2 bipod. The early version was marked U.S. 7790833, originally manufactured without a sling swivel. Problems were encountered with the jaws coming loose on the early bipod. A later version, marked U.S. 7790688, had the jaws redesigned to provide a more secure grip on the gas cylinder. Like the earlier version, the redesigned bipod originally lacked a sling swivel, but later versions did have a sling swivel added at the factory. An Army Modification Work Order was issued with instructions to refit the older model bipods by installing a sling swivel and longer yoke pin.

The bipod without the sling swivel was suitable for use with the M14 M (modified) that did not have a foregrip assembly. The swivel was added for use with the M14A1 stock that had a foregrip to assist in controlling the weapon during automatic fire. The swivel on the bipod was used in conjunction with the provision for attaching a sling on the foregrip.

A circa 1986 advertisement offering M14 part sets. (Courtesy of Donald G. Thomas)

NEW ORIGINAL U.S. G.I. PARTS IN ORIGINAL WRAPPERS!

Part	Price	Part	Price	Part	Price
TRW Bolt and Roller Assembly	**35.95**	Flash Suppressor Wrench	**6.95**	Safety Spring	**1.95**
Bolt Stop	**4.95**	Front Band	**3.95**	Sight, Front	**5.95**
Bolt Stop Pin	**.40**	Gas Cylinder	**15.95**	Sight, Rear Aperture	**3.95**
Bolt Stop Spring	**1.40**	Gas Cylinder Lock	**5.95**	Sight, Rear Base	**5.95**
Butt Plate Assembly, Hinged	**21.95**	Gas Cylinder Plug	**6.95**	Sight, Rear Cover	**2.95**
Cartridge Clip Guide	**6.95**	Gas Piston	**6.95**	Sight, Rear Elevation Knob and Spindle	**8.95**
Cartridge Extractor	**4.95**	Hammer	**9.95**	Sight, Rear Windage Knob	**4.95**
Chamber Cleaning Brush	**1.95**	Handguard Assembly	**4.95**	Trigger Group Assembly, Complete	**59.95**
Combination Tool	**4.95**	Magazine Latch	**5.95**	Trigger Guard	**4.95**
Firing Pin	**3.95**	Operating Rod Guide	**6.95**	Trigger Housing Assembly	**21.95**
Flash Suppressor	**16.95**	Operating Rod Guide Pin	**1.95**	Trigger and Sear Assembly	**4.95**
Flash Suppressor Nut	**1.95**	Operating Rod Spring	**6.95**		
Flash Suppressor Set Screw	**1.95**	Operating Rod Spring Guide	**6.95**		
		Safety	**6.95**		

Another 1980 era advertisement of M14 parts. Some of the parts listed, like flash suppressors, have increased disproportionately in price when compared to other parts. (Don Thomas)

The sling was adjusted so that there was tension between the bipod and foregrip. This was necessary to keep the rearward pressure from operator's hand on the foregrip from breaking the grip or stock where the grip attached. Original GI bipods were assembled by brazing the yoke bracket pieces together. The bipod attaches to the gas cylinder and is removed or installed with the combination tool.

Aftermarket M2 Bipods

There were a number of commercial copies of the M2 bipod manufactured overseas for the civilian market. Many of the commercial M2 bipods were made in Taiwan or China. Most of the foreign copies have the same or similar markings as those found on original GI bipods.

Some of the aftermarket bipods are satisfactory for civilian use, while others were very poorly constructed. The brackets for the attachment jaws on original GI issue bipods were assembled by brazing and can be identified by a bronze color in between the joints. Af-

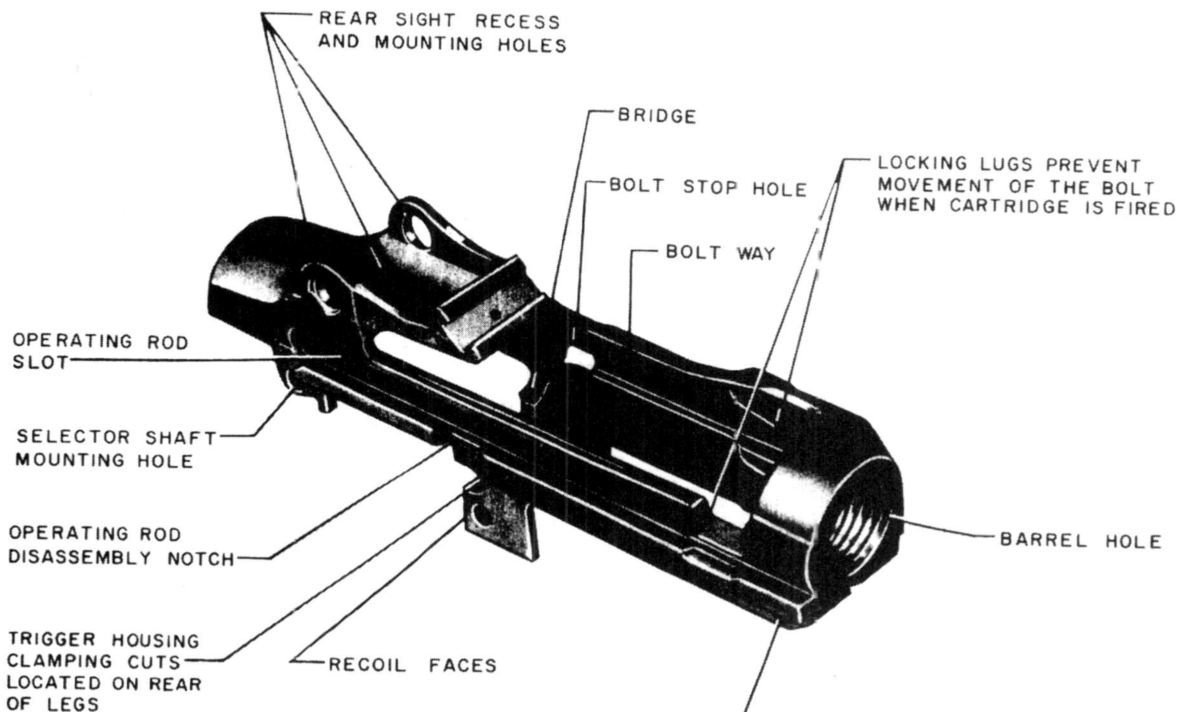

The M14 receiver.

termarket bipods were primarily assembled by welding of the component parts. Some foreign made bipods can be identified by the markings; W M I stamped on the yoke, some copies use a hexagon set-screw that requires the use of an Allen wrench for adjusting the attaching jaws. The attachment and securing of an original GI bipod only requires the wrench of the combination tool.

Parts

Parts for the M14 rifle were once very common on the surplus market, due in part to the sudden cancellation of the M14 contracts back in 1963. At that time the primary contractors had large stockpiles of M14 parts waiting to be used in assembling new rifles. With no use for the parts the contractors sold them off at bargain prices. Over the years a few of the surplus parts were slowly being sold to collectors and shooters. Companies that were assembling new M14 type rifles for the civilian market however, purchased the vast majority of the GI parts. As the companies making M14 type rifles proliferated, the parts began to dry up quickly, forcing the companies to manufacture whatever parts were needed. Many parts were subcontracted to companies in foreign countries, the Wayne Machine Company located in Taiwan being one of the largest suppliers. For a brief time period that spanned from 1996 to 2003 the Civilian Marksmanship Program (CMP) offered surplus GI parts and magazines at very reasonable prices. Reportedly the recent re-issue of the M14 rifle for use in Iraq and Afghanistan ended the further release of surplus GI M14 parts. The parts are now being used to support M14 rifles recalled to duty. Currently GI barrels, flash suppressors, gas cylinders and operating rods are becoming quite scarce and accordingly expensive. The only good news is that prices of GI M14 magazines have dropped dramatically since the 1994 ban expired.

M1 Rifle Parts the Same, but Different...

There are quite a few parts that will interchange between the M1 and M14 rifles. There are also a few that will interchange but are different. One is the sear, the M14 sear has a small tang protruding from the side. Although the M1 sear will fit in an M14, the automatic fire feature will be disabled. Another part is the sight elevation pinion, although they will

A standard configuration M14 barrel manufactured in August, 1967 by the government Springfield Armory. The markings were located on the left side of the barrel, and could not be seen by retracting the operating-rod.

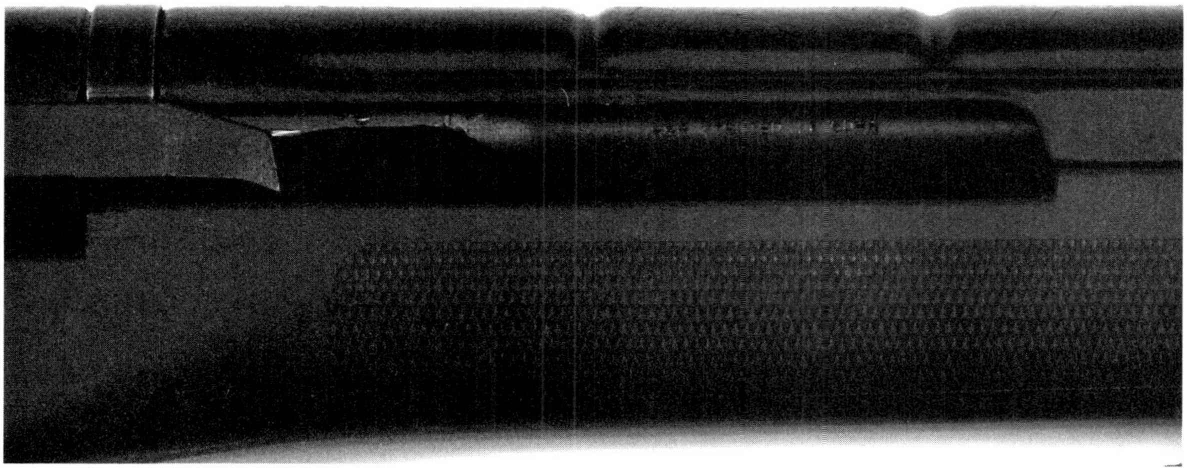

A Harrington & Richardson barrel with a manufacture date of November, 1960. (Jim Pecori)

interchange between the M1 and M14 rifles they are marked differently. The M1 sight elevation pinion is calibrated in yards with a maximum range of 1200. The M14 sight pinion is calibrated in meters, out to a distance of 1100. The M14 pinion is marked with a letter M to designate meters.

Receivers

M14 rifle receivers were manufactured by all four of the primary contractors. The receivers were made by the drop-forging method from 8620 steel, after all of the machining was completed the receiver was heat-treated. The heel of each receiver was marked with the primary contractor's name or logo, a serial number and the caliber marking 7.62 MM. The receivers were made to strict manufacturing tolerances. The left side of the M14 receiver was marked with its Ordnance drawing number 7790189.

Barrels

Government contract barrels made for the M14 rifle were manufactured from premium

A November 1962 dated Winchester barrel. (Bill Ricca)

A TRW barrel dated 8-63. (Jim Pecori)

chromium molybdenum vanadium or resulpherized 4150 steel blanks. The Springfield Armory used the same barrel blanks for both the M1 and M14 rifle barrels. The overall length of the standard issue M14 barrel was twenty-two inches, with a twist rate of 1 turn in 12 inches. All standard barrels and their chambers were chromium lined to extend service life and facilitate easier cleaning in the field. There were a number of companies contracted to manufacture barrels for the M14 to include; The Springfield Armory; Winchester; Harrington and Richardson; Thompson Wooldridge Ramo. These were the prime contractors. Other firms that manufactured barrels were Saco-Lowell and Canadian Arsenals. Contractors were held to strict tolerances on barrels. If even a few barrels from a lot were gaged and found to be substandard the entire lot would be scrapped. Barrels were generally marked on the right-side near the breech. The barrel markings, generally marked in the breech area of the barrel, included the DOD eagle stamp, the manufacture's name or code and the date of manufacture. A letter P indicating the barrel had been proof tested. The steel heat-lot code was identified by an alphabetical or alpha-numerical designation. The barrel markings can normally be viewed by retracting the operating rod.

During an arsenal rebuild used barrels were inspected for serviceability while remaining attached to the receiver. A visual inspection was done with attention being given to the chamfer and face of the muzzle. The gas port hole was to be clean and free of carbon or corrosion. The straightness of the barrel was checked by dropping a special gage into the bore. If the gage were unable to drop completely through the bore of the barrel under its own weight, the barrel would be considered bent and thus unserviceable. The bore was to be inspected for pitting, flaking of the chromium and sharp lands and grooves. If excess throat erosion, pitting, bulges or corrosion was present the barrel was subject to rejection. Barrels with a slight loss of chromium plating at the origin of the rifling within 1/2 inch; fine or scattered pitting; cloudy or hazy film from phosphating; faint rings; tool marks or scratches were deemed acceptable.

National Match Barrels

Government contract National Match barrels

GAS CYLINDER PLUG TIPS

What do you figure causes the most damage to the handguard? You're wrong. The real damage comes when you hold on to the handguard while you're tightening or loosening the gas cylinder plug.

The right way to do it is to hold on to the barrel and gas cylinder or, if the bipod's attached, grab that. If the barrel and cylinder are hot . . . you can't wait for them to cool off . . . and the bipod's not attached, stick a screwdriver between the barrel and cylinder for leverage.

By the incidentally . . . that combination tool, FSN 4933-768-0211, is the only thing you want to use for removing and installing the gas cylinder plug. Use the combination tool to tighten the plug as tight as you can get it. Don't—double don't—use any extension on the tool for added leverage when you tighten the plug. The only time you might need extra leverage on the tool is when you go to remove a carboned-up plug.

Speaking about tightening again . . . you can see that the cylinder and plug threads are fine, so you have to be careful or you'll cross-thread them.

It's worth remembering that the only stuff you use to clean the gas piston, gas cylinder bore and gas cylinder plug is bore cleaner. The thing to do is soak the piston and plug in bore cleaner and then wipe them with patches or a clean cloth soaked with bore cleaner. You don't have to put the gas cylinder in bore cleaner, but you do clean the bore with patches or a clean cloth loaded with bore cleaner.

Don't use any kind of abrasive or something sharp for scraping. The piston, f'rinstance, is made to mighty fine tolerances . . . and trying to clean it with something like crocus cloth, steel wool or sand sure can mess up things.

The inside of the piston, tho, can be cleaned with your bore brush and bore cleaner. The same goes for the gas cylinder plug.

The deal that is like to tear most guys out of the frame is the way components of the gas system change color once the rifle is fired. They figure they have to get the parts looking like new—polished to a shine. So they give the parts a good going over—short of sand blasting.

Army PS manual gas cylinder maintenance tips from Connie Rodd. (Courtesy of John M. Miller)

Just remember: That change of color is caused by gas and heat. And it doesn't have anything to do with the way the rifle works. So, skip trying to polish up those parts.

Another thing . . . seeing's how the components of the gas system are made of corrosion resisting steel, they're not going to rust. So you assemble them dry. That's right . . . no oil. Only the exterior surfaces get a light coat of oil.

There're two places you don't want to forget to put oil. Those're the inside of the operating rod tube and the spring and plunger for the connector assembly. **You get at the inside of the tube with an oily patch and your cleaning rod.**

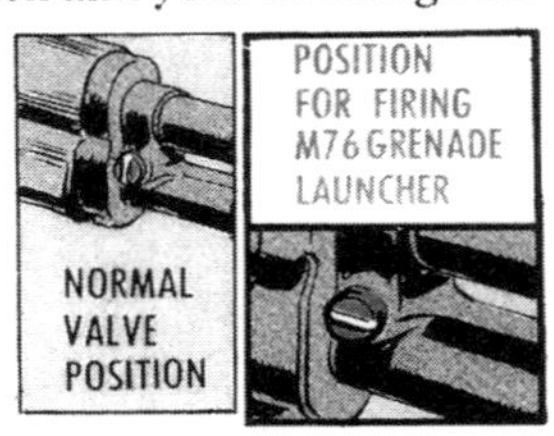

You can run into a rough situation when you get carbon between the spindle valve and gas cylinder. When this happens, it's a real job to rotate the valve. 'Course . . . the only time you rotate the valve from vertical to horizontal for firing is when you use the M76 grenade launcher. You want to push it in and rotate it once in awhile just to break loose and get rid of excess carbon.

Anyway . . . if you can't push in the valve so you can rotate it, put a block of wood against it and give the wood a few raps until you're able to move the valve in and out. Then you should be able to turn the valve. Don't try to take it apart. That's a job for Ordnance.

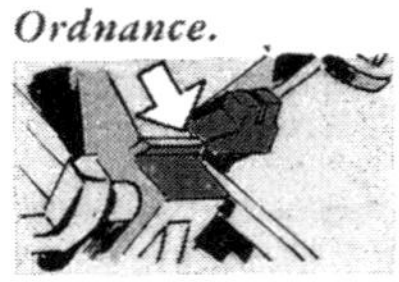

Now for some quick scoop. But it's important. When the firing mechanism's installed in the receiver and the selector is set for automatic fire, make sure the sear release is touching the sear.

Something else that needs checking — before and after firing — is the flash suppressor and front sight. Give the suppressor and sight a coupla shakes. And if either moves at all, ask your armorer to see if he can spot the trouble.

More maintenance tips from Miss Connie Rodd taken from an Army PS manual. (Courtesy of John M. Miller)

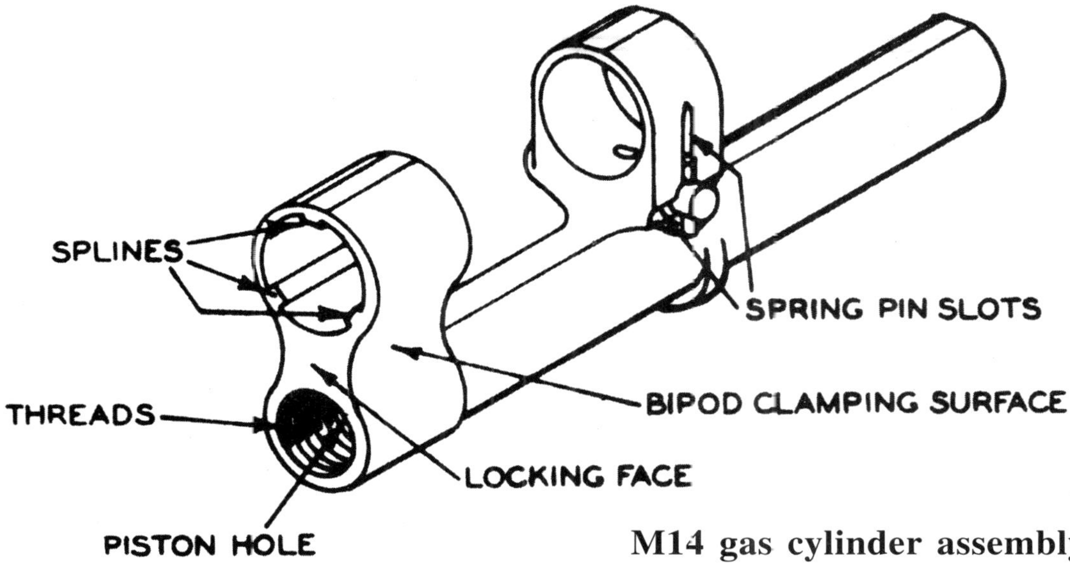

M14 gas cylinder assembly

were manufactured in light, medium and heavy weight configurations. The barrels were held to much closer tolerances than the standard issue M14 barrels. Many of the barrels were made by commercial firms. The barrels were not chromium lined as were standard issue barrels, the barrels were marked NM on the barrel shaft near the muzzle. National Match barrels were produced by Gene Barnett, Springfield Armory; Thompson Wooldridge Ramo; Canadian Arsenals LTD; Hart; Douglas; Kreiger; Saco-Lowell; SGW; Nomura. National Match as well as standard grade M14 barrels continue to be manufactured by modern firms to keep up with current demand.

Gas Cylinder

The gas cylinder of the M14 rifle was manufactured from 416 stainless steel. The gas cylinder housed the gas piston and a spindle valve. The spindle valve was used to divert

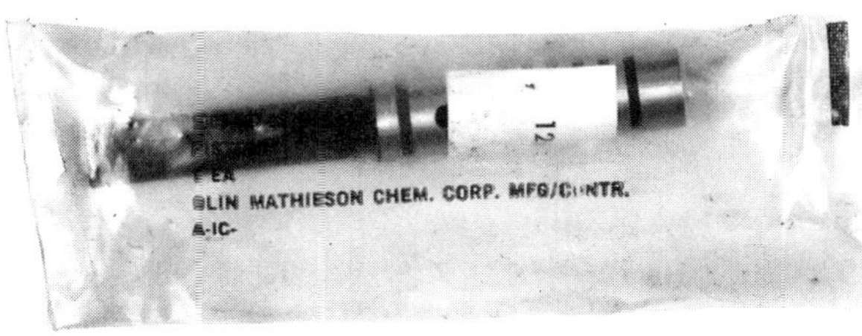

A new in-the-wrap Winchester gas piston. (Courtesy of Bill Ricca)

gas pressure for the launching of rifle grenades. The gas cylinder tapped propellant gas from a gas port at the bottom of the barrel after a fired projectile passed over the port. The gas was used to move the gas piston, which impinged on the face of the operating rod, moving both components rearward. Through a camming action the operating rod turned the breech bolt unlocking it and moving it rearward to cock the hammer. At the end of its travel operating rod would be propelled forward by the energy of the compressed recoil spring. As the bolt was moved forward by the operating rod, a fresh cartridge would be stripped from the magazine and guided into the barrel's chamber. The bolt would seat against the cartridge as the bolt was locked into position, and the cycle repeated after the cartridge was fired.

The gas cylinder, being made of stainless steel, was finished by the black oxide process.

The end of the gas cylinder was fitted with a threaded stainless steel plug to remove the piston and for cleaning of the inside of the cylinder. The plug could be removed with the combination tool. Troops often cross-threaded the fine threads of the plug rendering the gas cylinder and plug unserviceable. Another problem often encountered was over-tightening of the gas plug, particularly on a hot barrel. The rifles usually had to be turned in to an armorer to loosen the plug. The gas cylinder and piston were to be completely dry and free of oil when assembled. The gas plug was inspected for serviceable sealing and pilot bearing surfaces. If the threads were damaged or chased the plug was to be discarded. The gas chamber hole was to be free of carbon, although discoloration of the hole from heat was acceptable. The hex head of the plug was checked for damaged flats that had been inflicted by the use of an incorrect tool, if damaged the plug was discarded.

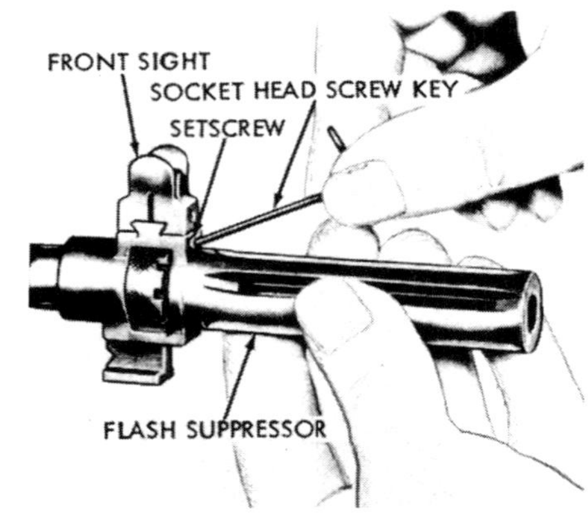

The first step in removing the flash-suppressor from the M14 rifle (after insuring the weapon is unloaded) is to remove the allen-head screw that locks the castle nut.

One of the most critical areas of the gas cylinder was the cylindrical hole for the gas piston. Gage 8440826 was used to check the cylinder for serviceability. The maximum dimension was .5015-inch. The gas spindle valve was also inspected for proper operation and correct assembly. When manually depressed the valve should easily rotate and be indexed to both positions. If the spindle valve was not rotated occasionally it was subject to carbon fouling and rotation was not possible.

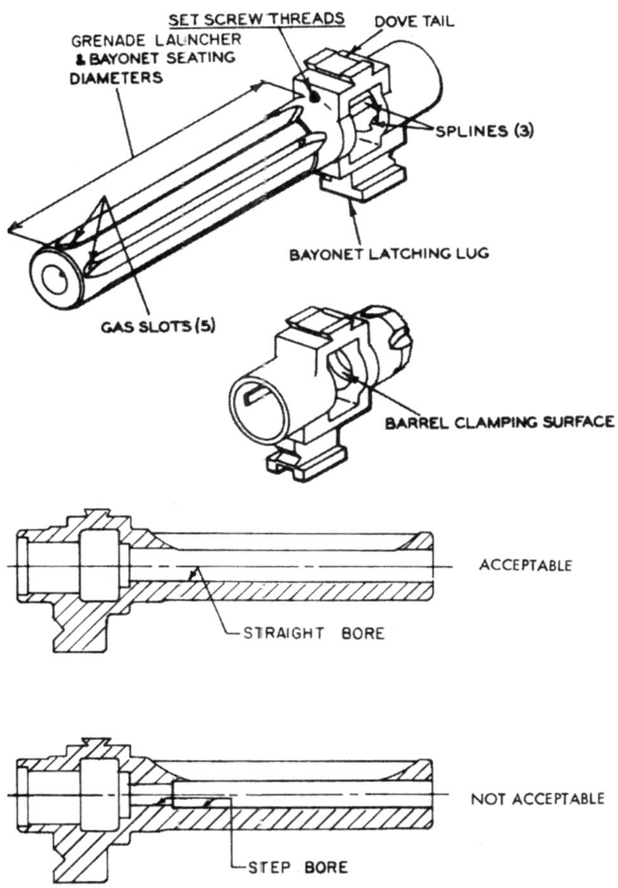

M14 flash suppressor

The Gas Piston

One of the features that differentiated the M14 from the M1 is the gas system. While the M1 operated by gas pressure concentrated directly on the tip of the operating rod, the M14 used a piston made from 416 stainless steel to actuate the operating rod. To operate properly and to prevent carbon build up, the piston and gas cylinder were to be kept free from oil and bore cleaner. Troops were trained (and warned) not to remove the piston as long as it would slide back and forth easily. Another directive was to turn the rifle upside down and turn the spindle valve so that the screwdriver slot was parallel to the barrel (grenade launching position) this would keep bore cleaner from seeping down into the gas cylinder. If cleaning of the gas piston and cylinder was necessary, the use of steel wool or scraping of carbon was prohibited. Only CR bore cleaner was to be used, if there was an excess of carbon build up, carbon removing compound P-C-111a (FSN 6850-620-0610) was authorized. The gas piston and cylinder were not to be lubed, and were to be reassembled completely dry. A threaded gas plug was placed at the forward end of the gas cylinder in order to remove the piston for cleaning. The threads of the gas plug were very fine and often were cross-threaded by careless troops. The threads on the plug were to be kept completely free of carbon. When being installed on the gas cylinder the plug was only to be turned finger tight followed by a slight "nudge" with the wrench on the

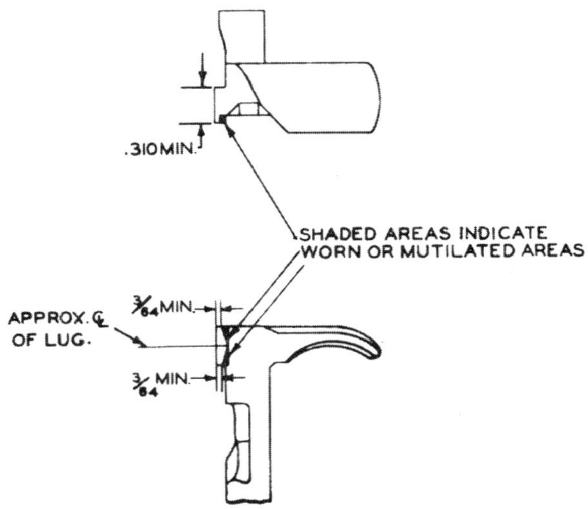

COMPONENTS WITH WEAR OR MUTILATIONS ON EITHER OR BOTH SIDES OF LUG WHICH ARE EQUAL TO OR BEYOND THE LIMITATIONS SHOWN ABOVE ARE NOT ACCEPTABLE

The numerical code 66118 indicates that this operating rod was manufactured by Winchester.

combination tool. Over tightening of the gas cylinder plug or tightening the plug on a hot barrel would make the plug very difficult or impossible to remove. The grenade launching spindle valve located on the gas cylinder was to be kept free by pushing it in and out and turning it on a daily basis, this would prevent carbon build up on the valve.

Flash Suppressor

The flash suppressor designed for the M14 rifle was the prong type with a closed end. The flash suppressor was originally designed as a casting, but this designed failed to hold up under sustained operation. The casting was replaced by a new suppressor machined from steel. The flash suppressor was machined with three splines for proper alignment onto the M14's barrel. The flash suppressor was secured to the barrel by a castle nut. The nut in turn was locked by a set-screw that fit into a threaded hole provided at the base of the front sight. There were five-gas slots machined into the sides of the flash suppressor. A flat web area was left at the bottom of the flash suppressor to offer limited muzzle control, and prevent escaping gases from kicking up dust possibly revealing the infantryman's position. All of the webs of the flash suppressor subtend at a 29-degree angle except for the wider bottom one that subtends at a 65-degree angle.

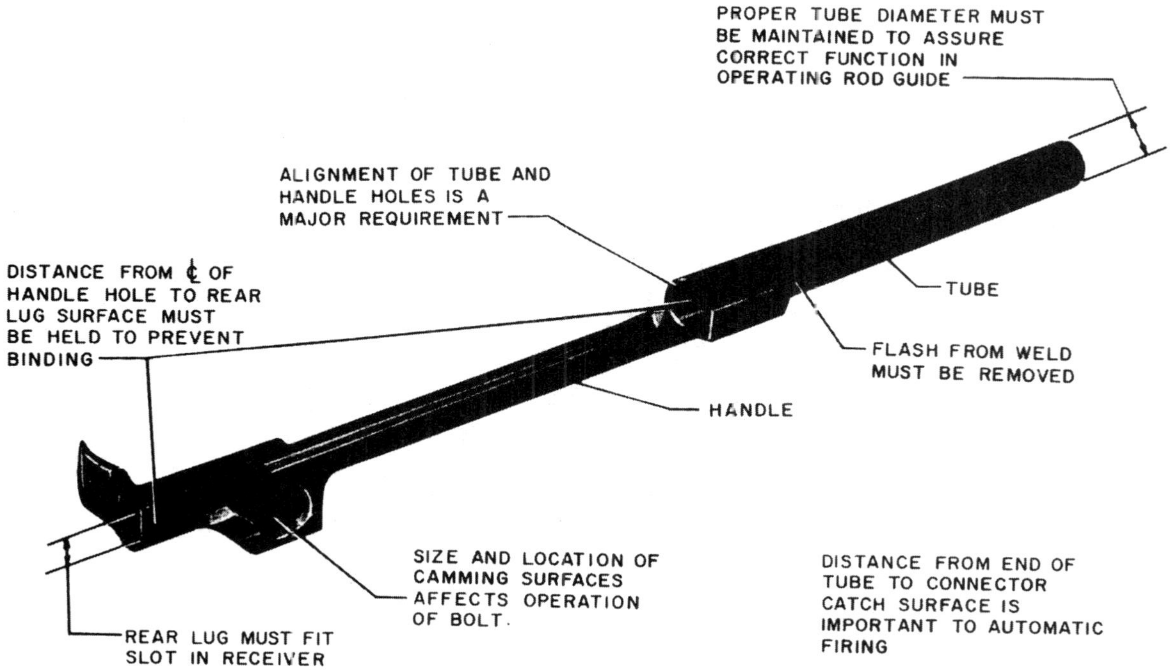

OPERATING ROD ASSEMBLY D7267064

Also incorporated into the flash suppressor was a dovetail for mounting the front sight assembly at the top and an integral lug at the bottom for attaching a bayonet, muzzle stabilizer, grenade launcher or a blank firing device.

The flash suppressor was inspected for burrs, cracks and mutilation with particular attention given to the bayonet latching lug, splines and set screw threads. The gas slots were inspected for burn-thru or cracks.

In addition to the early cast flash suppressor there was another design that had a step incorporated inside of the bore. The area inside of the suppressor closest to the muzzle was smaller than the rest of the bore through the device. Both the cast and stepped-type of flash suppressors were deemed obsolete and unacceptable.

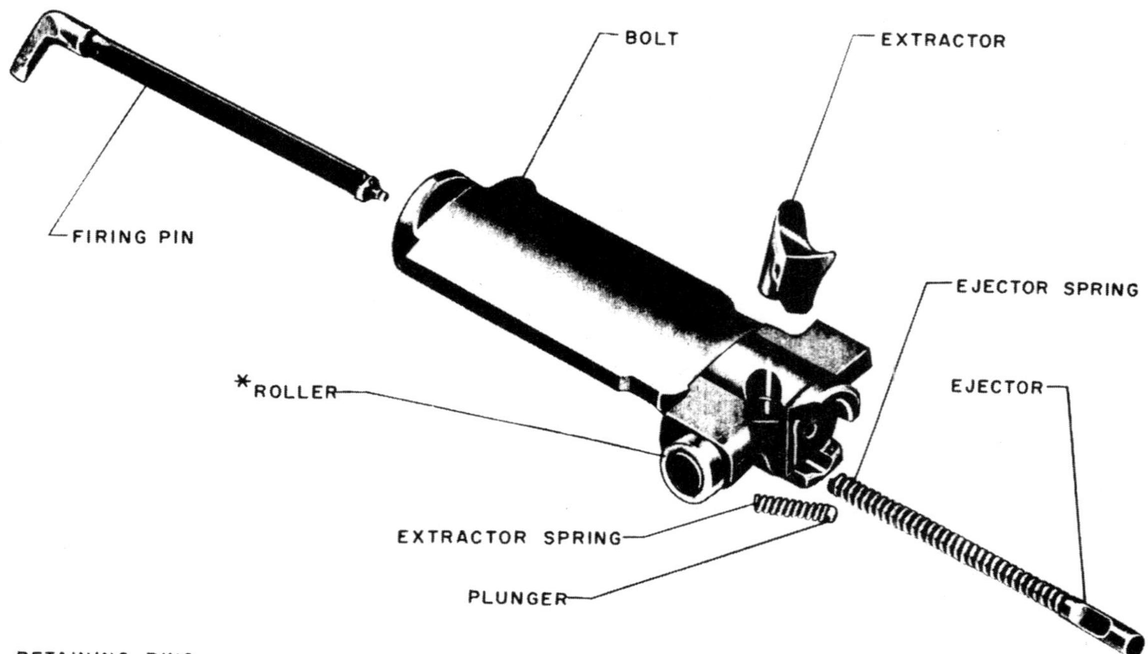

RETAINING RING THAT HOLDS ROLLER TO BOLT NOT SEEN

M14 bolt assembly

During the poor accuracy investigation of the M14 rifle during 1962 it was noted that the flash suppressor had a negative effect on accuracy. Noted in the report was that; the flash suppressor increases muzzle blast; the device has a tendency to loosen during firing; and that the tightness varies relative to the position of the notch on the castle nut in regard to the set-screw that is used to secure the nut; because the front sight is mounted on the flash suppressor accuracy suffers when the device is loose; the flash suppressor unnecessarily increases the overall length of the weapon.

Operating Rod

Operating rods were manufactured by all of the four primary contractors of the M14 rifle. All GI operating rods, except for those produced by TRW, were manufactured in two separate pieces that were welded together, using a method originally developed by the Remington Arms Inc. TRW operating rods were constructed in one-piece. The original Ordnance Corps drawing D7267064 designated that a 0.077 + 0.006 -inch hole be located in the face-end of the tube that contacts the gas piston, although a number of operating rods were manufactured without the hole. The M14 Depot Overhaul Manual published by the Rock Island Arsenal stated that operating rods both with and without the hole were acceptable. The Mercury Tool and Machine Company manufactured replacement operating rods during 1986.

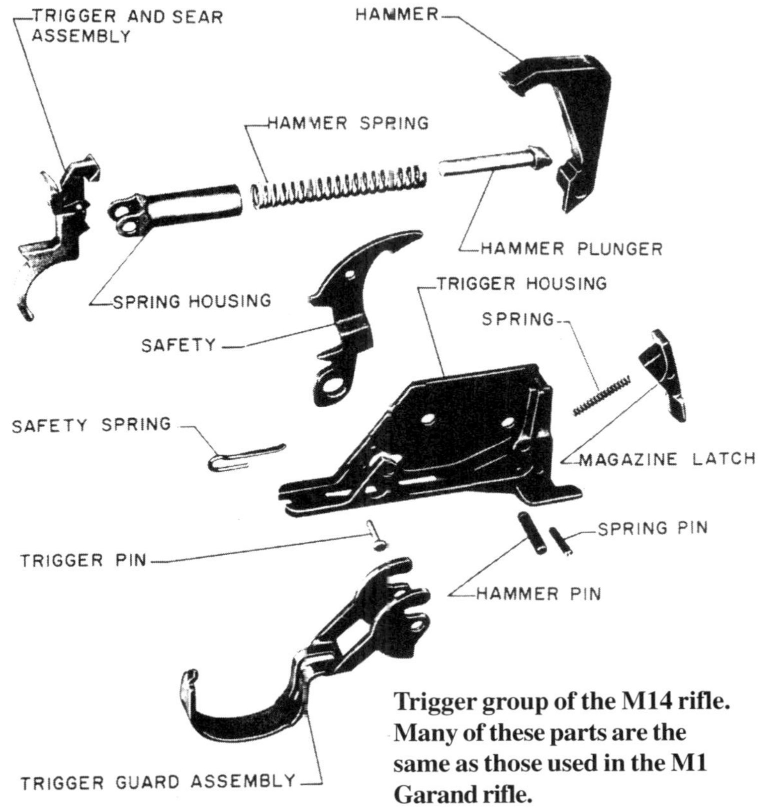

Trigger group of the M14 rifle. Many of these parts are the same as those used in the M1 Garand rifle.

The markings to identify the manufacturer of the operating rods and the reference part number were marked on the right side on the web that connects the handle to the tube. Operating rods were made of 8645 forged steel. For lower echelon maintenance the inspection requirement was a visual inspection and that the operating rod had free unrestricted movement with the operating spring guide.

At a higher level of maintenance the operating rod guide lug, cam area and receiver contact area were to undergo a visual inspection for missing or defective finish, burrs, cracks or mutilation. The guide lug height was .065-inch with a minimum dimension of .310-inch. There was no requirement for a magnetic particle inspection of the camming slot.

There were a large number of GI operating rods destroyed by the demilitarizing process, in other words cut into pieces. A few enterprising individuals reclaimed demilled operat-

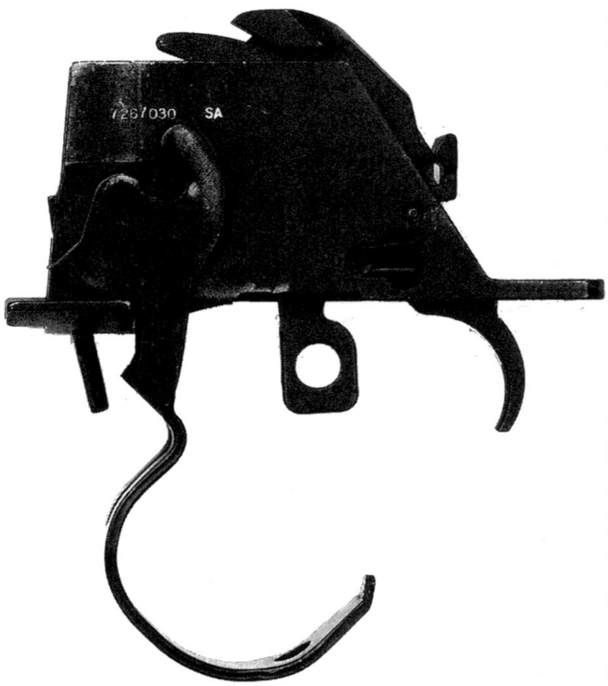

Springfield Armory (SA) trigger group.

ing rods by fitting and welding them back together. The quality would be as good as the individual who performed the work, and there have been some that were well done and serviceable. It would be something to be aware of when considering the purchase of a replacement or spare M14 operating rod.

Bolt, Breech

Breech bolts for the M14 rifle were constructed from 8620 steel. GI bolts were marked on the top with the manufacturer's name and part number. Other bolt markings included the heat treatment and steel lot used in the manufacture of the bolt. A prick punch mark designates that the weapon was proof tested. A letter M was inscribed on the bottom of bolts that had under gone a magnaflux inspection for cracks. The locking lugs of the bolt were the most susceptible area for developing stress cracks.

During depot maintenance the hammer cam, locking surfaces, firing pin slot, bolt face, firing pin hole were visually inspected for erosion or damage. The diameter of the firing pin hole was measured with a NO-GO gage 7458398, maximum dimension was 0.083-inch. Pits or dents on the bolt face from careless use of the cleaning rod were not considered reason for rejection. Washout or erosion at the firing pin hole was cause for rejection. Erosion up to 1/16-inch in width or length and not more 1/32-inch in depth was considered acceptable on other areas of the bolt face. The bolt roller was checked for rotation in both directions while exerting a push-pull force on the roller.

All bolts that were not cracked or having excessively eroded firing pin holes were to be reclaimed. All rollers and retention clips removed from used bolts were scrapped. The bolt was to be refinished with Phosphate Type Z or Type M Class 1 Mil-P-16232. Thompson Wooldridge Ramo (TRW) manufactured

MWO 9-1005-223-30/1

DEPARTMENT OF THE ARMY MODIFICATION WORK ORDER

RIFLE, 7.62-MM, M14A1:
MODIFYING STABILIZER ASSEMBLY
TO IMPROVE LOCKING RETENTION CAPABILITIES
Ref: TM 9-1005-223-20
TM 9-1005-223-35

Headquarters, Department of the Army, Washington, D. C.
8 April 1968

1. **Application.**

a. Category of Maintenance. Direct support.

b. Applied By. Small Arms Repairman (MOS 45B20).

c. Time Required. 0.6 man-hour.

d. Remarks. None

2. **Supply of Parts Required.**

Note. Until six months from publication date the Army will provide the necessary modification kits on a free-issue basis when the MWO is cited as requisitioning authority. Requisitions received subsequent to the above prescribed date will be supplied as free-issue items only if assembled kits initially procured are available in stock. Otherwise, the requisitions will be canceled and the customer advised that the items required to accomplish modification will be issued on a reimbursable basis from the stock fund upon receipt of requisitions for individual line items.

a. New Parts and Material Required.

FSN	*Description*	*Part No. and mfg code*	*Qty*	*Fig.*
1005-933-4321	MODIFICATION KIT, MUZZLE STABILIZER ASSEMBLY:	5910816 (19204)	1	
	Consisting of:			
5215-929-0862	PIN, SHOULDER, HEADED	7791664 (19205)	1	2
5305-956-3127	SCREW, CAP, HEXAGON HEAD:	11686519 (19205)	1	2
1005-951-3056	STOP, MUZZLE STABILIZER:	11686518 (19205)	1	2

the majority of replacement M14 bolts. After bolt replacement a headspace check was mandatory. Production bolts could vary up to .004-inch in length.

Firing Pins

There were several firing pin designs for the M14 rifle. The earliest variation were Parkerized, the second variation were Parkerized with a chrome tip. The third variation were completed chrome plated. The pins were inspected with particular attention given to the striker tip and the tang areas. If any burrs, cracks, sharp edges or chips were present the pin was to be replaced. For replacement only the chrome plated firing pins were to be used.

M14E2 and M14A1

There were several parts specifically for the M14E2 and M14A1 rifle variations. The most obvious was the stock.

Muzzle Stabilizer

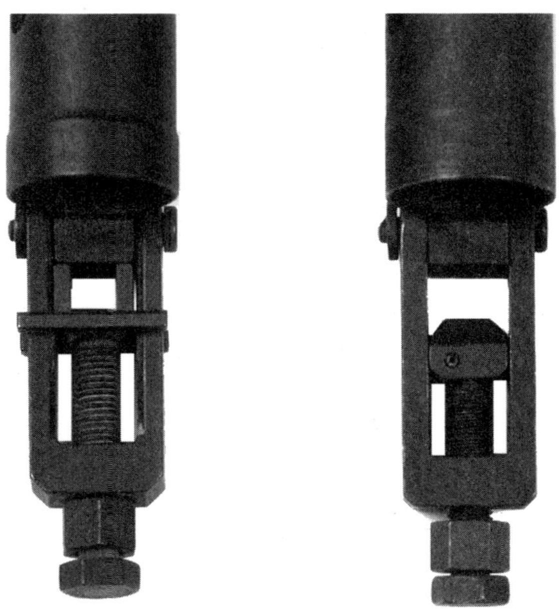

One of the problems reported from the field with the M14E2 rifle was that the muzzle stabilizers were coming loose. The Springfield Armory studied the problem and designed a few new parts to solve the problem. The stabilizer on the left has been modified, the one on the right is in the original configuration. (Bill Ricca)

Part number of a stabilizer with the new locking mechanism (top) and the older version (below). (Bill Ricca)

The muzzle stabilizer was designed for taming the muzzle rise of the M14 rifle when in the Automatic Rifle configuration. The muzzle stabilizer was basically a steel tube with holes drilled into it to direct muzzle gases in the opposite direction that the muzzle had a tendency to pull, upward and to the right. The original prototype stabilizer completely replaced the M14's flash suppressor, and was attached using the same castle nut. The new redesigned production stabilizer slipped over the flash hider and was attached to the bayonet lug by a hinged clamp. The unit was secured with a bolt and jam nut to keep the bolt from backing off. The head of both the bolt and the nut were the same diameter as the nut on the gas cylinder plug, and the locking bolt on the M2 bipod. Each device could be loosened of secured with the wrench on the M14's combination tool.

One of the problems reported from the field was that the muzzle stabilizers were coming loose and continually had to be retightened.

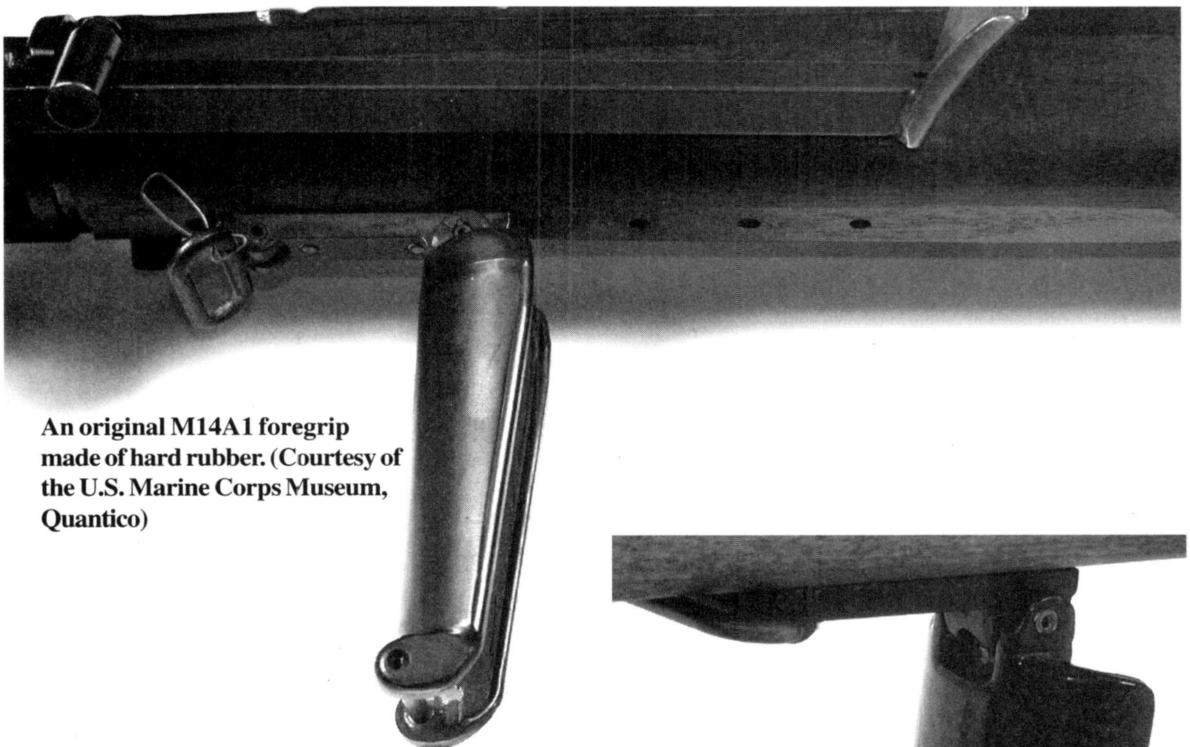

An original M14A1 foregrip made of hard rubber. (Courtesy of the U.S. Marine Corps Museum, Quantico)

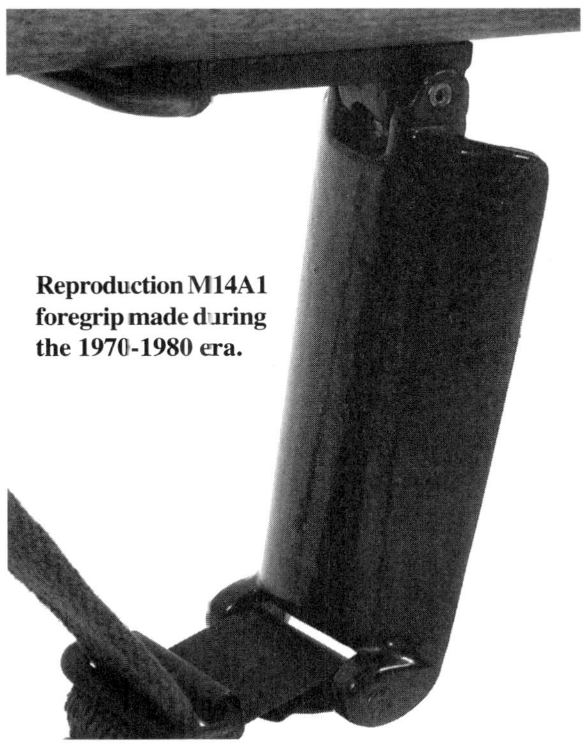

Reproduction M14A1 foregrip made during the 1970-1980 era.

The Springfield Armory studied the problem and redesigned and manufactured a few new parts to solve the problem. A Modification Work Order was issued on updating existing stabilizers in the field, while new manufacture stabilizers would be made to the new specifications. This resulted in three variations of muzzle stabilizers; the original unmodified version; the field-updated variation and the new stabilizers built with the improved locking mechanism.

Scarcity of original GI issue M14A1 muzzle

NORMAL

MWO 9-1005-223-40/1

DEPARTMENT OF THE ARMY MODIFICATION WORK ORDER

RIFLE, 7.62-MM, M14A1:
REINFORCING SHOULDER GUN STOCK
Re: TM 9-1005-223-20
TM 9-1005-223-35

Headquarters, Department of the Army, Washington, D. C.
18 June 1968

1. Application.

a. Category of Maintenance. General support.

b. Applied By. Small Arms Repairman (MOS 45B20).

c. Time Required. 0.6 man-hour.

d. Remarks. None.

2. Supply of Parts Required

Note. Until one year from publication date, the Army will provide the necessary modification kits on free-issue basis when the MWO is cited as requisitioning authority. Requisitions received subsequent to the above-prescribed date will be supplied as free-issue items only if assembled kits initially procured are available in stock. Otherwise, the requisitions will be cancelled and the customer advised that the items required to accomplish modification will be issued on a reimbursable basis from the stock fund upon receipt of requisitions for individual line items.

a. New Parts and Material Required.

FSN	Description	Part No. and Mfg Code	Qty	Fig
1005-933-4320	MODIFICATION KIT, STOCK ASSEMBLY, GUN SHOULDER:	5910815 (19204)	1	1
	Consisting of:			
5325-904-9303	GROMMET, RUBBER:	11686524 (19205)	4	
1005-951-3254	PLATE, BACKING, HANDGRIP:	11686522 (19205)	1	
5305-956-3401	SCREW, MACHINE:	11686523 (19205)	2	
5310-194-9209	WASHER, LOCK, COUNTER-SUNK:	MS-35336-21 (96906)	2	

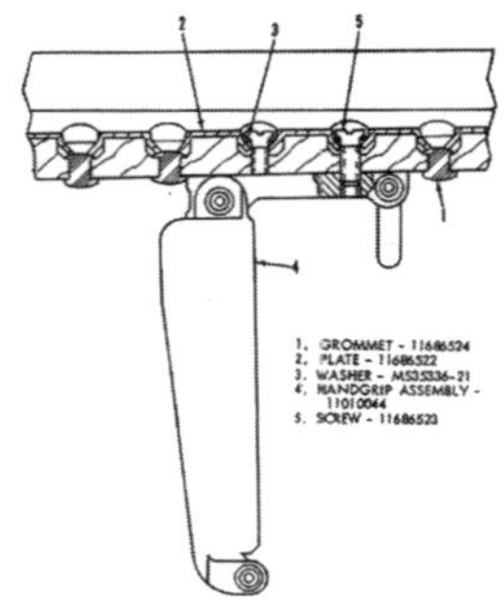

Figure 1. Modification of shoulder gun stock

b. Source of Supply. To be requisitioned through normal supply channels.

c. Date Available. Now.

d. Weight and Cubage of Parts and Materials. Not pertinent.

3. Purpose of Modification. To provide a means of strengthening the shoulder gun stock by reinforcing the front handgrip area.

4. Priority Classification. This modification work order is classified in the NORMAL group and will be applied to all issued equipment in active Army using units, maintenance float, and station stock as is consistent with the needs of the service, but in all cases within six months of publication date, and to all other Army-owned equipment (except unserviceable, economically repairable items) within 12 months of publication date.

5. Major Item Affected.

Description	Federal stock No.	TM No.
RIFLE, 7.62-MM M14A1	1005-072-5011	9-1005-223-35

6. Assembly or Component Affected.

Description	Federal stock No.	Drawing No.	Date
STOCK ASSEMBLY	....	F7791671	23 Oct 63

7. Parts Affected

a. Part to be Modified.

Description	Part No.	Federal stock No.	Qty
STOCK, GUN, SHOULDER	7791679	1005-072-5391	1

b. Excess Parts.

Description	Part No.	Federal stock No.	Qty
BUMPER, RUBBER	11010048	1005-072-5380	4
BURR, STOCK ASSEMBLY:	7790474	1005-072-5381	2
SCREW, HANDGRIP ASSEMBLY:	7791675	1005-072-5387	2
WASHER, LOCK:	MS-35335-32	5310-596-7691	2

c. Disposal of Excess Parts. Parts will be disposed of as scrap.

8. Drawings Required to Apply Modification. None.

A Modification Work Order for addressing the stock breakage problems encountered with the M14E2/A1 rifles. (Courtesy of the Rock Island Arsenal Museum)

stabilizers has spawned a few reproductions, some very good while others are cast and poorly made.

Front Hand Grip

A forward handgrip was designed for the M14E2 and M14A1 to aid the infantryman in controlling the weapon during automatic fire. General Tire was one of the original contractors that worked in concert with the Springfield Armory to design and develop the folding foregrip. As originally designed the foregrips proved fragile and underwent a design change to improve their reliability. In addition to General Tire original GI M14E2 and M14A1 foregrips were manufactured by NE Inc. and later by the Do-Well Manufacturing Co.

When using an original M14 rifle stock on a semiautomatic M1A™ or civilian M14 type rifle, the slot cut into the wood for the full automatic selector shaft is present. This hole gives an unfinished look to the rifle. To address the problem and to provide more of an authentic look, non-functioning selector kits are available. The nonfunctional shaft can be fitted with either a GI selector switch or lock. (Photos courtesy of Tony "Hawk" Diaz)

All of the rifles on this page have been fitted with faux selector shaft kits. The rifle to the right has a selector switch, while the 3 rifles below have selector locks on them.

After the breaking foregrip problem was seemingly solved, another problem began to surface, the M14A1 stocks began to crack and break in the area where the foregrip was attached. The redesigned foregrips were now transferring the stress to the foregrip of the stock. In addition to the stock breakage, the hinges of the foregrips began to experience breakage. To solve the two problems a new sling swivel was added to the M2 bipod, the swivel was mounted on the bipod's yoke pin. The primary purpose of the additional swivel was to attach the sling in a fashion that would maintain sling pressure on the foregrip so that the rearward pressure from the shooter's weak hand when firing the weapon would be absorbed by the sling and not the stock itself. The swivel was similar to that mounted on the rear of the M14 stock, but the hole was drilled slightly larger and was not tapped. Breakage problems still continued to occur when the sling was installed incorrectly. The Army issued several illustrated instruction bulletins on how to install and adjust the sling on the M14A1 rifle.

Original hard rubber GI foregrips for the M14E2 and M14A1 rifle have always been, and remain, quite rare. Back in the late 1970s a surplus dealer made replacements made of rubber coated aluminum castings to fulfill a civilian market need. Unfortunately today the rubber-coated foregrips are often bought and sold as original U.S. GI issue.

Chapter 12

Stocks for the M14 Rifle

Wooden Stocks

Generally speaking when wood was used as the primary material for the stocks and foregrips of U.S. military rifles, black walnut was always the first choice. However, often due to shortages of walnut, substitute hardwoods were used in the place of walnut.

Birch wood when used as a material for rifle stocks has quite a few advantages over walnut. Birch is naturally more dense than walnut, therefore stronger. Birch is also more resistance to cracking and denting than walnut. When the use of birch wood was originally attempted as a material for the M14 rifle stocks, the linseed oil normally used to treat rifle stocks failed to penetrate the wood resulting in excessive residue, causing rejection of the stocks. After an investigation it was discovered that birch wood has a different grain than walnut and has a slower capillary action in absorbing oil. The standards used in treating the birch wood stocks were the same as formerly used for the walnut stocks. A new procedure was established to dip the birch stocks twice into class II oil with a four hour drying time between immersions. This was followed by a seven day drain and drying time

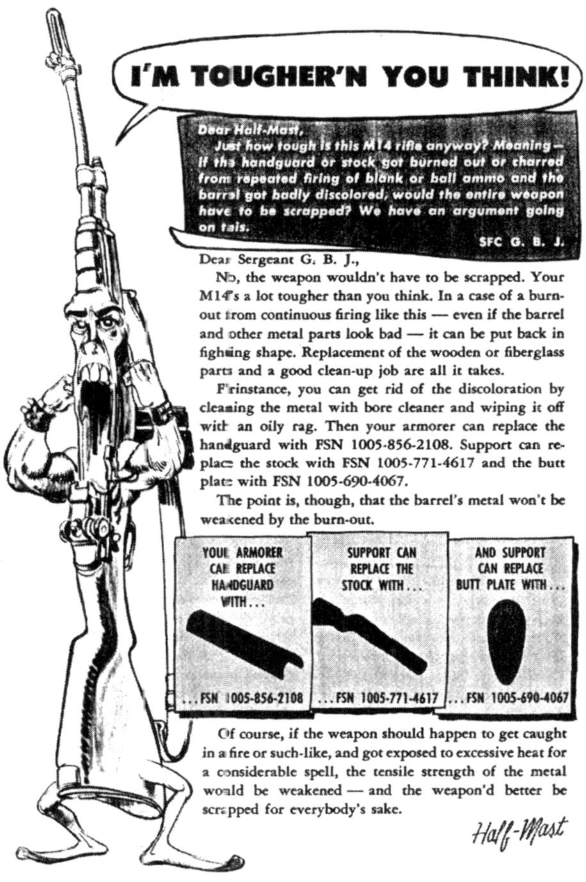

Tips from the U.S. Army PS manual. (John M. Miller)

An early M14 rifle with the flat M1-type buttplate and wooden handguard. (Courtesy of the U.S. Marine Corps Museum, Quantico)

prior to testing for smoke and water resistance. The seven day time period proved to be too brief resulting in the need to artificially dry the wood. Double dipping of the birch stock was done to darken the color of the wood to resemble the darker shade of the walnut stocks. Despite the attempts to dry the birch wood, excessive oil was still present. Double dipping of the birch stocks in oil was abandoned, and a single dip procedure was adopted, resulting in the birch stocks being lighter in color than the walnut versions. It was deemed that the color would be acceptable. Considerations were given to the commercial procedures used on birch wood to darken it. That procedure involved the staining of the wood before dipping it once in class I oil.

Both birch and walnut stocks for the M14 were made at the Springfield Armory. Commercial contractors used to produce the wooden stocks were the primary M14 contractors; Winchester, Harrington and Richardson. Stock makers S.C. Overton & Company and Sykes Manufacturing also produced wooden stocks for the M14. The manufacturer of M14 stocks can be easily identified by the contractor's initials stamped into the butt of the stock behind the buttplate.

A brief test of laminated stocks for the T44 rifle took place at Yale University during 1953-1954. The vertically laminated stocks were made from alternating layers of walnut and maple wood, were tested for shear strength and percentages of failures. The most vulner-

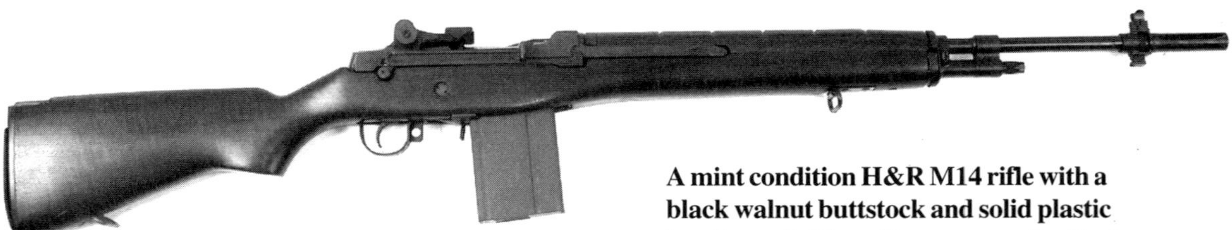

A mint condition H&R M14 rifle with a black walnut buttstock and solid plastic handguard. (Steve Brousseau collection)

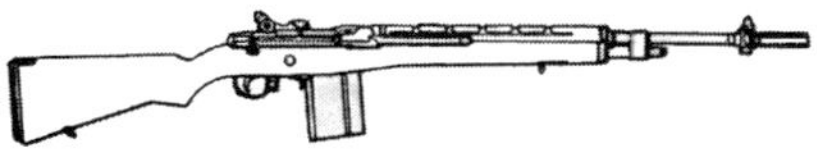

An M14 rifle with a birch-wood buttstock and slotted plastic handguard. Birch wood is generally lighter in color than walnut. The flip-up buttplate is shown with the shoulder rest is in the up position.

able area of the M14 stock was the wrist area that when subjected to rough use, such as airborne parachute jumps, would crack or break making the M14 rifle rather ineffective for dealing with the enemy.

Markings present on most U.S. GI M14 wood stocks were the DOD or Department of Defense eagle acceptance stamp located on the left side of the stock under the heel of the receiver. The DOD stamp represented that the rifle had been accepted as U.S. Government property. The other mark is a letter P generally inside of a circle. The presence of this mark indicated that the weapon has been subjected to and passed a proof firing test. This mark is located on the inside surface of the hand grip area of the stock. Replacement and synthetic stocks were usually not marked as described.

Synthetic Stocks

Military rifles are often subjected to the elements for long periods of time. Humidity, rain, extreme heat, cold and extended exposure can be very hard on weapons. When rifles were fitted with wooden stocks the problems that were encountered were usually magnified.

Depending on such variables as temperature and humidity, wood is subject to shrinking or swelling. When a wooden stock is used on a rifle this process can greatly affect the weapon's ability to deliver accurate fire. This would be of particular importance on a sniper's weapon. Another disadvantage of using wood on military rifles is that the wood will crack or break rather easily under rough use in the field. Because of hard use, constant cleaning and lubricating, wood stocks would become oil soaked. During extended periods of both semiautomatic and automatic fire of the M14 rifle their wooden hand guards, and even sometimes the stocks, would often char, smolder or even catch on fire. Wood fiber is also subject to compression, which in the case of the M14 rifle is under constant pressure when the trigger group is assembled to the rifle.

Due to the many disadvantages of wood, the use of a suitable synthetic stock on a military rifle would be considered quite advantageous. However, the successful fielding of a suitable synthetic stock took a longer time than anyone had originally anticipated. New advancements made in plastic manufacturing techniques were finally achieved in making a durable stock that could stand up to rigorous military requirements.

The Springfield Armory experimented with a number of materials from rubber-coated wood to aluminum in its quest to increase durability,

Soon after the M14 rifle was in production a redesigned buttstock was introduced to utilize the hinged "flip-up" type of buttplate with a shoulder rest. A problem obtaining the new buttplates from the contractor necessitated a rubber spacer (part number 7790914) to be temporarily used on the new stocks, in order to use the older style M1 buttplate. The rubber spacers were used until the new style buttplates could be delivered. (Courtesy of Bill Ricca)

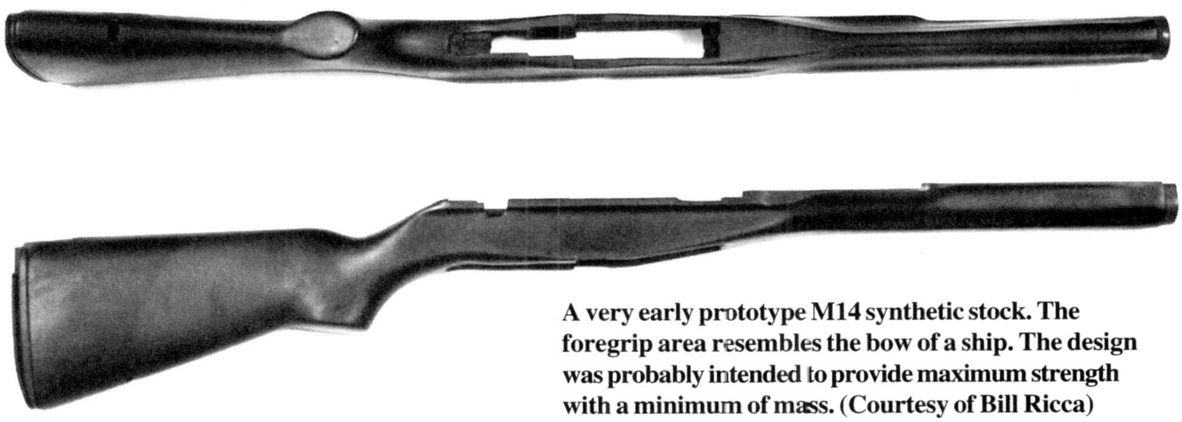

A very early prototype M14 synthetic stock. The foregrip area resembles the bow of a ship. The design was probably intended to provide maximum strength with a minimum of mass. (Courtesy of Bill Ricca)

and to seek a suitable material for replacement of the standard wood stocks and handguards.

The synthetic stock as adopted was a polyester plastic, glass fiber filled. A huge advantage from the use of a synthetic stock was that they could be made to closer tolerances than their wooden counterparts. The fiberglass reinforced plastic stocks did not require the metal stock liners as used in the wood stocks. The average weight of the M14 synthetic stock was 34 to 36 ounces, compared to wood stocks that averaged 34 to 38 ounces in weight. The first samples of the fiberglass M14 stocks were delivered in the summer of 1960. One of the first designs had an odd shaped foregrip area that resembled the bow of a boat. One of the primary problems of this earlier design was a lack of material to provide adequate strength for the front sling swivel.

A substantial amount of time and effort was expended on perfecting a fiberglass reinforced

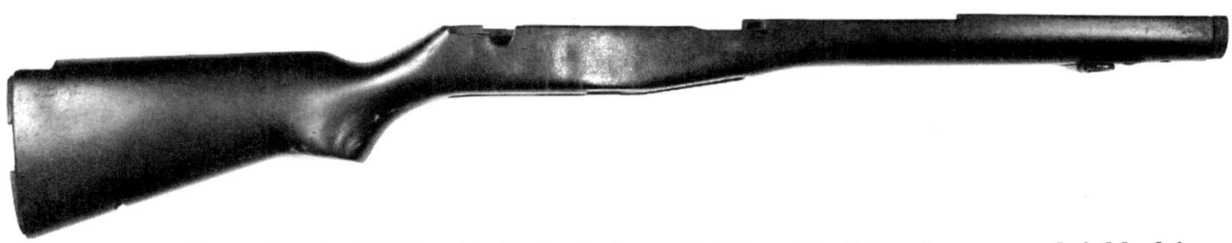

An early production M14 synthetic stock, the material used in this unique example is black in color. Note the lack of checkering on the grip surfaces. (Courtesy of Bill Ricca)

Another example of the early "smooth" style synthetic M14 stock, this stock is chocolate brown in color. The early uncheckered stocks were very slippery when wet from sweat or rainfall. (Courtesy of Bill Ricca)

plastic stock for the M14. The General Tire & Rubber Company working with the U.S. Ordnance Department Research and Development Division was credited with compiling drawings and a Technical Data Package for production of a synthetic stock. In October of 1962 the Springfield Armory was authorized to procure 10,000 pilot line synthetic stocks from General Tire by the Springfield Armory Engineering Division.

Developing a rifle stock made of synthetic materials had long been a goal of the Springfield Armory, a few prototypes were made for the M1 rifle and M1 carbine. When the M14 rifle was adopted in 1957 a program was launched to provide a stock made of a synthetic plastic material. By the time a suitable synthetic stock was perfected the M14 was near the end of its production run, cut short by the Secretary of Defense.

On 29 September 1961 a letter was received from the U.S. Army Infantry Board indicating that the Engineering test of the synthetic stock had been successful. At this time the Industrial Engineering phase was accelerated for incorporating the synthetic stock into the production of the M14 rifle. Ninety pre-produc-

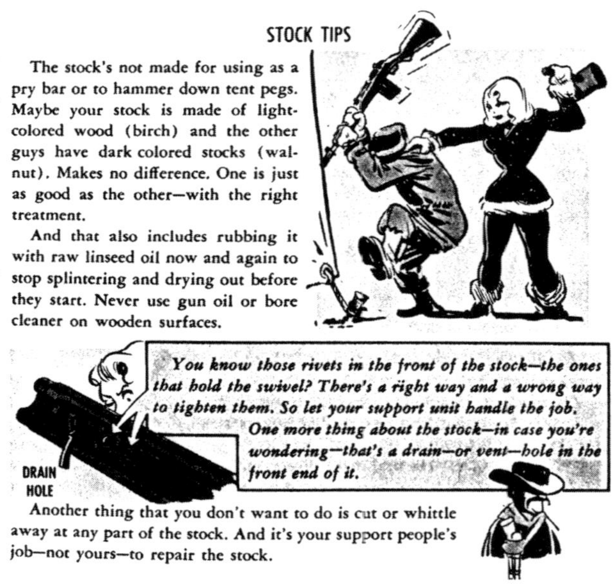

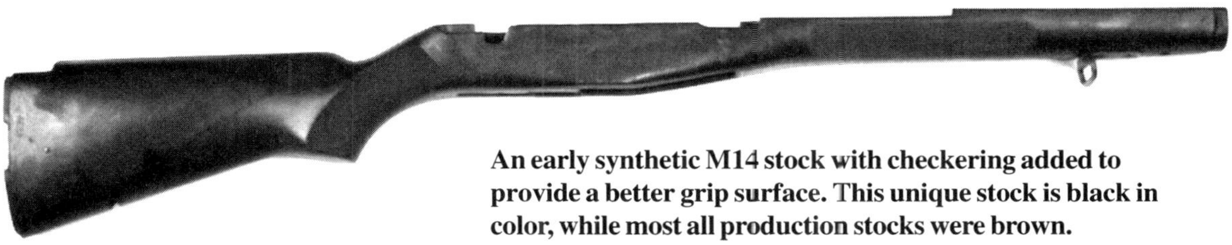

An early synthetic M14 stock with checkering added to provide a better grip surface. This unique stock is black in color, while most all production stocks were brown. (Courtesy of Bill Ricca)

tion samples, along with ten left and ten right stock halves were expected to be delivered and inspected on 24 April 1962. However difficulties with the supplier of the molds delayed delivery until the end of June. Evaluation of the samples was eventually completed on 30 June 1962. During the fourth production order for 49,000 M14 rifles a provision was added that 10,000 of the rifles be furnished with the new fiberglass reinforced plastic stocks. An engineering study was authorized for the development of a plastic stock repair kit, with instructions, intended for use in the field.

Early plastic stocks had no checkering on the grip areas them making them very slippery when exposed to rain, sweat or any other form of moisture. Most of the early stocks were a chocolate-brown color. Other colors that have been noted were black, olive drab and arctic white. These stocks were not painted, but the color was impregnated right into the material the stock was made from. A second problem discovered with the early synthetic M14 stocks was that of the top buttplate screw. The stock was molded around a small piece of wood that was provided for the top buttplate wood screw to thread into. The wood allowed just a few threads to engage, and was easily stripped. The stock was redesigned to provide an inleted area for a machine-screw, retainer and nut. Another redesign of the M14 synthetic stock was the addition of "checkering" to the wrist and foregrip areas of the stock. This alleviated much of the "slippery when wet" condition experienced with the earlier design.

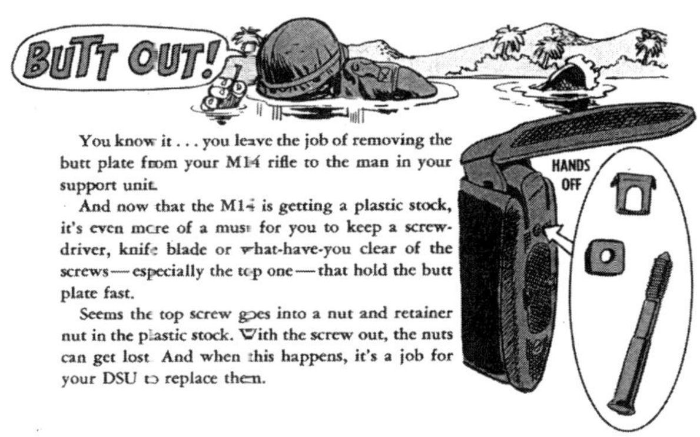

Prototype heavy fiberglass McMillan sniper stock for the M14 rifle. The buttstock is adjustable for length. The bipod is from an H&K G3 rifle. (Courtesy of John M. Miller)

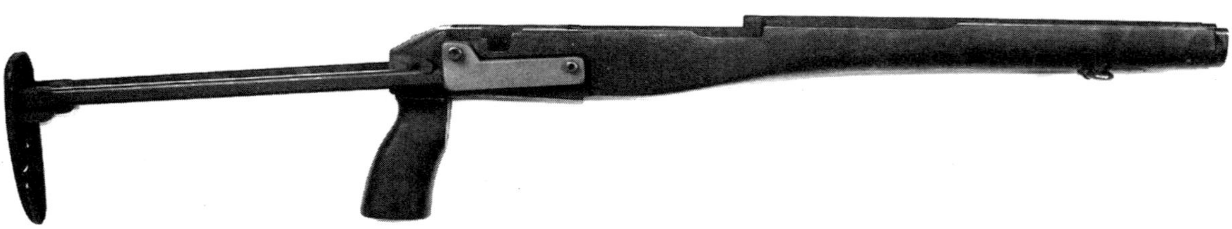

A experimental skeletonized M14 stock. The extension on this stock is fixed in place and does not fold. The design was probably intended to reduce the weight of the M14 rifle. (Courtesy of Bill Ricca)

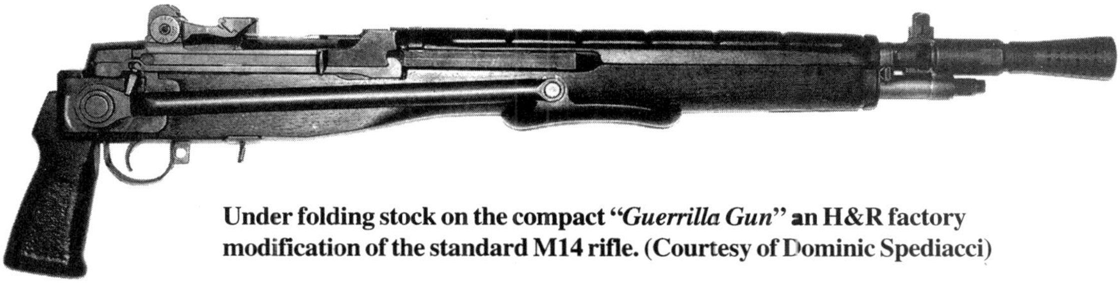

Under folding stock on the compact *"Guerrilla Gun"* an H&R factory modification of the standard M14 rifle. (Courtesy of Dominic Spediacci)

In July of 1968 the Army announced that they had awarded a contract to the General Tire and Rubber Company for manufacture of 500,000 of the improved synthetic stocks for the M14 rifle. The Sheffield Corporation also manufactured a number of synthetic stocks for the M14 rifle. The stocks would be used as replacements for unserviceable wood stocks.

Folding Stocks

One of the roles that the M14 rifle was intended to fulfill was that of the U.S. submachine gun, the M3A1. The problem was that the overall length of the M14 rifle simply made it too long to successfully serve as a compact weapon like the M3A1. In 1961 design studies were launched at the Springfield Armory for the development of a folding stock version of the M14 rifle. The role of the folding-stock M14 rifle was to arm tank crews, paratroopers and vehicle operators; all those that would normally be issued a submachine gun. There were a total of five different designs of folding stocks developed and tested by the Springfield Armory; all had pistol-style rear grips.

A fiberglass M14 stock with a side folding stock and Beretta BM59 plastic pistol grip, this stock was modified by a Navy SEAL team. (Courtesy of the U.S. Navy Historical Center)

One of the roles that the M14 rifle was intended to fulfill was that of the U.S. submachine gun. The problem was that the overall length of the M14 rifle simply made it too long to successfully serve as a compact weapon. In 1961 design studies were launched at the Springfield Armory for the development of a folding stock. There were a total of five different designs of folding stocks developed and tested by the Springfield Armory, lack of funding ended the program. (Springfield Armory Museum)

Another type of experimental stock was this telescoping design, similar to that of the M3/M3A1 submachine gun. This is one of the prototypes that led to the modern Sage EBR M14 stock. (Courtesy of John M. Miller)

During 1962 there was a maximum effort on the concept and feasibility relevant to the development of a folding stock for the M14 rifle.

The first prototype folding stock was a pantograph type. The stock was fitted to an otherwise standard M14 for demonstration to ORDTS on 5 September 1961. During early 1962 a second folding stock was presented. The stock was an improved pantograph design, but its thin struts could not withstand rough field use. In October of 1962 a third design designated as the Type III, was an exact copy of the Russian folding stock used on the AK47 rifle. The Russian stock itself was an improved version of an earlier German design that originated during World War II on the MP38 and MP40 machine pistols. The type IV was similar to the type III, but had a flip-up shoulder rest and a folding front foregrip.

The final M14 folding rifle stock design from the Springfield Armory was the Type V. During January of 1963 a Type V side-folding stock for the M14 was fabricated by the Springfield Armory. The stock's arms, buttplate and swivel bracket were made from aluminum bar stock. When not in use the stock folded to the left side of the rifle. The Armory planned a second model of a similar design, but with forged aluminum components. The response from the using services was enthusiastic about the proposed Type V stock. A few suggestions were offered that included a request to change the design so that the stock would fold to the right instead of left, and that the foregrip fold to the rear rather than forward. An M14 rifle with the type V stock was forwarded to ATAC for the study of mounting the weapon for storage in vehicles. The folding type V stock was estimated to cost from $37.00 to $57.00 to manufacture.

On 27 June 1963 a report on the "Folding Stock M14 Rifle" was received from the U.S. Army Test and Evaluation Command. The re-

Top view of a Springfield Armory folding stock design. This stock is very similar to that of the H&R *Guerrilla Gun* folding stock. (U.S. Marine Corps Museum)

The Department of Defense (DOD) eagle acceptance stamp was impressed on many walnut and birch M14 stocks.

port stated that the 7.62 mm rifle has the potential when suitably modified to be an acceptable replacement for the M3A1 submachine gun. The folding stock version of the M14 was to be designated as the M14E1 rifle. By the end of 1963 no guidance or recommendations had been received from the Headquarters of Army Weapons Command. Due in part to lack of funding and waning interest in the M14 rifle, no folding stock models were ever Standardized.

M14 Rifle Handguards

The original upper handguards used on the T44

Detail of an experimental M14 underfolding stock and plastic pistol grip. (Courtesy of the U.S. Marine Corps Museum Quantico, Virginia)

and early production M14 rifles were made of walnut matching the stocks of the weapons. Due to the M14 rifle's ability to fire rapidly in semiautomatic, and at a rate of 700 rounds per minute during full-automatic fire, problems began to occur with the use of wooden handguards during extended fire situations. The problem was that the handguards would char, smolder and even catch fire from the heat generated from a hot barrel.

Sample plastic handguards were first received from the contractor General Tire, in June of 1960. The handguards were forwarded to the Engineering Division for study. The Springfield Armory Operations Divisions were preparing for in-house manufacture of the plastic handguards. Liaison was maintained with the commercial contractor of the component to learn as much as possible about the process as possible. It was noted in the initial report that this was a new process and that considerable work may be required to perfect it.

The Springfield Armory acquired the necessary tooling, equipment and supplies to begin the compression-molding production of the newest plastic handguard. The process under

A ventilated fiberglass M14 handguard with a manufacturer stamp RPC 10. The significance of the stamp in unknown. This is a very early manufacture handguard that lacks the silver heat reflecting paint on the inside surface. (Courtesy of Bill Ricca)

development at the Armory involved the use of resin impregnated fiberglass mat in contrast to the wet lay-up method used in the commercial industry. By the end of the summer of 1960 the production engineering work for the M14 fiberglass-reinforced hand guard was completed. After testing the performance of the plastic handguard was considered superior to the original wooden design and the cost per piece lower.

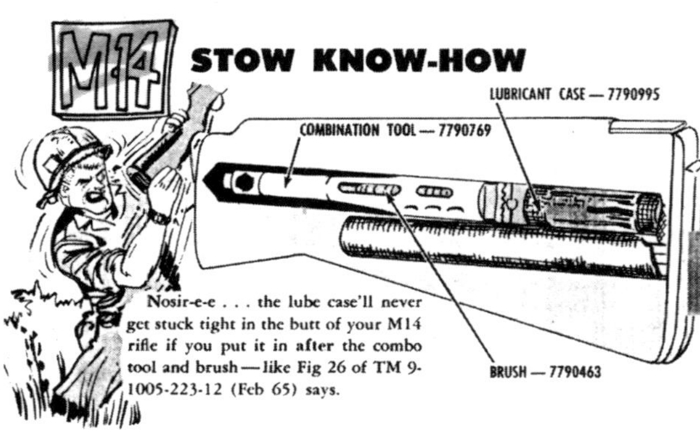

If the M14 maintenance kit was improperly stored inside the buttstock compartment, it could be very difficult to remove. (John M. Miller)

As Armory personnel became better oriented in the process, Ordnance officials visited the General Tire and Rubber Company Plant in Marion, Indiana to assist them with production problems being encountered with manufacture of the plastic handguard at that facility. The first production plastic handguards were molded in April of 1961. The initial trial run showed that the dies would produce acceptable dimensions when using General Tire's methods, but when using the Armory's method of using pre-impregnated material, acceptable dimensions could not be maintained. Work with the Engineering Depart-

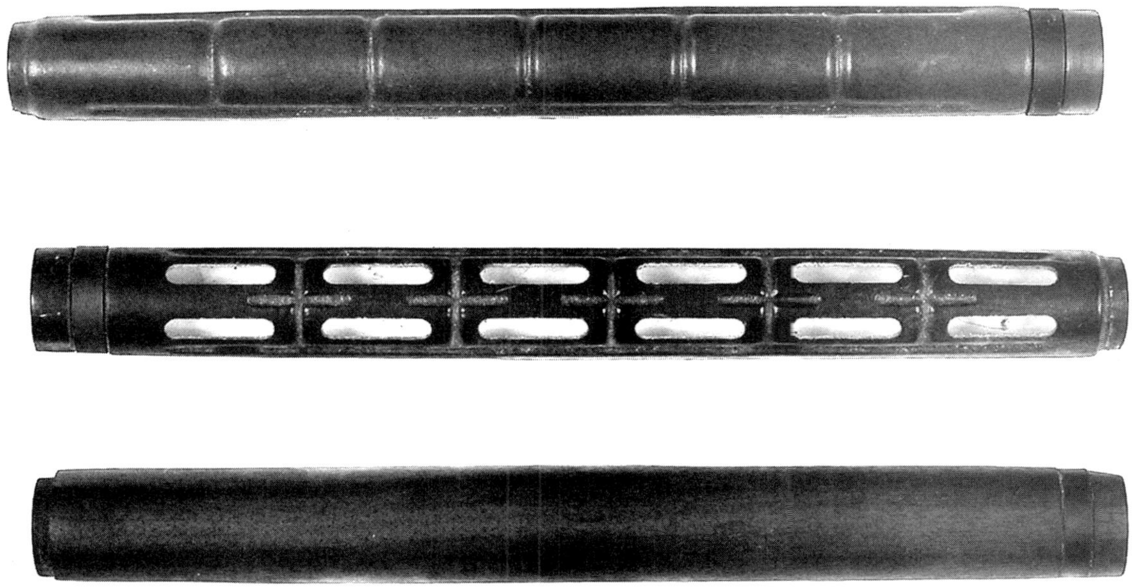

There were three primary types of hand guards used on the M14 rifle. The earliest was made of walnut (bottom). The next type was the fiberglass ventilated style (middle), and finally the solid fiberglass type (top). The fiberglass handguards varied in color from light to very dark brown.

ment continued in order to develop the ultimate product for strength and durability. The Armory was using one four-cavity mold and one single-cavity mold. With this equipment the Armory estimated a production capacity of 5,000 handguards per month.

During the Arctic test of the M14 during January 1960 it was reported that the fiberglass-reinforced plastic handguards became chipped around the retaining band and lip. It was determined that the vibration between the handguard and the front band was caused the damage.

There were three separate types of upper handguards undergoing study by the Springfield Armory. Type A had no ventilation holes, Type B had 38 round holes and Type C had 14 slotted holes. A report submitted by the Infantry Board dated 10 April 1959 recommended that the for the Infantry rifle role the M14 be fitted with a Type C handguard (with 14 slotted holes) and the hinged buttplate. In December of 1959 the heavy barrel M15 rifle was dropped, being replaced by the M14 M (Modified) with the slotted handguard, Type III bipod and the aluminum folding buttplate.

Dear Half-Mast,

Is there anything in the regulations that would keep us from painting identifying numbers or initials on the stocks of our rifles?

Sgt. R. C.

Dear Sergeant R. C.,

You won't find anything official that says you can't—but it's not a good idea. Paint's too hard to get off once a weapon's been turned in for re-issue.

If local SOP says you're to number or initial or otherwise label your rifle, it's better to use masking tape. This comes off easy and'll leave the stock good as new.

A good out-of-the-way place to put tape is in front of the front swivel. But be sure you do a neat job.

To get paint off your stock (whether it's made of wood or plastic), use a few dabs of paint remover. However, after you wipe the paint off a wooden stock, apply a good dose of raw linseed oil to keep the wood healthy.

But, whatever you do, never use sandpaper or any other abrasive to get the paint off. You'll do more harm than good.

Half-Mast

The M14 M rifle (Modified) was Standardized in December 1959.

The development of a suitable fiberglass hand guard proved more difficult than originally planned. The first production fiberglass hand guard was made with fourteen cooling slots. Two problems were encountered with this design. One was that a mirage generated by heat from the barrel would obscure the shooter's view of the front sight post. The second problem was that the handguards were weakened by the presence of the cooling slots, causing them to be easily broken.

The next production handguards were made of the same material, but there were no cooling slots incorporated into the design. The fiberglass handguards were made in a variety colors from a very dark to light brown. Some of the darker handguards were almost black in color. Although they proved somewhat more durable than the earlier slotted style of hand guards the first solid handguards also proved to be quite easily broken. The engineers then proceeded to increase the thickness of the handguard from .070 of an inch to .090 of an inch while this solved the problem it required new molding dies to be obtained. To aid in dissipating heat, the inside of the later production handguards were coated with silver paint. The silver color was used to reflect the heat from the barrel helping to keep the handguard cooler.

Chapter 13

Experimental Items

Multilite Sights

Special experimental night sights were designed for use with M14 rifle. The sights were designated by the manufacturer as Multilite sights. The sights were tested under varied field conditions. Initial testing conducted at Fort Benning by the Infantry Board were favorable, except for the luminous material that pulled away from the sights posts when exposed to heat. To address the problem new adhesives and material were used for future testing. During Arctic testing the luminous material on the sights began developing cracks after ten minutes of sustained firing. During the Arctic testing it was discovered that the sights offered no significant improvement in accuracy at night over the M14 standard sights. However, the test results from the Infantry Board had shown a considerable improvement in accuracy at night. It was determined that a soldier needed to become accustomed to using the Multilite sights before his shooting skills would improve. The conclusion reached by the Arctic Board was that the use of the night sights were not suitable under Arctic conditions.

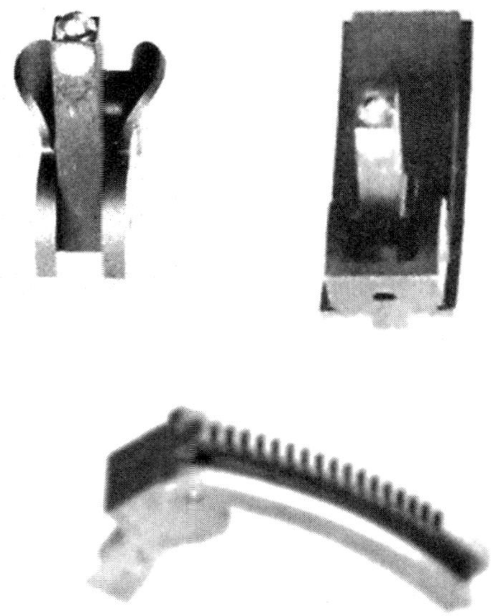

Flip-up Multilite night sights. There were experimental Multilite Sights produced for a number of U.S. small arms, but none were adopted as Standard. (NARA)

Subsequent work on the Multilite sight system improved the resistance of the luminous material to damage from extreme hot and cold temperatures. The night sights were redesigned to automatically adjust themselves when the weapon's standard sights were adjusted for targets at average battle ranges. Multilite sights were intended for use on several weapons to enhance their effectiveness during low-light conditions and at night. Sights were designed

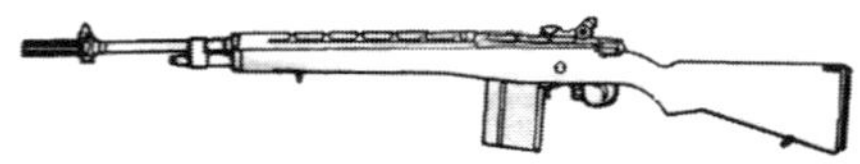

During 1960 a prototype 40-mm breech-loading grenade launcher, the X-1 was built. The launcher was designed to be attached to the M14 rifle's gas cylinder and bayonet lug. (Springfield Armory Museum)

for the M14 rifle, M60 machine gun, 3.5-inch Rocket Launcher and the 106 mm Recoilless Rifle. During October 1961 six additional sets of sights were procured and shipped to the Radium Chemical Company for luminous marking. A problem was encountered, resulting in the delivery being delayed until January 1962. Fifteen M14 Multilite sights were returned from the Infantry Board on 4 May 1962 for corrective action, specifically the lowering of the illumination element. The sights were modified and returned to the Infantry Board on 4 June 1962. Additional testing found that the sights were not fully developed enough for use on military rifles.

The Harrington and Richardson Arms Company also experimented with promethium night sights for the M14 rifle. The front and rear night sights were designed to normally be in folded position and easily flipped up into po-

The Springfield Armory designed X-1 rifle mounted grenade launcher. (Springfield Armory Museum)

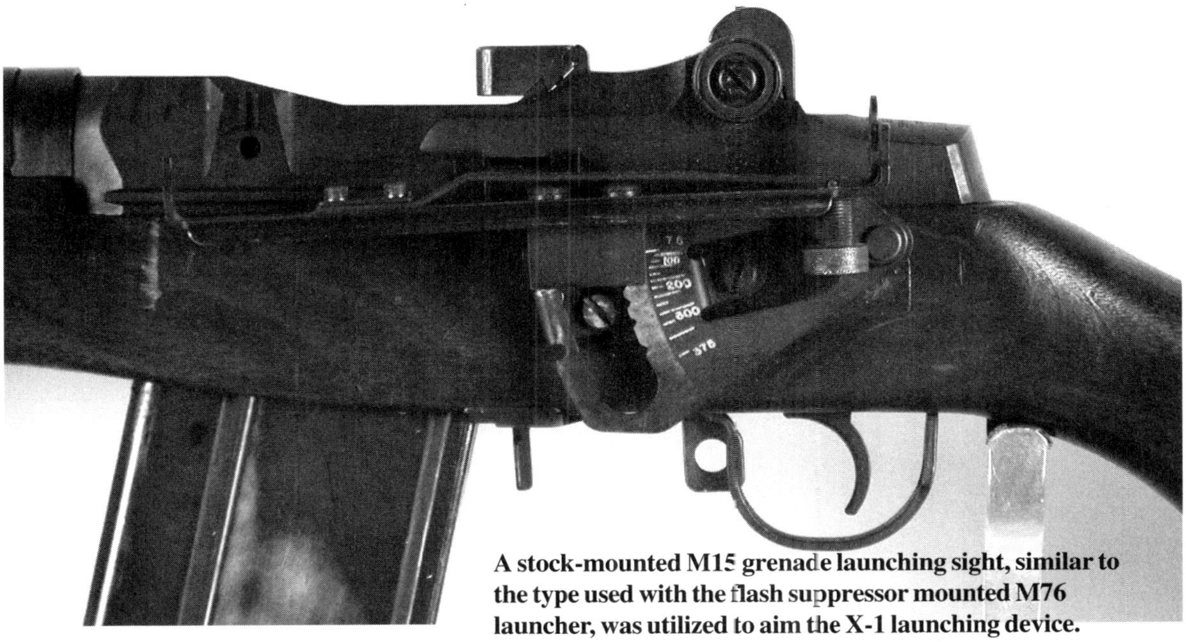

A stock-mounted M15 grenade launching sight, similar to the type used with the flash suppressor mounted M76 launcher, was utilized to aim the X-1 launching device. (Courtesy of the Springfield Armory Museum)

sition when required. The sights were never adopted.

Grenade Launcher X-1

During 1960 a prototype 40-mm breech-loading grenade launcher, the X-1 was built. The launcher was designed to be attached to the M14 rifle's gas cylinder and bayonet lug. The standard stock-mounted M15 grenade launching sight, sililar to that used with the muzzle style M76 launcher, was used to aim the X-1 device. As the M79 Grenade Launcher and Law Rockets became widely issued there was no longer a need for a rifle mounted launcher, especially with the already heavy M14 rifle. The concept was later used on the lighter M16 rifle.

Caseless Ammunition

One of the primary missions of the United States Ordnance Corps was to provide the infantryman with efficient modern equipment, and to increase the efficiency of the infantry rifleman. One such project was converting the M14 rifle to fire caseless ammunition. The

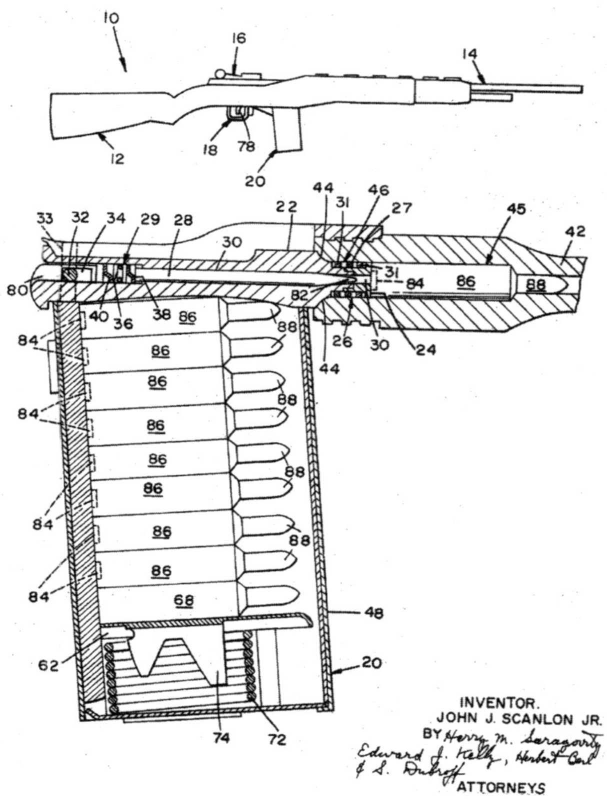

Patent drawing for the caseless ammunition system designed for the U.S. 7.62 mm M14 rifle. (Courtesy of Donald G. Thomas)

objective of the program was to modify the M14 rifle to function successfully with caseless ammunition in both semiautomatic and full automatic modes of operation.

When conventional design rifles are fired the breech is sealed by the expansion of the brass cartridge case. The case expansion is caused by the gas pressure of the rapidly burning gun powder inside the case. With caseless ammunition, as the name implies, there is no cartridge case. Replacing the brass case to seal the chamber is an obturator, or sealing plug, added to the face of the breechbolt.

The Armory designed an obturation device intended for incorporation into the M14 rifles to fire prototype samples of caseless ammunition. Initial testing of the caseless ammunition revealed that there were deficiencies in the design of the obturation cup. The M14 rifle served as the prototype weapon for the development of caseless ammunition. During September 1961 the Springfield Armory completed the development of two M14 rifles modified to fire caseless ammunition and shipped them to the Frankford Arsenal. Accompanying the weapons were spare obturation cups and special tooling for fabricating the cups.

During the 1960s period the Springfield Armory designed an improved obturator device that was incorporated into the M14 rifle to fire a new type of caseless ammunition under development at the Frankford Arsenal. During initial testing there were certain deficiencies revealed in the original design of the obturation device. During 1961 the obturator was redesigned and the development of the modified M14 rifle satisfactorily progressed.

The advent of caseless ammunition offered many advantages over standard brass cased ammunition. However there were several obstacles encountered adapting the caseless ammunition for military use, the primary problems involved heat and cartridge feeding. Another problem was the rapid rate in which infantry rifles are fired. Excessive heat from the flame produced in the barrel's chamber from the use of caseless ammunition caused damage to the o-ring used for obturation.

The high-temperature problem was addressed by the implementation of a firing pin obturator located at the rear of the firing pin's shaft. This effectively reduced the temperature of the firing pin. In addition specific areas of the receiver, and firing pin were covered with a protective coating of a heat-resistant material such as chrome or tantalum alloy. The obturator o-ring was protected on both sides by split rings. The o-ring was made of Teflon or similar substance to better endure high temperatures.

Feeding Problems

Caseless ammunition when compared to conventional brass cased cartridges is quite fragile. The inherent fragility of caseless cartridges causes a problem in a rapid feeding military rifle. In the M14 rifle prototype a special magazine was developed to carefully position the cartridges into the receiver.

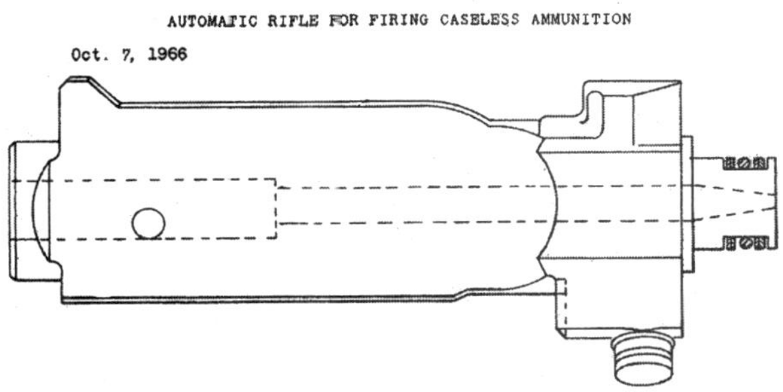

With caseless ammunition, as the name implies, there is no cartridge case. Replacing the brass case to seal the chamber is an obturator, or sealing plug, added to the face of the breech-bolt. (Courtesy of Donald G. Thomas)

The barrel and receiver assembly had a special feed ramp leading the chamber of the barrel. The radius of the ramp was of such magnitude that it could facilitate the insertion of the bolt's obturator. Adjacent to the feed ramp was a sealing area for the chamber to prevent propellant gases from escaping. The entire length of the barrel was coated with special chrome plating.

The modified magazine had a capacity of nineteen rounds of 7.62 mm caseless cartridges. The cartridge follower plate that was used to raise the ammunition had a slot located at the rear to permit positioning of a bar that prevented the cartridges from moving laterally in the magazine during the feeding cycle. On the top of the follower was a block shaped like a cartridge used to facilitate stacking of the

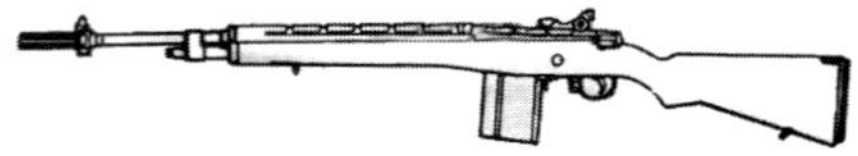

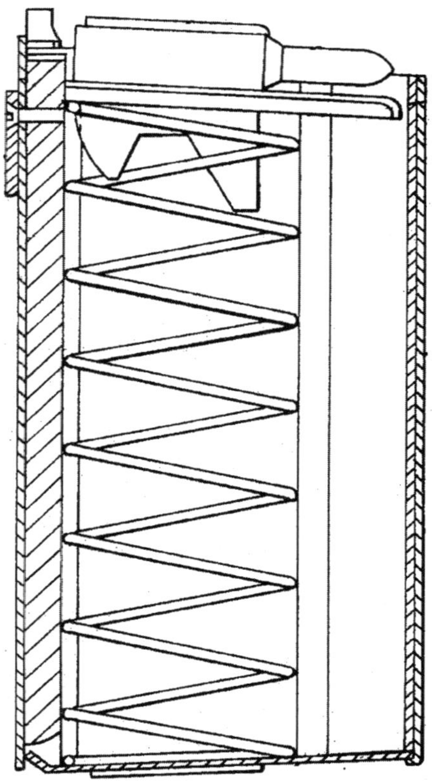

Caseless ammunition when compared to conventional brass cased cartridges is quite fragile. The inherent fragility of caseless cartridges causes a problem in a rapid feeding military rifle. In the M14 rifle prototype a special magazine was developed to carefully position the cartridges into the receiver. (Don Thomas)

rounds in the magazine. A bolt stop device was also provided to retain the bolt in the open position after the last round was stripped from the magazine. This feature facilitated rapid reloading of the weapon.

Operation Principle

Pulling the trigger causes the hammer under spring force to fall forward. The hammer strikes the firing pin that moves forward a distance of approximately .060 to .065 inch until its point strikes the combustible primer located in the rear of the caseless round. This action ignites the caseless round which then burns rapidly to evolve gas pressure that forces the projectile out of the barrel at a velocity of 2800 feet per second. The same gas pressure is exerted against the face of the bolt and obturator. The high pressure gases also expand through the firing pin passage way in the bolt and act against the firing pin obturator. As the projectile and gas pressure travel through the barrel gas pressure behind the projectile is tapped through a port in the barrel entering the gas cylinder actuating a gas piston in the normal manner and cycling the action for the next round to be fired.

Support of Frankford Arsenal's caseless ammunition program was continued by the Armory's manufacture and delivery of M14 weapons, barrels and magazines modified for the caseless rounds. During May of 1964 a meeting was held between the Frankford Arsenal and the Springfield Armory. The purpose of the meeting was to outline a future program in which the Frankford facility and the Armory would work together on the caseless ammunition project. The plan was to continue development of the caseless M14 rifle as well as choosing a belt-fed weapon for similar development. Studies of new weapons designed specifically for caseless ammunition were also

manufacturer has had a program to develop caseless ammunition, today the quest for suitable caseless ammunition continues.

The T44E6 Lightweight Rifle

All during the T44 program finding ways to

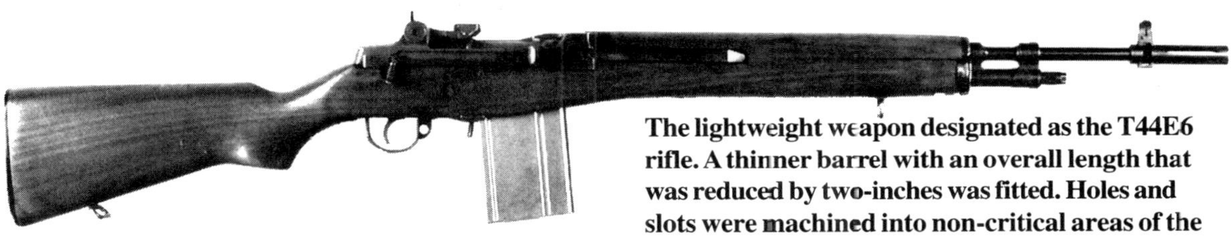

The lightweight weapon designated as the T44E6 rifle. A thinner barrel with an overall length that was reduced by two-inches was fitted. Holes and slots were machined into non-critical areas of the receiver. (Springfield Armory Museum)

to be developed. Funds for the program were granted in June 1964. No progress was achieved by the time the M14 rifle was being phased out in favor of the M16A1 rifle, and the caseless ammunition program for the M14 was terminated.

Just about every large commercial firearms

reduce to the total weight of the rifle was a priority. Under the SALVO program the Army relaxed the requirement of a barrel heavy enough to withstand the launching of anti-tank grenades and aggressive bayonet maneuvers. The grenade-launching requirement was waived due to the recent development of the new Light Antitank Weapon that fired a 66

An experimental short-barrel M14 rifle was used to study the effects of barrel length on projectile velocity. (U.S. Marine Corps Museum)

There were a few attempts to design and adopt a folding stock for the M14 rifle. (Springfield Armory Museum)

mm Heat Rocket, and the 40 mm M79 grenade launcher.

The lightweight weapon was designated as the T44E6 rifle. A thinner barrel with an overall length that was reduced by two-inches was fitted. Holes and slots were machined into non-critical areas of the receiver. Plastic stocks were planned but failed to reduce the weight and thus were not adopted. The steel

A .22 caliber M14 training rifle designed by the Remington Arms Company (Don Thomas)

H&R maintains a competent staff in its development, design and engineering departments. Most recent efforts have resulted in the development of the M-14 Simulator, a training rifle that looks and has the feel of the M-14 rifle, caliber 7.62mm. Where the 7.62 mm cartridge for the M-14 costs about 10 cents each, the caliber .22 ammunition used by the Simulator costs less than one cent each—.0016 cents to be exact. With trigger pull, weight and configuration of the Simulator conforming to the M-14 rifle in every way, the cost of training and improved markmanship is drastically reduced, while utilizing the many small bore firing ranges available for year round training.

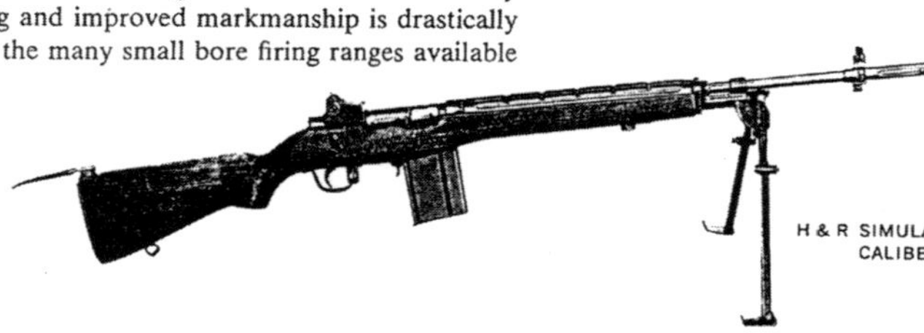

H & R SIMULATOR M-14
CALIBER .22

The .22 caliber rimfire M14 *Simulator* rifle was designed by the H&R Company in hopes of selling them to the government for training purposes.

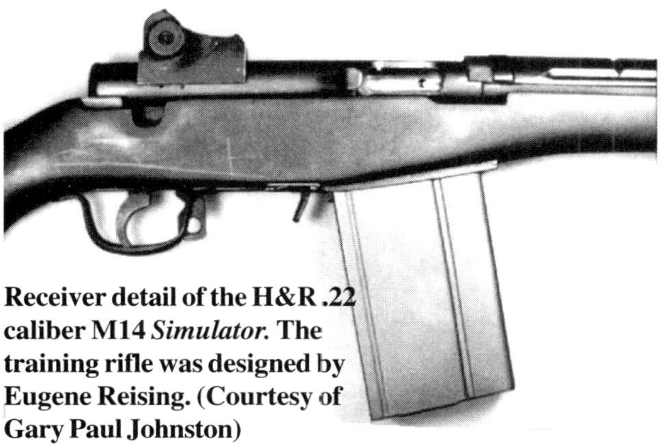

Receiver detail of the H&R .22 caliber M14 *Simulator.* The training rifle was designed by Eugene Reising. (Courtesy of Gary Paul Johnston)

buttplate was replaced with one fabricated from aluminum. An aluminum magazine that was substantially lighter than the steel version was used reducing the weight by 4.3 ounces. An aluminum stripper clip guide was also fitted. The manually operated gas valve, normally part of the gas cylinder for launching grenades was removed. The cumulative weight reduction of the weapon with an empty magazine was .863 pounds. The weight of the T44E6 with an empty magazine was 7.395 pounds. The T44E6 was tested at the Aberdeen Proving Ground from May to July of 1958. The T44E6 was tested and compared with the Armalite AR-15 (which would later be adopted as the M16) and the Winchester Lightweight Rifle.

The T44E6 was primarily fired in the semiautomatic mode.

Conclusions of the testing were:

1. The T44E6 was not as accurate as the T44E4 because of the shorter and lighter barrel.

2. The short barrel generated a greater flash and report with lowered projectile velocity.

3. The light barrel lowered the temperature at which rounds would cook-off in an overheated barrel.

4. The reduced weight resulted in an increase of felt recoil and instability during automatic fire.

The Board recommended that the T44E6

Short barrel H&R *Guerrilla Gun* with full wooden stock. (Springfield Armory Museum)

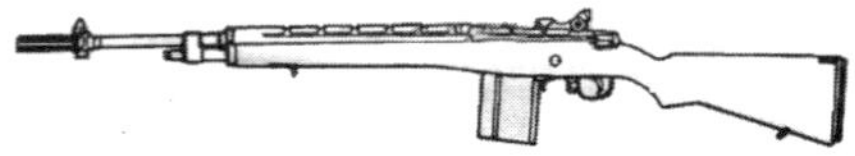

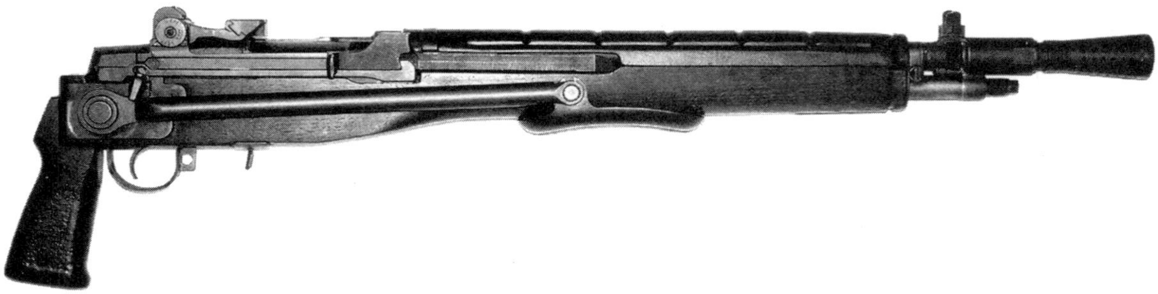

H&R *Guerrilla Gun* with folding stock. This weapon was designed as a lightweight, short-barreled version of the M14. The H&R designed rifle was intended for jungle, paratrooper and tank crew use. The weapon uses the same 7.62 cartridge as the standard M14 rifle. The *Guerrilla Gun* has a unique gas system and a cone-style flash hider. The stock was very well made of forged components. A few of these unique rifles were sold at the H&R bankruptcy auction in 1986. The fully transferable rifles are highly desirable collector's items. (Photos courtesy of Dominic Spediacci)

weapon be turned over to the Army Ground Field Forces, for field testing, to determine if the weight reduction was worth the loss in performance. The Ground Forces Report was unfavorable toward the T44E6, as a result all efforts on the T44E6 rifle were terminated in January of 1959.

Disposable Magazines

The standard steel 20-round box magazines issued with the M14 rifle were to be retained after expending the cartridges in them. The empty magazines were to be reloaded by the infantryman when time permitted. Stripper clips and guides were provided to expedite the hand reloading of magazines. The previous infantry rifle, the M1 used 8-round steel clips that were considered expendable in combat, although during training exercises they were to picked up and saved for later reloading. During the development of the M14 rifle there was a requirement to develop inexpensive magazines that could be discarded after use.

The Remington Arms Company Inc. was enlisted to perform a feasibility study to develop a 20-round magazine for the M14 rifle at a cost sufficiently low enough that the magazine could be considered expendable.

The primary requirements set by the Ordnance Corps for a disposable magazine and a 10-

Most of the Harrington & Richardson *Guerrilla Guns* had serial numbers with an X prefix. The receivers were marked R14 instead of M14. (Courtesy of Dominic Spediacci)

round charger were: fit in the present rifle; a shelf life of three years; withstand rough field use and be the same weight as the current steel magazine. A 10-round charger clip was also requested that would be low enough in cost to be disposable and enable the infantryman to reload the magazine through the top of the receiver if necessary.

Remington experimented with several different ideas to reduce the cost of a 20 round magazine for the M14 rifle.

The initial study conducted by Remington

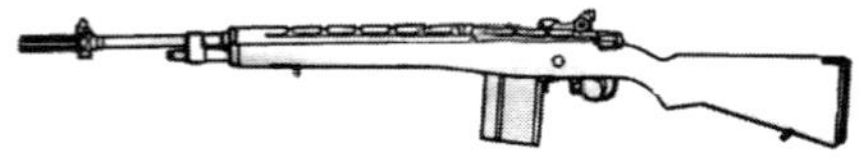

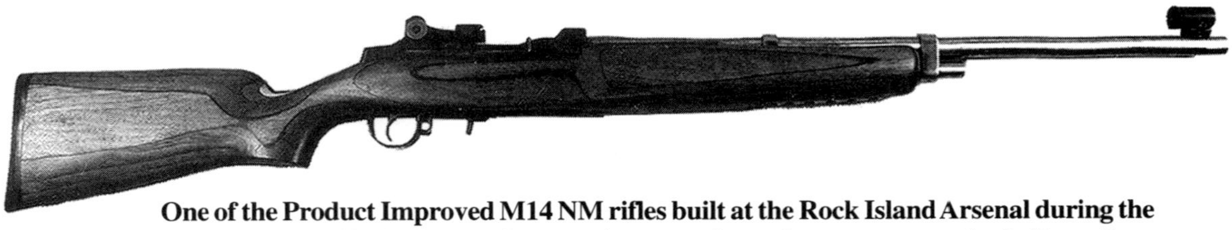

One of the Product Improved M14 NM rifles built at the Rock Island Arsenal during the 1970 era. This rifle had many features integrated to enhance accuracy including a free-floating stainless steel barrel and a heavy reinforced stock. (Courtesy of the Rock Island Arsenal Museum)

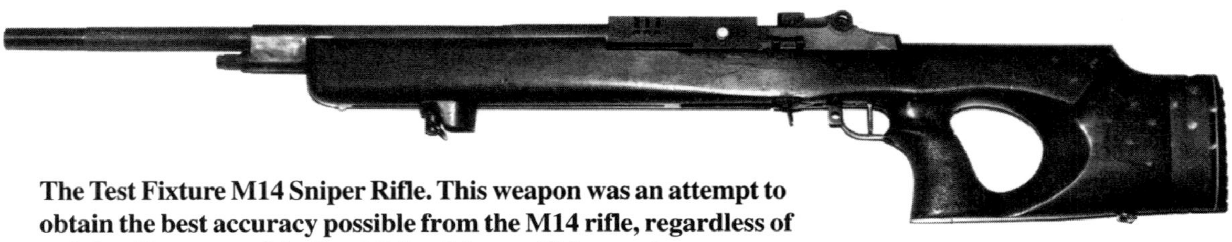

The Test Fixture M14 Sniper Rifle. This weapon was an attempt to obtain the best accuracy possible from the M14 rifle, regardless of weight. (Courtesy of the Rock Island Arsenal Museum)

During the mid-1970 era the Rock Island Arsenal undertook a Product Improvement Program to enhance the accuracy of the M14 National Match (sniper) rifle. Funds for the program were provided through the U.S. Army Material Command. There were a total of four different weapons built each in a different configuration. The program was headed by a civilian employee of the Arsenal, Mr. R. Snodgrass. The rifles that were built, all produced at least a sub 2-inch MOA at 100 meters with M118 ammunition. The rifles featured hi-quality rotary-forged heavy 24-inch barrels, and laminated wood stocks. Another variation featured a thumbhole style target stock with a heavy stainless steel barrel. After the war in Vietnam ended the funding was cut, because there was a no longer a requirement for such a weapon.

Arms Inc. indicated that some cost reductions could be realized by simplifying the standard issue steel magazine. Each individual magazine component was analyzed to determine if it could be produced at a lower cost. Based on quotes offered by several contractors, it was estimated that a suitable magazine could be fabricated at an approximate cost of .60 per piece. A newly designed magazine made from a combination of plastic and metal was estimated to offer even greater savings. Based on the quotes from contractors, indicated that such a magazine could be produced at a cost of .40 per piece; initial tooling was estimated to cost $20,500.00

Magazine Follower

An injection molded plastic magazine follower was made of a polyethylene called "Alathon 37". The plastic follower was basically the same as the metal follower except that the plastic stop and follower could be easily molded as a single unit. Cost reduction was estimated at 80-percent over the price of the standard follower. A new spring would be required, because the spring fastening lugs used on the steel follower could not be reproduced on the plastic version. A new spring was designed that would work with the plastic follower and it could be made from the existing spring, although no cost estimates were obtained for the conversion.

The plastic SKD-463 magazine was designed to function in the same manner as the standard steel magazine. The magazine was constructed by bonding a plastic magazine body to a steel band that formed the feeding lips. This design was abandoned due to its high estimated cost of construction. (Courtesy of Bill Ricca)

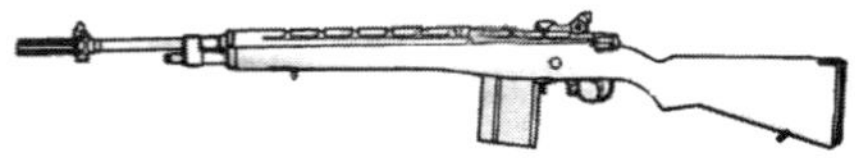

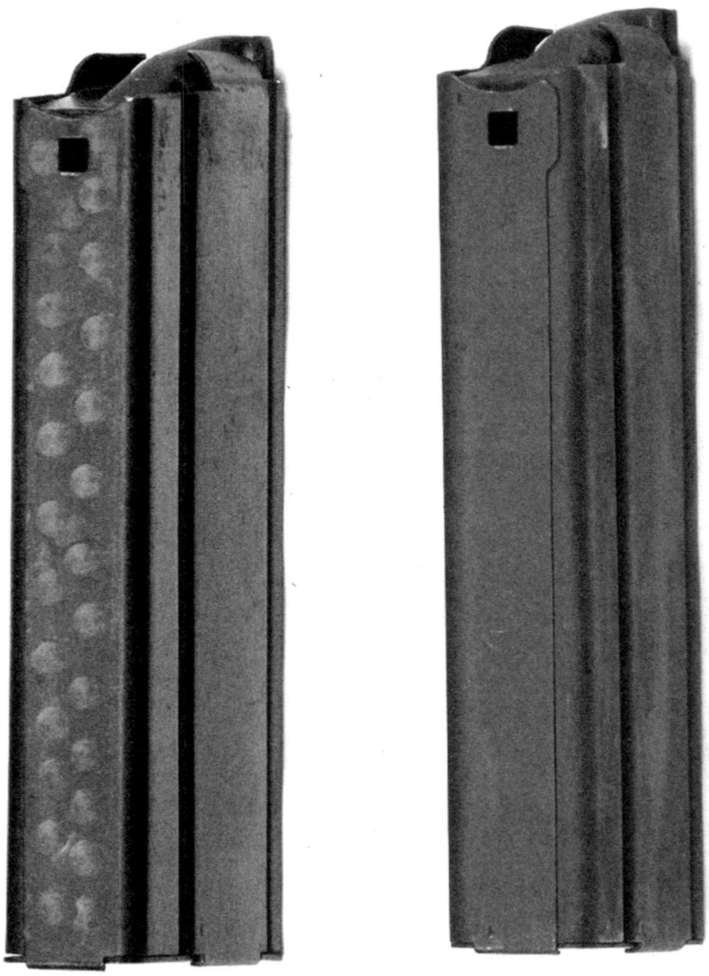

Prototype M14 aluminum magazine (left) is displayed next to a production steel magazine. (U.S. Marine Corps Museum)

Plastic and Steel Magazine SKD-463

This magazine was designed to function in the same manner as the standard steel magazine. The magazine was constructed by bonding a plastic magazine body to a steel band that formed the feeding lips. This design was abandoned due to its high estimated cost of construction.

Aluminum Magazine SKD-511

This magazine was the same as the steel magazines except it was designed to be formed from 2024-T4 aluminum.

One Piece Magazine SKD-416

This design was basically the same as the original 20-round M14 magazine except for the fabrication methods used. The SKD-416 magazine was formed from a single piece of 2024-T4 aluminum or FS1050 heat-treated steel. The magazine was to be formed with the back left open for assembly purposes. After the spring and follower were installed, the back would be closed and retained by four snap detents on the magazine walls. Twenty 7.62 mm cartridges could be loaded using the same method used on the standard issue magazine. The apparent advantages of this magazine over the current design were that it eliminated the need of a latch plate and spot-welding of the plate to the magazine body.

Disposable Magazine Housing

Another approach to the design of a practical disposable magazine was the idea that if the essential close tolerances and feeding lips of the magazine could be incorporated into the rifle itself, a low cost container could be designed to accommodate the spring and follower (which would be considered expendable) and twenty rounds of ammunition.

A solution was proposed in the form of a magazine housing assembly designated as the SKD-443. The magazine housing assembly was composed of two components. One was the housing constructed of FS1050 heat-treated steel. The housing was fabricated in the same manner as the standard magazine tube. The housing had all of the same dimensions as the standard issue magazine except for its length. The overall length was two-inches shorter than a standard magazine, and the bottom portion was open with leading surfaces formed by flaring of the opening. Six slots were located to provide a locking arrangement and permit functioning of the disposable magazine. The sec-

Plastic M14 magazine. The Remington Arms Company Inc. was enlisted to perform a feasibility study to develop a 20-round magazine for the M14 rifle at a cost sufficiently low enough that the magazine could be considered expendable. (Bill Ricca)

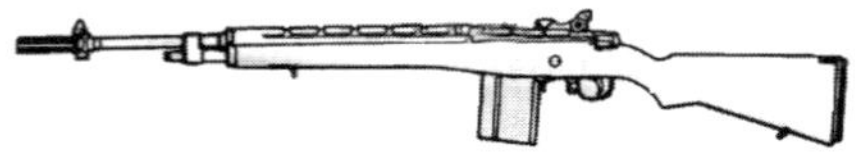

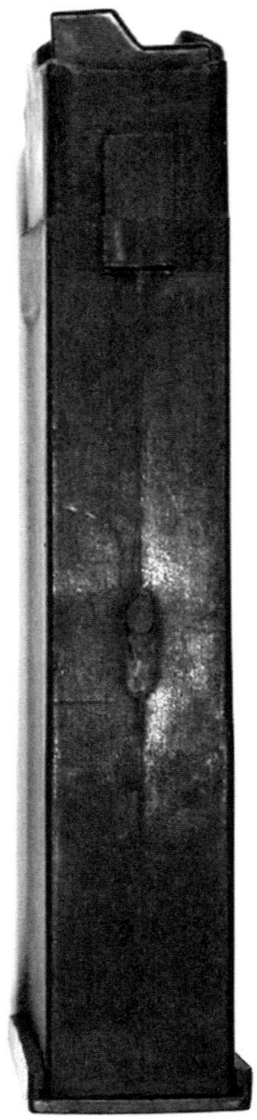

Rear view of the experimental plastic magazine for the M14 rifle. (Courtesy of Bill Ricca)

ond component was a magazine latch plate, which was the same as on the standard issue steel magazine. The latch plate was to be spot-welded to the back of the magazine housing.

The SKD-443 magazine housing offered greater stability, and was designed to be a semi-permanent component of the M14 rifle. The housing was installed in the same manner as the standard magazine using the original latching method. The housing could be easily removed for cleaning or replacement. When assembled to the rifle the housing provided a method of fastening the disposable magazine to the weapon and controlled the feeding of the cartridges into the chamber.

The Disposable Magazine

The disposable magazine assembly No. 3, SKD-461 was preceded by three earlier designs. The Plastic disposable magazine assembly SKD-510, disposable magazine No. 1, SKD-509 and disposable magazine No.2, SKD-508.

The No. 3, SKD-461 disposable magazine was constructed using 3S-0 aluminum to form a U-shaped magazine tube. The tube was designed to contain a spring, follower and twenty cartridges. Two locking fingers located on the sides of the tube were provided as a means of latching it to the magazine housing. Two retaining lips were located at the top portion of the tube to keep the cartridges contained in the tube. The follower was made of plastic. The retaining collar was manufactured from injection molded "Alathon 37" polyethylene plastic.

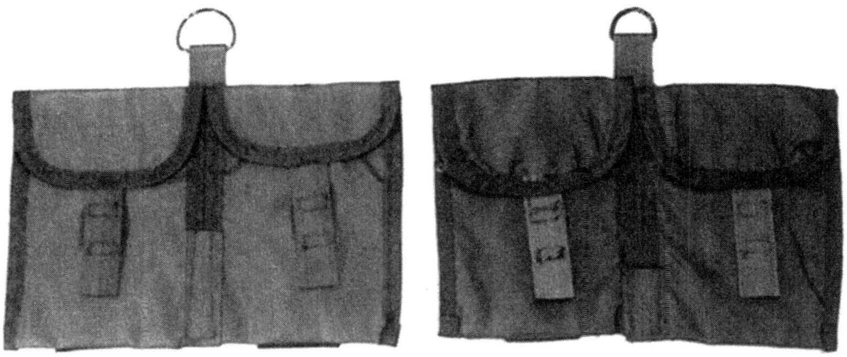

Experimental M14 nylon magazine pouches designed by the Marine Corps. (Courtesy of Craig Pickrall)

The retaining collar was assembled over the tube and positioned below the retaining lips of the tube. The spring, follower and twenty cartridges were then installed. The retaining collar was then slid up the tube and positioned to support the retaining lips. This kept the cartridges in place during storage and handling. When the magazine assembly was inserted into the magazine housing of the rifle, the retaining collar contacted the housing and was stopped. The magazine tube continued its upward travel into the housing. The retaining lips would leave the collar and then be supported by the walls of the magazine housing. Once the magazine was inserted to the correct depth, the locking fingers on the magazine tube engaged the slots in the magazine housing. With the magazine tube in this position, the retaining lips were not supported due to the slots in the housing. The cartridges then forced the retaining lips back through the slots. The cartridges were then able to feed into the rifle's chamber. To remove the magazine from the magazine housing the retaining collar was firmly grasped and pulled downward. In doing so the retaining collar passed over the locking fingers compressing them. The locking fingers are then disengaged permitting the downward motion to remove the disposable magazine assembly from the magazine housing.

The disposable magazine was estimated to cost approximately $.40 with an assembly cost of $.04.

The plastic follower was estimated to cost $.02 each while the tooling would cost $4900.00. The quote was provided by the Bridgeport Molded Products, Inc.

The retaining collar was estimated to cost $.0275 each with estimated tooling costs of $9100.00. The quote was provided by the Bridgeport Molded Products, Inc.

The magazine tube was estimated to cost approximately $.157 each, with the necessary tooling costing $6500.00. The price quote was provided by the Crosby Company of New York. The Crosby Company had manufactured drum and box magazines for the Thompson Submachine Gun during World War II.

Charging Clips

A plastic charger would be able to reduce long term costs even further. Two 10-round charging clips for loading through the top of the receiver could be produced at a cost .18 to .20. Initial tooling costs were estimated at $15,500.00 Two steel chargers of the same type cost .35 with a tooling charge of $5000.00.

M14 rifle serial number 8788 was supplied to Remington Arms Inc. along with 6 standard issue magazines for use in the magazine development program.

The Caliber 5.56mm M14 Rifle

During the initial procurement of the 85,000 AR-15 (M16) rifles from Colt there was thought that this would be merely a one-time purchase until the long-awaited Special Purpose Individual Weapon (SPIW) was fully developed. Once the SPIW was developed it was anticipated that all previous infantry rifles would be rendered obsolete. After it was announced that production of the M14 rifle was being terminated and replaced by the 5.56 mm M16 rifle, the Thompson Ramo Wooldridge Company attempted to develop a version of the M14 rifle chambered for the same 5.56 mm cartridge that was used in the M16 weapon. The purpose was to salvage what they could out of the canceled M14 project. Production numbers of the 7.62mm M14 had fallen short of the weapons needed to replace the M1 rifle. TRW had hoped that the 5.56 mm M14 would be considered and production would continue at least until the time that the SPIW was ready. The first contracts for the SPIW prototypes were awarded to Harrington and Richardson and Winchester. Eventually the SPIW concept proved to be a failure and the AR-15 would be standardized as the M16 rifle and despite early problems in the field, would be procured in vast numbers as the "Standard A" infantry rifle.

Chapter 14

The 7.62 MM NATO Cartridge

During World War II there were two basic cartridges used. The full-power rifle round effective out to 500 meters, and the much smaller pistol cartridge that was used in handguns and submachine guns. When used in a submachine gun the maximum effect range of the pistol round was approximately 150 meters. Weapons using the aforementioned cartridges each had its own disadvantages, one was too powerful for close-in fighting, and the other was too underpowered for long range fire.

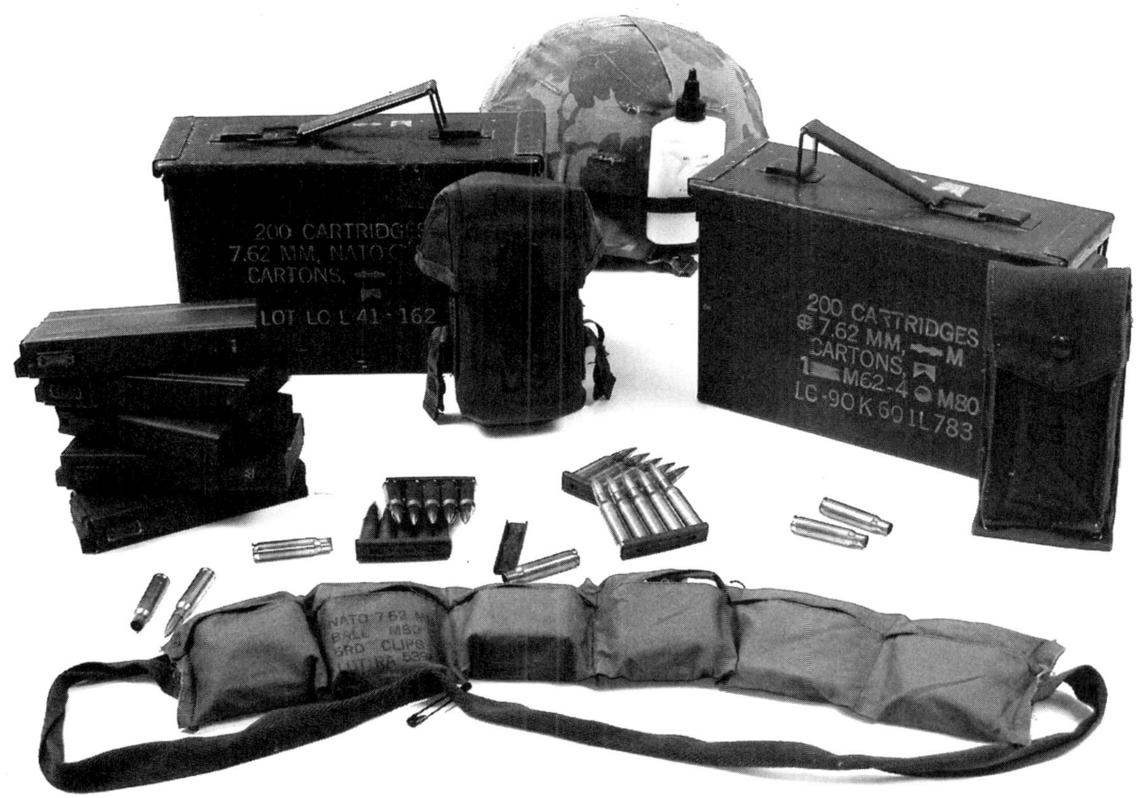

U.S. Rifle 7.62 mm M14

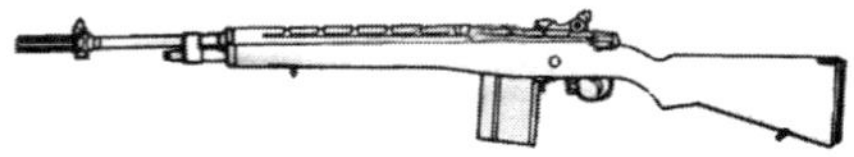

In a hypothetical scenario, a squad of soldiers armed only with submachine guns could be pinned down by an enemy using battle rifles from a distance that the submachine guns could not return effective fire. On the other hand a squad of soldiers armed with bolt-action or even semiautomatic battle rifles would be no match for an enemy armed with quick-firing submachine guns in an ambush or house to house type of engagement. Ideally a weapon was required that would allow the soldier to be prepared for either scenario. Studies had shown that soldiers rarely exposed themselves in combat. At distances exceeding 300 meters they would probably not be able to be seen, particularly while wearing camouflaged uniforms that blended in with the surrounding landscape. According to a number of studies, a full powered rifle cartridge was essentially not needed by the infantryman on the modern battlefield.

A U.S. Army soldier fires a German *Strumgewehr* captured in Vietnam. The World War II weapon fired the midrange 7.92x33 mm *kurz* (short) round.

Prior to the United States' entry into World War II the Ordnance Department had been busy developing and testing a "light rifle" to arm soldiers whose primary duties did not include daily front line engagements with the enemy. Destined to be armed with the new weapon were artillerymen, mechanics, military police and other rear echelon troops. The increasing use of paratroopers by the enemy was likely to expose those in the rear to attacks. The light rifle, while not overburdening them, would allow these soldiers to defend themselves in the event such an attack occurred. The weapon that emerged from the development program was the .30 caliber M1 carbine. The carbine was very popular, it was light in weight, had a 15 round detachable magazine and could provide somewhat effective fire out to 300 yards. The biggest problem was the limited effect of the carbines small .30 caliber (7.62x33mm) round. The carbine was the closest thing to a midrange weapon the U.S. military would field during the war. Had the carbine been designed to use a slightly more powerful cartridge it may have beat the Germans to the modern mid-range assault rifle concept. During 1945 the M2 version of the U.S. carbine was introduced. The M2 had select-fire capability and used a large capacity 30-round magazine.

During the Second World War the Germans had been developing a new intermediate rifle cartridge and a new weapon for the round. Both the *Sturmgewehr* weapon and its midrange cartridge started a worldwide revolution in small arms design. The German weapon used the 7.92x33 mm *kurz* (short) round. The cartridge provided controllable full-automatic fire for close range engagements and accurate fire out to approximately 300 meters. Besides being an effective combat arm there were quite a few other advantages as well. The smaller cartridges were lighter, allowing the infantryman to carry a larger ammunition load. The ammunition was also cheaper to produce using fewer critical materials; this was a very important feature during wartime. Probably the most important characteristic of the midrange round was the ability to design and build lighter weapons to fire the cartridge.

The U.S. M2 select-fire carbine with muzzle brake and 30-round magazine. The weapon fired a short, low power .30 caliber (7.62x33 mm) cartridge.

Post war American interest in developing a new rifle cartridge was initially generated by both the German's successes with their *7.92 kurz round* and the new gunpowder that was developed by the Olin Chemical Corporation. Olin's new ball powder enabled designers to retain the ballistics of the 30'06 cartridges in a shorter and subsequently lighter cartridge case. Savings in both weight and cost inspired the U.S. to develop a new short rifle cartridge of their own. Initial study and work on the new cartridge was begun at the Small Arms Development Branch. John C. Gray, a civilian engineer, was placed in charge of the program. Final development was done by Ordnance personnel at the U.S. Frankford Arsenal working with engineers from Olin's Western Cartridge Division. The new cartridge case was designated as the T65, during the development process the case was modified a number of times. The final design had a longer neck and sharper shoulder radius than the first cases developed. The final design submitted for adoption was designated as the T65E3 case. The new cartridge was somewhat intermediate in size, but was as powerful as the older 30'06 round it was intended to replace, in a smaller package.

dardize all of their weapons and cartridges so in the next conflict they could easily assist each other, and reduce the potpourri of small arms and their parts as well as ammunition. Although the theory was sound any reasonable implementing of the plan would not be easy. Each country had its own idea of what was best. National pride was also a problem. It seemed

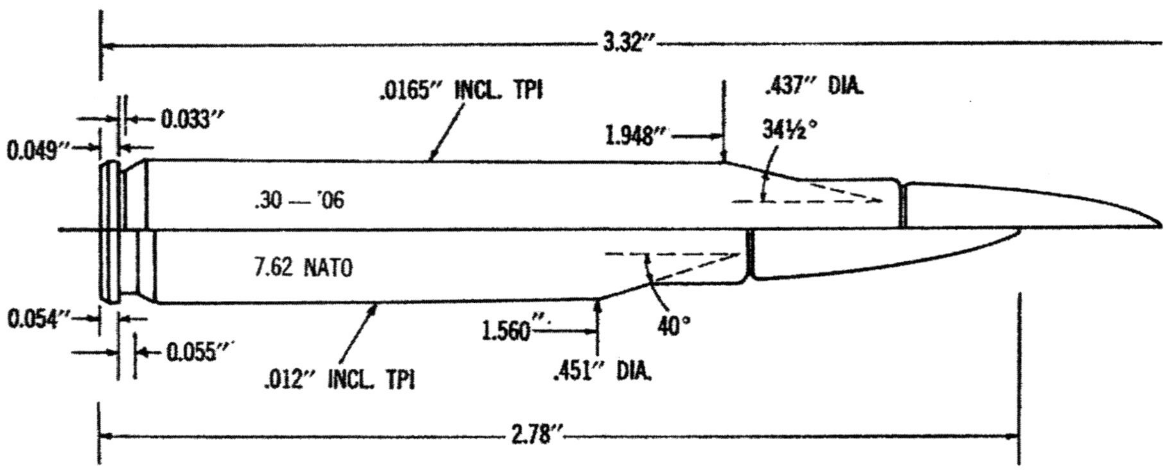

This unique drawing compares the older 30'06 M2 cartridge (top) with the new T65 7.62 mm round.

During World War II the United States had a number of allies; all of them shared a common problem. They all used a diverse number of different weapons and cartridges. This situation caused a supply and logistics nightmare. After the war a number of the allied countries formed the North Atlantic Treaty Organization (NATO) and agreed that they should stan-

that no one wanted to adopt an infantry rifle that was designed by another country.

The origins of the 7.62x51 mm cartridge date as far back as 1944. The Chief of Ordnance had directed the U.S. Frankford Arsenal to investigate the feasibility of a new .30 caliber that would be shorter in length than the current issue 30'06 round. The development of a shorter service cartridge would coincide with the development of a new infantry rifle, using a shorter round, the weapon itself could be reduced in size and weight. The key to a smaller cartridge case, while retaining the same basic ballistics as the 30'06 round, was the recent developments of improved gun powders. The primary goal was to reduce the length of the old M2 round (30'06) by .5-inch.

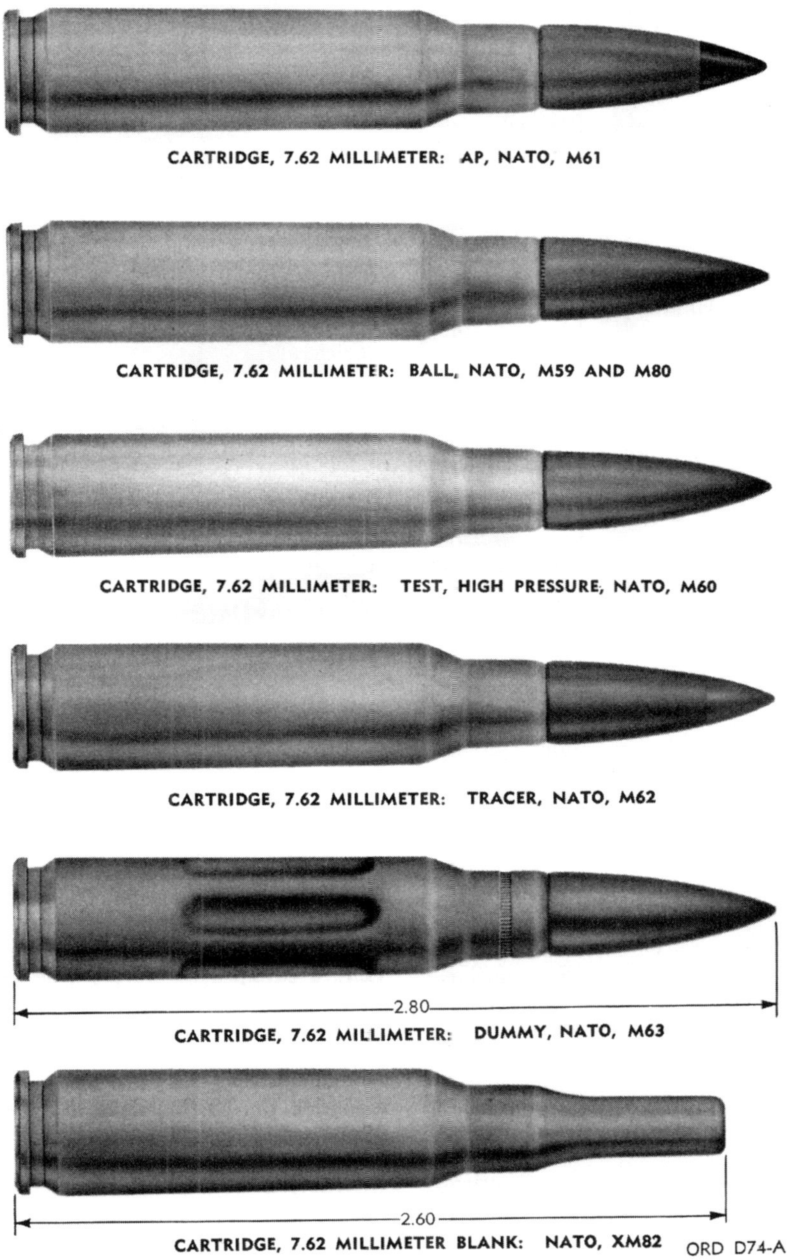

Development initially began with the commercial .300 Savage cartridge cases using military projectiles and a variety of new powders. The new rounds were initially called the .30 caliber short. The original .300 Savage cases that were used slowly evolved into the T65 case that had a reduced shoulder angle, a longer neck and less case taper. A thicker case rim was designed to withstand the rigors of extracting rounds

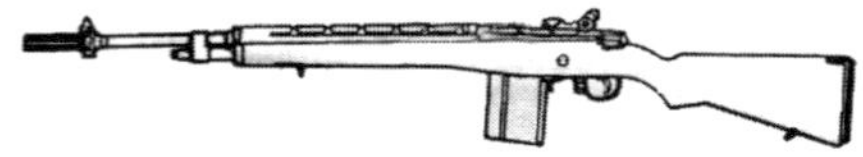

from overheated and dirty chambers as well as tentative use in machine guns. Prior to adoption of the T65 round the Winchester Company saw the sporting potential of the new round, and had asked the Ordnance Department for permission to market the cartridge commercially as the .308 Winchester.

The overall length of the T65 cartridge is 2.84 inches as compared to the 30'06 M1 rifle cartridges 3.34 inches length. Despite a .5-inch shorter case, the T65 retains nearly the same ballistics as the older 30'06 round. The shorter T65 case is made possible by use of an improved ball gun powder, requiring less cartridge space. During early testing a T65 case loaded with a 150-grain projectile recorded an average muzzle velocity of 2,812 feet per second. T65 cases loaded with 165-grain armor piercing bullets easily penetrated .25-inch steel plates at a range exceeding 200 yards.

Although wood is seldom used as armor on a modern battlefield, penetration testing using 1-inch pine boards was conducted. A block of 6) 1-inch pine boards were placed at a distance of 2,000 yards from the firing line. The T65 cartridge using a 150-grain lead projectile penetrated five of the wood boards and lodged in the sixth one. A standard GI issue steel helmet could be penetrated at a distance of 1,200 yards.

A British experimental .280 caliber cartridge (shown left) is compared with the U.S. T65E3 7.62 mm cartridge (right). Courtesy of John M. Miller

The British .280 Caliber Round

The British like the Russians were quick to grasp the German's midrange cartridge concept, and began development of their own midrange round. The British began their program by establishing the Small Arms Ideal Calibre Panel, which first met in 1945. Research for the determination of the ideal cartridge for an infantry rifle was undertaken by Dr. Richard Beeching, Deputy Chief Engineer of the British Armaments Design Establishment. Dr. Beeching and his staff undertook an unprecedented study of projectile shape, weigh and ballistics to determine the optimum combination needed for a maximum range of 549 meters or 600 yards. After reviewing the report generated from the extensive scientific study, the Panel proposed a .276 caliber round that had a 140 grain projectile with a muzzle velocity of 2300 feet per second. Although the actual diameter of the projectile was .276 caliber it was designated by the British as the .280 caliber round.

After the British designed what they felt was the optimum infantry round, they proposed their concept to the U.S. Ordnance Department. Upon presenting their proposal to the U.S. authorities the British discovered for the first time that the United States had been developing their own midsize cartridge (the T65) and new lightweight rifle to fire it. The U.S. after only a brief test rejected the British cartridge as having less power than desired in a service rifle round. While the U.S. authorities would later agree that the British .280 cartridge would be adequate in a weapon to replace the carbine and submachine gun, it would not be suitable for use in battle rifles or machine guns. The United States simply refused to accept any cartridge that had less wounding ability, penetration and range than the old World War II 30'06 M2 round.

During a comparison test of the British .280 cartridge and the United States' T65, the more powerful U.S. developed round proved to have

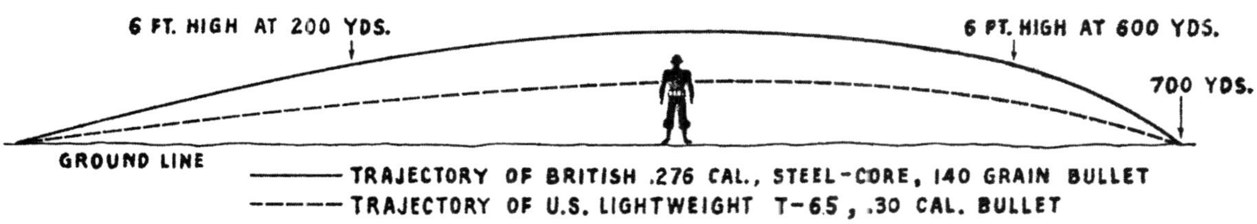

One of the concerns the Ordnance Corps had with the British .280 caliber round was referred to as the "danger space". When firing at a target at a range of 700 yards, the adjusted point of aim using the British cartridge the projectile would be about rise in an arc up to 8-feet above the target during its travel, decreasing the "danger space" between the shooter and his target. Firing the T65 under the same scenario the trajectory to the target was much flatter and would pass through the chest of a six-foot man where the British projectile would pass two-feet over his head.

superior penetration and range. The United States Ordnance Department had a number of other concerns with the less-powerful British cartridge. One of the concerns was with the trajectory of the .280 round. When firing at a target at a range of 700 yards, the adjusted point of aim using the British cartridge, the projectile would rise in an arc up to 8-feet above the target during its travel, decreasing the "danger space" between the shooter and his target. Firing the T65 under the same scenario the trajectory to the target was much flatter and would pass through the chest of a six-foot man where the British projectile would pass two-feet over his head. Another objection that was raised was the effect extreme cold weather had on projectile velocity. The U.S. Ordnance Department cited that under Arctic temperatures the velocity of the British cartridge could be reduced as much as 200 feet per second. In addition to reducing the lethality, the cold might cause the rifle to have functioning problems from the reduced power of the round.

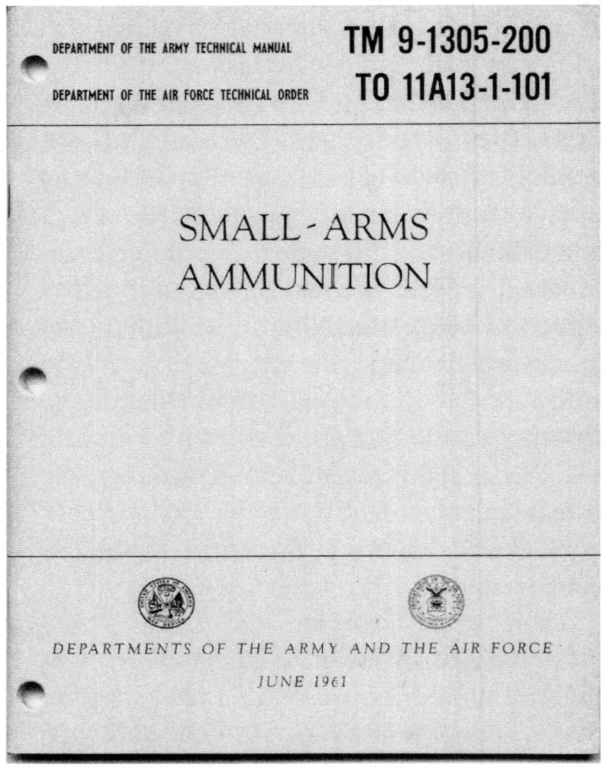

The British were quite disappointed at the United States Ordnance Department's blunt rejection of their new cartridge. The British dismissed the U.S. "danger space" scenario as simply a U.S. concocted theory. Upon their return to England the British team went to work to increase the projectile velocity of their .280 cartridge. To help solve the problem the British substituted a lead core projectile in place of the former steel core bullet. This allowed them to pack more powder into the cartridge case. The new improved Mk1Z .280 round achieved 2,595 feet per second. Another British development was the 7mm HV cartridge that had an average velocity 2,670 feet per second. The British in a futile attempt to make their newly designed round more adaptable to U.S. small arms, redesigned the base of the case to have the same dimensions as the U.S. .30 caliber cartridge, the cartridge was named the .280/30.

The U.S. Army issued an official statement in 1951 stating their position on the adoption of a new cartridge. "The Army is firmly opposed to the adoption of any less effective smaller caliber cartridge for use in either its present rifle or in new weapons being developed. Any new rifle cartridge must have wounding power, penetration performance and ballistics at least equal to that in use today. Battle experience has proven beyond question the effectiveness of the present rifle and ammunition, and there have been no changes in combat tactics that would justify a reduction of rifle caliber and power."

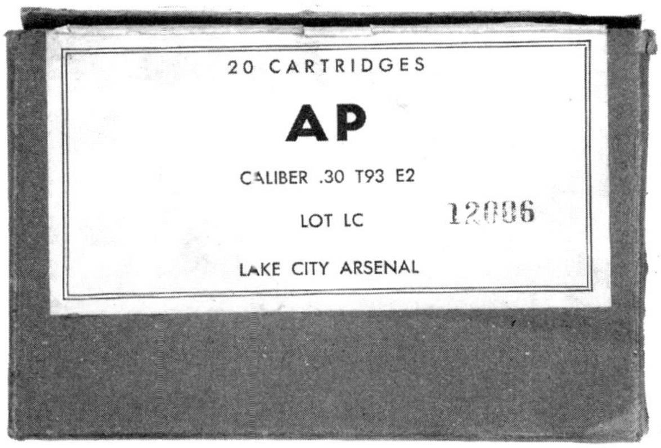

The T93E2 Armor Piercing round was adopted as the 7.62 mm M61 AP cartridge. (Courtesy of John M. Miller)

During the original testing and development phase of the 7.62 mm NATO cartridge, the biggest threat to the NATO alliance was the Soviet Union. If and when the next war came it was expected to be fought in Eastern Europe, much as World War II had been. At the time no one anticipated a guerilla type of conflict like would take place years later in Southeast Asia.

The United States was thinking along the lines of a universal cartridge and weapon to replace the M1 rifle, the BAR, carbine and the submachine gun. The same cartridge was also to be used in the general purpose machine gun. The U.S. officials felt that their T65 cartridge would be suitable for machinegun use, but not too powerful for use in shoulder fired weapons. What the United States was looking for was a "magic" bullet, one that could defy the basic laws of physics.

Few countries could agree on which new cartridge to adopt as the "standard" for all of the

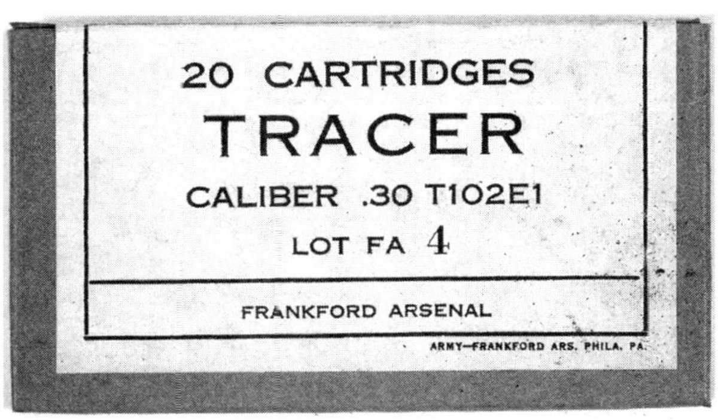

A box of experimental .30 T102E1 Tracer rounds (7.62 mm). The cartridges were adopted as the 7.62mm M62 Tracer cartridge. (Courtesy of John M. Miller)

A crate of U.S. 7.62 mm ammunition contained four ammunition cans.

NATO members. To come to a decision it was determined that a comprehensive set of ballistic, engineering and field studies would be conducted to choose the best overall cartridge. The test results would be compared directly against a group of preset characteristics desired in a military round. The test site that was chosen for the research was the Pendine Experimental Establishment located in Great Britain. Each country interested in submitting a candidate cartridge would be required to send a technical representative to the facility for the trials. The Pendine testing was quite comprehensive and included; velocity, time of projectile flight for various ranges, maximum effective ranges, penetration capabilities in armor plate, steel helmets, wood and sand. At the test's conclusion there were several candidates that had proven capable of doing the job. Ultimately the United States' entry was chosen. The T65E3 cartridge was adopted as the 7.62 mm NATO. It was the very first standard NATO round to be adopted.

The North Atlantic Treaty Organization (NATO) officially announced on 15 December 1953 the adoption of the T65E3 cartridge as developed by the United States. In February of 1954 the NATO alliance agreed on detailed specifications so that the cartridges, regardless of where they were manufactured would be compatible with NATO standard weapons. To positively identify the rounds as NATO standard, a small encircled cross was to be stamped on the base of all 7.62 mm cartridges, by all countries producing the ammunition. At a meeting of the Ordnance Committee during August of 1954 the new cartridge was formally adopted as the Cartridge, NATO Caliber 7.62 mm. The standard NATO ball cartridge had a 150-grain full metal jacketed lead projectile with an average muzzle velocity of approximately 2900 feet per second.

The United Kingdom, Belgium and Canada standardized lead-core ball, while the United States standardized steel-core ball ammunition. The U.S. saw no point in standardizing the type of ammunition manufactured during peacetime, while using another (steel-core) in times of war. Caliber .30 steel-core armor-piercing ammunition was standard issue during World War II.

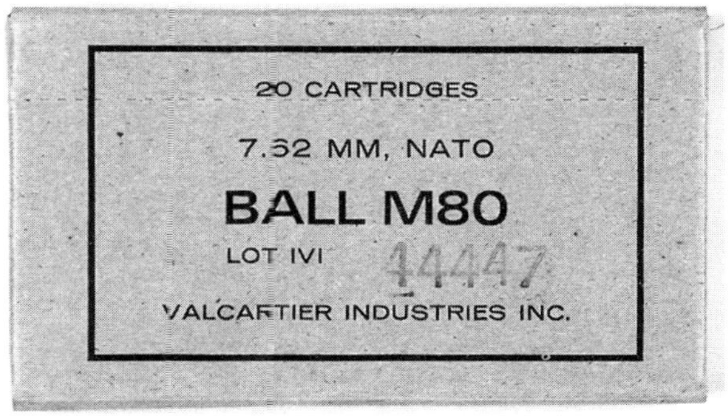

The M80 is a standard cartridge with a lead core projectile covered by a gilding-metal steel jacket. The round is used for training, against light material and enemy personnel.

Special high-pressure test cartridges for proof firing under controlled conditions. These rounds can be very dangerous and should never be fired by unqualified personnel. (Courtesy of John M. Miller)

7.62 Cartridges

7.62 mm Ball, M59

This cartridge was intended for use against personnel and unarmored targets. The projectile has a mild steel core. Projectile weight 150.5 grains, muzzle velocity is 2,750 feet per second. The projectile is not painted.

7.62 mm Ball, M80

The M80 is a standard cartridge with a lead core projectile covered by a gilding-metal steel jacket, it has a weight of 146 grains. The projectile is not marked. The round is used for training, against light material and enemy personnel.

U.S. Rifle 7.62 mm M14

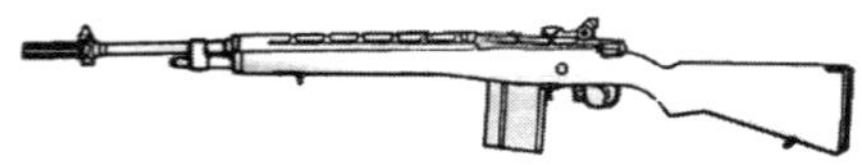

An M14 rifle is displayed with crates of 7.62 mm ammunition that would be used for an endurance test of the weapon. (Courtesy of the Springfield Armory Museum)

7.62 mm Ball, M60

This ammunition is used for high-pressure proof testing of weapons during manufacture or repair. It is not for general use and should never be fired by unauthorized personnel. The cartridge case is silver in color.

7.62 mm Ball, M61 AP

The M61 AP cartridge originated as the T93E1 armor-piercing round with a flat-base projectile, manufactured at the Frankford Arsenal in 1951 for testing purposes. The projectile was similar to that of the .30 caliber AP M2 (30'06) except for its knurling.

The M61 is armor piercing ammunition for use against personnel, light armored or unarmored vehicles, concrete shelters or similar cover. Projectile weight is 150.5 grains, muzzle velocity 2,750 feet per second. The tip of the projectile is painted black. This ammunition is not authorized for training purposes.

7.62 mm Ball, M62 Tracer

For observation of fire, signaling and incendiary effects. Projectile weight 146 grains, velocity 2,750 feet per second. The tip of the projectile is painted orange. For machinegun use, a ratio of four ball rounds for each tracer is used.

7.62 mm Ball, M276 Dim Tracer

Combat cartridge issued to troops using night

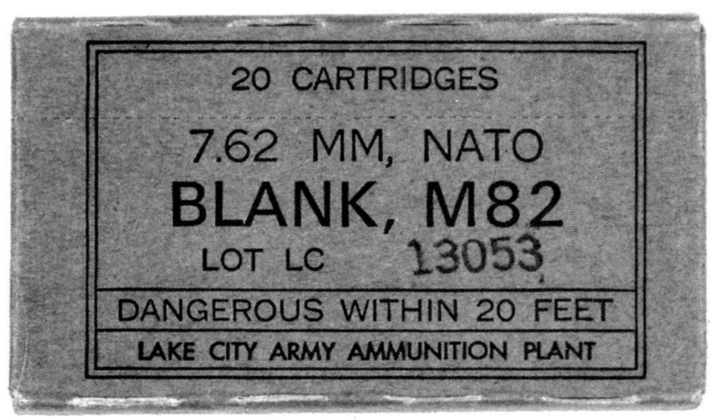

The blank training 7.62 mm cartridge for U.S. weapons is the M82. The blank cartridge was Standardized on 21 September 1961.

vision equipment, which is sensitive to bright light.

7.62 mm M993 Armor Piercing

This round is intended for used against light armored targets. The M993 is more effective than previous issue AP ammunition. The 126.6 grain projectile has a tungsten steel penetrator contained inside of an aluminum cup, and jacketed in tombac.

7.62 mm M82 Blank Cartridge

The blank training 7.62 mm cartridge for U.S. weapons is the M82. The blank cartridge was Standardized on 21 September 1961. To allow for reliable feeding in semiautomatic and full automatic weapons the blank cartridge case was formed with a long neck and formed into a shape that approximates that of a normal projectile. The shape is distinctly different so that blanks are not confused with live cartridges.

For blank cartridges a powder that burns faster than normal is required. Because there is insufficient pressure to operate the breech mechanism of a weapon a blank adapter must be used. The adapter restricts the gas leaving the muzzle of the weapon and instead directs it to operate the gas system. Blank adapters were produced for both the M14 rifle and M60 machine gun. Powder from blank cartridges should never be used for reloading purposes.

The 7.62 mm M198 Duplex Round contained two 80-grain full metal jacketed projectiles. The base of the front projectile was partially hollow to allow the tip of the rearmost projectile to fit into it. A cut-away M198 cartridge is shown on the right. (John M. Miller)

7.62mm Overhead Fire Ammunition

During the early 1960 period special cartridges were designed and fielded for firing over the

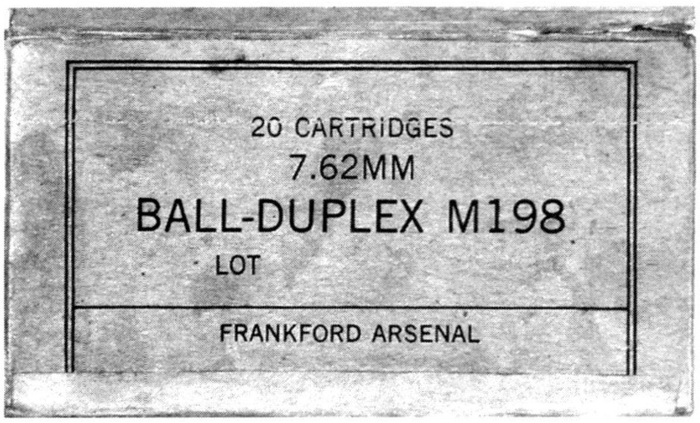

The M198 Duplex Round contained two 80-grain full metal jacketed projectiles. The base of the front projectile was partially hollow to allow the tip of the rearmost projectile to fit into it. The base of the rear projectile had a 9-degree angle to give it the desired dispersion pattern once it left the muzzle. (John M. Miller)

heads of troops during training exercises. Special projectiles were used to eliminate injuries from fragments. To allow this feature the bullets had solid gilding metal without sealing disks. There were several variations of Overhead Fire ammunition produced. XM178 was a ball round that weighed 149.3 grains. XM179 was an Overhead Fire tracer round with solid gilding with a cavity in the base of the projectile for the tracer compound. The tracer bullet weighed 142.6 grains. The XM179 rounds were identified by a red painted tip covering 5/16 inch of the projectile's tip. The last Overhead Fire round made was the XM180. The projectile of this round had a smaller tracer cavity in the projectile's base. The tip of the XM180's bullet had a 1/16 inch red band painted on the tip. The tracer rounds were not designed to trace as far as standard issue tracers. Problems were encountered with retaining the compound in the projectile. Around 1963 it was decided that there was no advantage to special overhead fire cartridges and the program was terminated.

7.62 mm Duplex M198 Round

The development of the duplex round was an attempt by the U.S. Army to improve the probability of hitting targets during combat. Although the round was standardized as the 7.62 Ball, Duplex, NATO M198 on 7 May 1964, the multiple projectile round concept was not a new one. Winchester-Western developed the U.S. M198 Duplex Round under contract with the U.S. Ordnance Department. The duplex round would work in any U.S. weapon chambered for the standard 7.62 NATO cartridge, although it proved to be most effective when used in machine guns. Testing conducted under a wide variety of conditions revealed that the duplex round was anywhere from 25% to 100% more effective than a cartridge firing a single projectile.

The M198 Duplex Round contained two 80-grain full metal jacketed projectiles. The base of the front projectile was partially hollow to allow the tip of the rearmost projectile to fit into it. The base of the rear projectile had a 9-degree angle to give it the desired dispersion

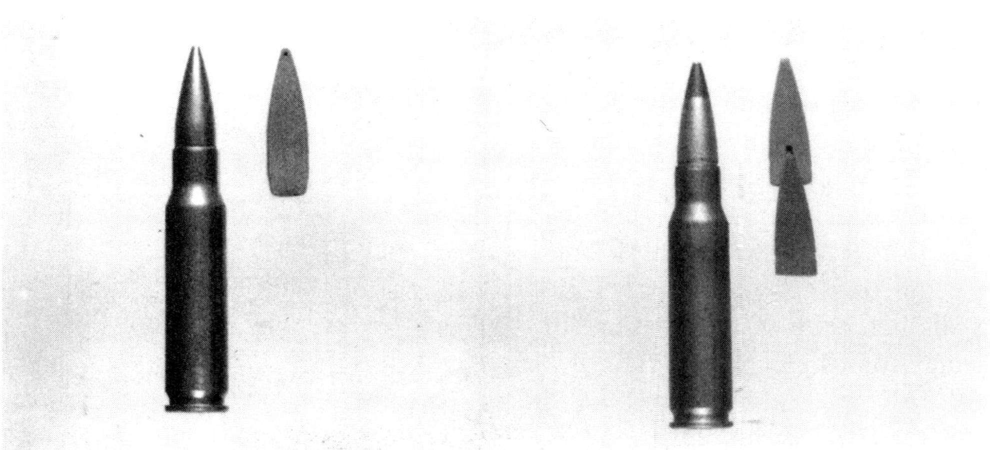

M80 (STANDARD) M198 (TEST)

CARTRIDGE, 7.62-MM, DUPLEX, BALL, M198

DATA

STANDARD: Cartridge, 7.62-MM: NATO, Ball, M80 (Ord Dwg No. C-8595543)
TEST: Cartridge, 7.62-MM: Duplex, Ball, M198 (Ord Dwg No. C-FC 8579)

	Standard	Test
Bullet(s)	M80, Wt 149 gr	FAT 92, Wt 82.5 gr FAT 93, Wt 82.5 gr
Case	Brass, FAT1-E3	Brass, FAT1-E3
Propellant	WC 846.4, 47.1 gr	WC 846.4, 47.6 gr
Primer	FA 36	Winchester 120 M
Instrumental Velocity at 78 feet, fps	2750 ± 30	FAT 92: 2750 ±30 FAT 93: 2000 Min
Chamber Pressure, psi	50,000 Max	52,000 Max.

View showing the M80 cartridge with a sectionalized bullet (standard) on the left and a M198 cartridge with sectionalized bullets (test) on the right.

S18-001-C-P386/AMC-65 ABERDEEN PROVING GROUND 11 February 1965 pattern once it left the muzzle. The average muzzle velocity of the front bullet was 2800 feet per second. The rear bullet was slightly slower at 2600 feet per second. The front projectile had approximately the same accuracy as a normal 7.62 round. The rearmost projectile had approximately 6-times as much dispersion as the front projectile. In other words if the front projectile shoots a 10 shot group 5-inches in diameter, the rear projectile will disperse as much as 40-inches. When fired at great distances from a helicopter mounted machine gun, the use of the duplex round allowed twice as many projectiles to be placed into a target area. Although the result was favorable, the 7.62 duplex cartridges were not widely used except for testing purposes.

7.62 mm M160 Frangible Cartridge

The M118 cartridges were commonly issued for use in the M21 Sniper rifles. The M118 round features a match-grade 173-grain full metal jacket, boattail projectile with a muzzle velocity of 2,550 feet per second.

A new cartridge has been introduced for the 7.62 mm Designated Marksman Rifle. The new round is the M118LR Special Ball, "LR" indicating Long Range.

The 7.62 mm frangible cartridges used the same 108-grain projectile as the earlier .30 caliber M22 frangible round. The projectile was constructed from a combination of 50% lead and 50% bakelite composite. The average muzzle velocity of the projectile is 1,360 feet per second. The primary purpose of frangible ammunition is to leave a mark on a target at the point of impact. The cartridges are generally used for firing single shots in machineguns for targeting purposes. The projectile has a green-painted tip with a white ring at the base of the green area.

7.62 mm Match Ammunition

During World War II there was no match grade or any other special type of ammunition available for snipers. The most common round used was the M2 AP armor-piercing

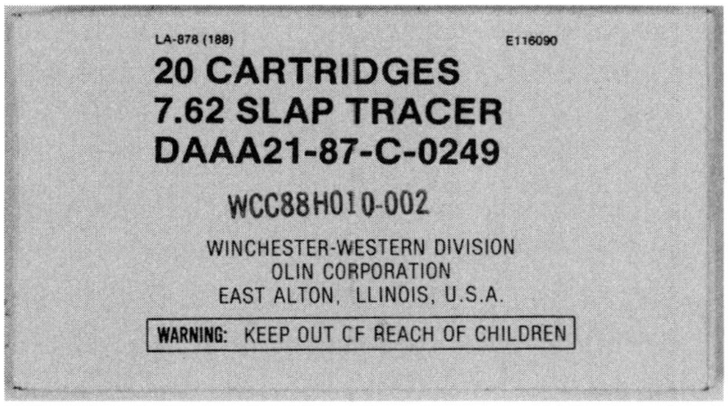

7.62 mm Saboted Light Armor Penetrating (SLAP) rounds. (Courtesy of John M. Miller)

.30 caliber. Prior to 1957 special ammunition was only procured in limited numbers and for special sanctioned matches, most of them international competition. Around 1957 there had been considerable interest generated in rifle competition at all inter-service levels as well as the larger matches. Shortly thereafter special .30 caliber ammunition was made available for both practice and competition use, although the best of the lot was reserved for the National Matches.

In 1962 Lake City began to manufacture both .30 caliber Match and National Match ammunition. National Match ammunition was loaded with carefully selected components and a specific powder charge for those components. In 1961 7.62 mm Match was made available followed by 7.62 National Match the following year. Match M118 is a Standard type always made to the same specifications. The M118 cartridges were commonly issued for use in the M21 Sniper rifles. The M118 round features a match-grade 173-grain full metal jacket, boattail projectile with a muzzle velocity of 2,550 feet per second, muzzle energy is 2,605 foot pounds.

A new cartridge has been introduced for the 7.62 mm Designated Marksman Rifle. The new round is the M118LR Special Ball, "LR" indicating Long Range. The M118LR uses a 175-grain match projectile with a velocity of 2580 feet per second, muzzle energy is 2,586 foot pounds.

7.62 mm M852 Match Ammunition

This ammunition is intended and specifically loaded for weapons designated as competitive rifles and for marksmanship training and

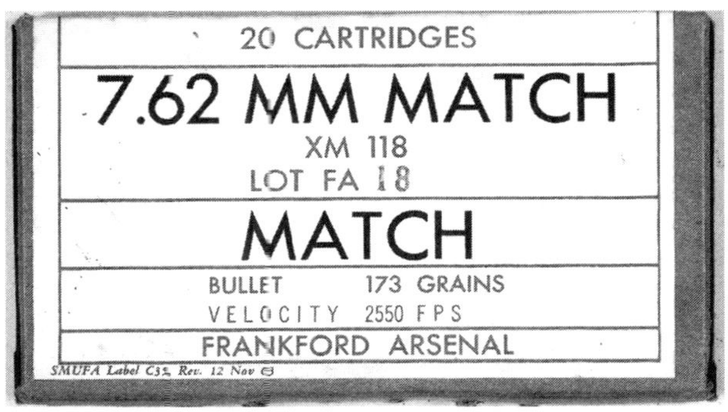

In 1962 Lake City began to manufacture both .30 caliber Match and National Match ammunition. National Match ammunition was loaded with carefully selected components and a specific powder charge for those components. (Courtesy of John M. Miller)

practice. The cartridges are not for combat use. The base of the cartridge case is stamped with the word MATCH. The 168 grain projectile is a hollow-point design.

7.62 mm T275 Match Ammunition

The T275 series of special Match ammunition was produced for Olympic and similar International 300 meter rifle competitions. The particular loadings of T275 cartridges varied from one year to the next. The T275 series went as high as the T275E5, but the cartridge was never adopted as Standard.

Accuracy tests conducting with the Harrington and Richardson T48 rifles indicated that the T93E1 and T93E2 (M61) ammunition were comparable at ranges at a distance of 600 yards, but the T93E1 round proved far superior at a 100 yard range.

The T275 series of special Match ammunition was produced for Olympic and similar International 300 meter rifle competitions. (Courtesy of John M. Miller)

Chapter 15
Military Publications

There were numerous technical (TM) and field manuals (FM) published for the M14 rifle. The original GI manuals have been copied and sold at gun shows by a number of individuals and companies. While the reproduction manuals are less expensive than originals the quality is inferior to the originals.

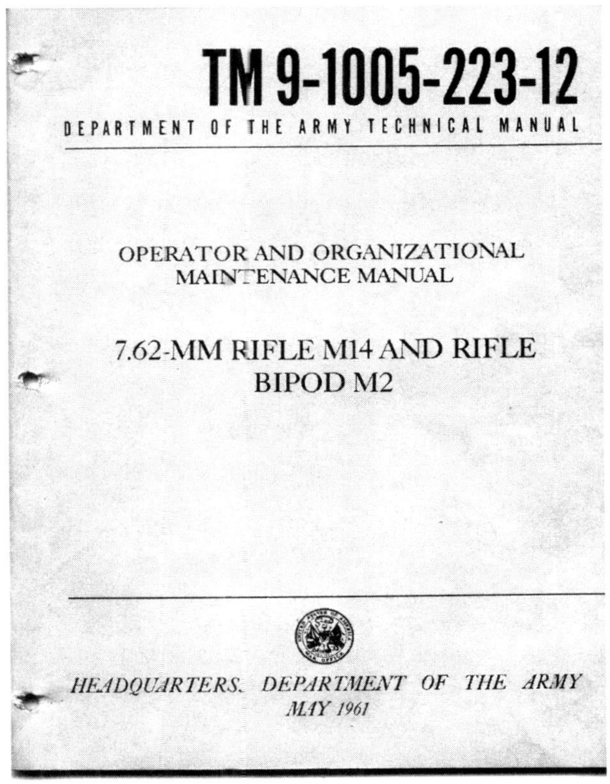

TM 9-1005-223-12

DEPARTMENT OF THE ARMY TECHNICAL MANUAL

OPERATOR AND ORGANIZATIONAL
MAINTENANCE MANUAL

7.62-MM RIFLE M14 AND RIFLE
BIPOD M2

HEADQUARTERS, DEPARTMENT OF THE ARMY
MAY 1961

The first letters TM indicate that the publication is a Technical Manual; The number 9 indicates that Ordnance is the responsible agency; The number 1005 designates that the series covers all weapons up to 30 mm inclusive; The number 223 is the item identifier, 223 for the M14 rifle. The number -12 designate the level of maintenance and repair the publication is intended for.

The numbering system used for Army technical manuals can be confusing, but after becoming familiar with the Army's system most will find that it is actually very simple.

The numbering system is as follows;

Example; Manual TM 9-1005-223-10

The first letters TM indicate that the publication is a Technical Manual;

The number 9 indicates that Ordnance is the agency responsible;

The number 1005 designates that the series covers all weapons up to 30 mm inclusive;

The number 223 is the item identifier, 223 was assigned exclusively to the M14 rifle. Each weapon was assigned a different number;

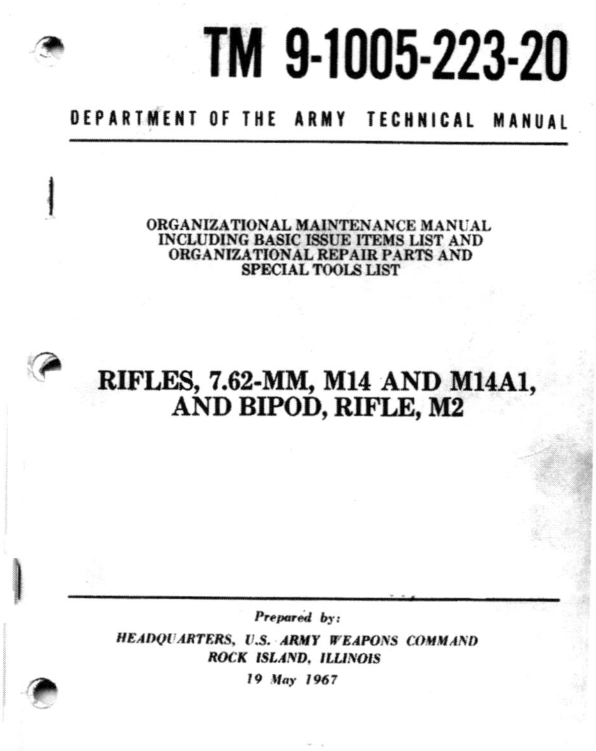

TM 9-1005-223-20

DEPARTMENT OF THE ARMY TECHNICAL MANUAL

ORGANIZATIONAL MAINTENANCE MANUAL
INCLUDING BASIC ISSUE ITEMS LIST AND
ORGANIZATIONAL REPAIR PARTS AND
SPECIAL TOOLS LIST

RIFLES, 7.62-MM, M14 AND M14A1,
AND BIPOD, RIFLE, M2

Prepared by:
HEADQUARTERS, U.S. ARMY WEAPONS COMMAND
ROCK ISLAND, ILLINOIS
19 May 1967

A manual with a title ending in a number 20 indicates that the publication was only intended for level 2 maintenance. Organization maintenance or the unit armorer.

The numbers 10-12-20-24 and 35 designate the level of maintenance and repair the publication is intended for. The maintenance levels are explained next.

Level 1- This would be an individual soldier or a crewmember. The individual that actually uses the equipment.

Level 2- Organization maintenance or the unit armorer. This is the next level above the user level. More involved disassembly and repairs are authorized at this level.

Level 3- Direct Support maintenance. This is generally a Maintenance Battalion within a Division. The Direct Support Unit travels to field areas within the combat division to perform repairs.

Level 4- General Support. This is generally the highest level of Maintenance assigned within a combat theatre. A General Support Unit is not assigned to any one particular combat unit. The General Support Unit is usually stationary and located in a rear area. General Support can usually perform any task that a Depot unit is capable of. General Support is also organized to repair more weapons within a given time period than Direct Support could.

Level 5- Depot Maintenance. This was the highest level of Maintenance within the Ordnance system during the 1960 era. Depot level maintenance was authorized and equipped for complete overhaul or rebuilding in large numbers. The Depot stocked large numbers of spare parts and assemblies to support lower levels of support. A Depot could be located overseas or in the United States. During the 1960s some manufacturers like the Springfield Armory and the Rock Island Arsenal also performed Depot Level maintenance functions in certain areas.

The last two digits used to describe a manual designate the maintenance level or levels to which the manual would apply. Usually the manuals applied to more than one level.

10- A manual with this number at the end of the title would designate that the publication was intended for level 1 maintenance, the operator or user of the equipment.

12- A manual with a title ending in a number 12 indicates that the publication was intended for both maintenance levels 1 and 2.

20- A manual with a title ending in a number 20 indicates that the publication was only intended for level 2 maintenance.

25- A manual with a title ending in a number 25 indicates that the publication was intended for maintenance levels 2, 3, 4 and 5.

34- A manual with a title ending in a number 34 indicates that the publication was intended for maintenance levels 3 and 4.

35- A manual with a title ending in a number 35 indicates that the publication was intended for levels 3, 4 and 5.

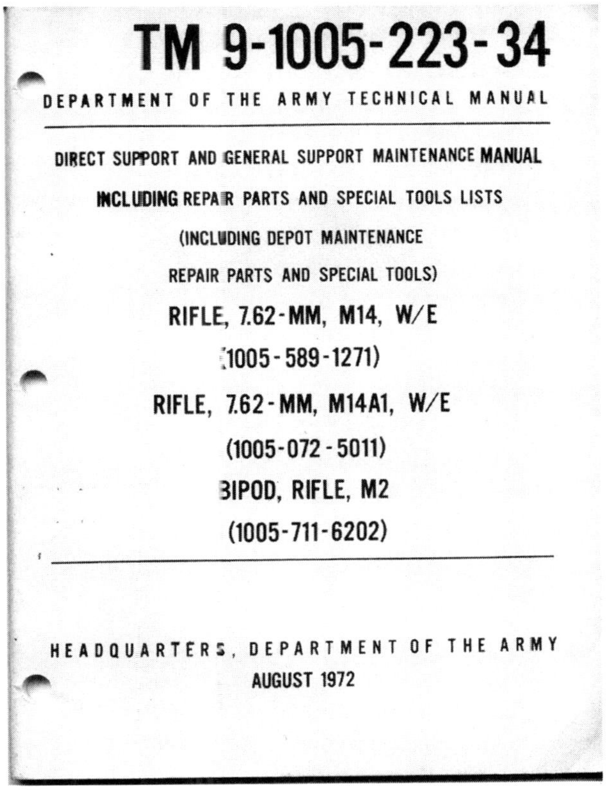

TM 9-1005-223-34

DEPARTMENT OF THE ARMY TECHNICAL MANUAL

DIRECT SUPPORT AND GENERAL SUPPORT MAINTENANCE MANUAL

INCLUDING REPAIR PARTS AND SPECIAL TOOLS LISTS

(INCLUDING DEPOT MAINTENANCE

REPAIR PARTS AND SPECIAL TOOLS)

RIFLE, 7.62-MM, M14, W/E

(1005-589-1271)

RIFLE, 7.62-MM, M14A1, W/E

(1005-072-5011)

BIPOD, RIFLE, M2

(1005-711-6202)

HEADQUARTERS, DEPARTMENT OF THE ARMY

AUGUST 1972

A manual with a title ending in a number 34 indicates that the publication was intended for maintenance levels 3 and 4; Direct Support and General Support.

Higher maintenance level manuals were generally multilevel publications.

Some manual title designations ended with a P, P/1 or P/2. The presence of the letter P was used to indicate that the manual contained an equipment parts listing.

31 July 1959 Manual 223-12P
12 October 1959 Manual 223-12
12 October 1959 (1960) Manual 223-12
26 May 1961 Manual 223-12
15 November 1961 Manual 223-12
16 March 1962 Manual 223-20P
22 January 1963 Manual 223-12
22 January 1964 Manual 223-12
8 February 1965 Manual 223-12
14 February 1966 Manual 223-35
19 May 1967 Manual 223-20
23 February 1968 (National Match Rifles)
1 July 1968 Manual 223-35
June 1969 Manual 223-35
July 1971 Manual 223-35
August 1972 Manual 223-34
10 March 1972, Manual 223-10
August 1976 Manual 223-34
August 1980 Manual 223-34

Known M14 Rifle FM Publication Dates
Often it would be stated in the newer manuals that that it superceded a previous manuals. The dates of the previous manuals being superceded would be listed.

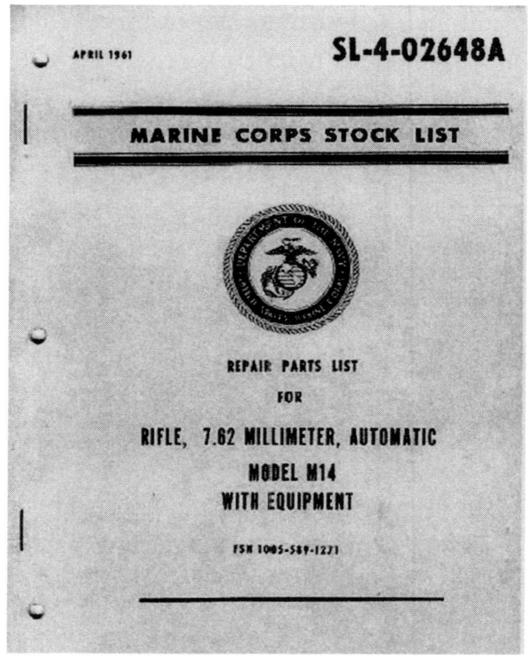
APRIL 1961

SL-4-02648A

MARINE CORPS STOCK LIST

REPAIR PARTS LIST
FOR
RIFLE, 7.62 MILLIMETER, AUTOMATIC
MODEL M14
WITH EQUIPMENT

A rare 1961 dated USMC parts manual. (Courtesy of David Albert)

Field Manuals (FM)

Field manuals generally contained information for the soldier actually using the equipment or weapons covered. Small Arms manuals were assigned number 23. The manuals were periodically updated but would retain the same designation. For example there were a number of FM-8 manuals printed for the M14 rifle with different titles and dates. Inside the cover would be the dates and names of similar, but earlier manuals the new manual superceded.

PS Magazine Be Your Own Inspector BYOI

The origins of the PS magazine dates back to World War II when the Ordnance Department had two regular publications, they were *Army Motors* and The *Ordnance Sergeant.* The *Ordnance Sergeant* magazine was printed and distributed monthly. The Magazine delved into high echelon changes that affected the entire Ordnance spectrum. The success of these two magazines paved the way

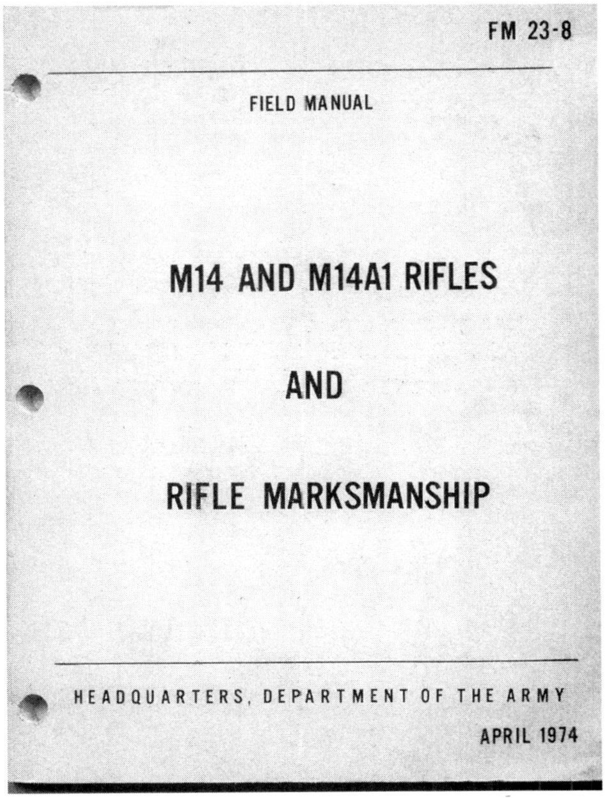

FM 23-8

FIELD MANUAL

M14 AND M14A1 RIFLES

AND

RIFLE MARKSMANSHIP

HEADQUARTERS, DEPARTMENT OF THE ARMY

APRIL 1974

An Army Field Manual (FM) for M14 and M14A1 marksmanship. (Donald G. Thomas)

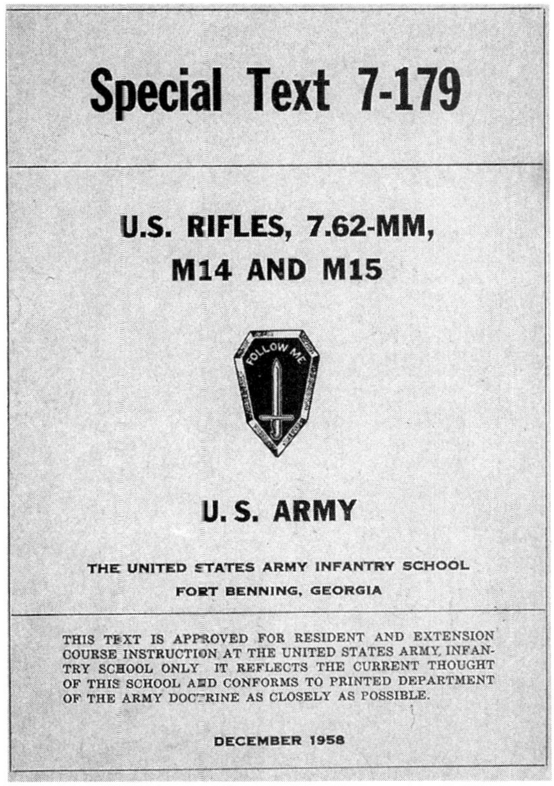

Special Text 7-179

U.S. RIFLES, 7.62-MM, M14 AND M15

U. S. ARMY

THE UNITED STATES ARMY INFANTRY SCHOOL
FORT BENNING, GEORGIA

THIS TEXT IS APPROVED FOR RESIDENT AND EXTENSION COURSE INSTRUCTION AT THE UNITED STATES ARMY INFANTRY SCHOOL ONLY. IT REFLECTS THE CURRENT THOUGHT OF THIS SCHOOL AND CONFORMS TO PRINTED DEPARTMENT OF THE ARMY DOCTRINE AS CLOSELY AS POSSIBLE.

DECEMBER 1958

A very early Army manual describing the M14 and M15 rifles. (Courtesy of Bill Ricca)

for PS magazine that was geared more toward the soldier in the field.

The first issue of PS magazine was published in June of 1951 during the Korean War. The purpose of the magazine was to get information out into the field as quickly as possible. The designation of PS was an abbreviation for Post Script to the various technical manuals and bulletins that often took a long period of time before getting published and distributed out into the field. The PS magazines usually contained information on maintenance for soldiers who were the end users of the equipment.

The PS magazines were designed to be informative and provide a small degree of enter-

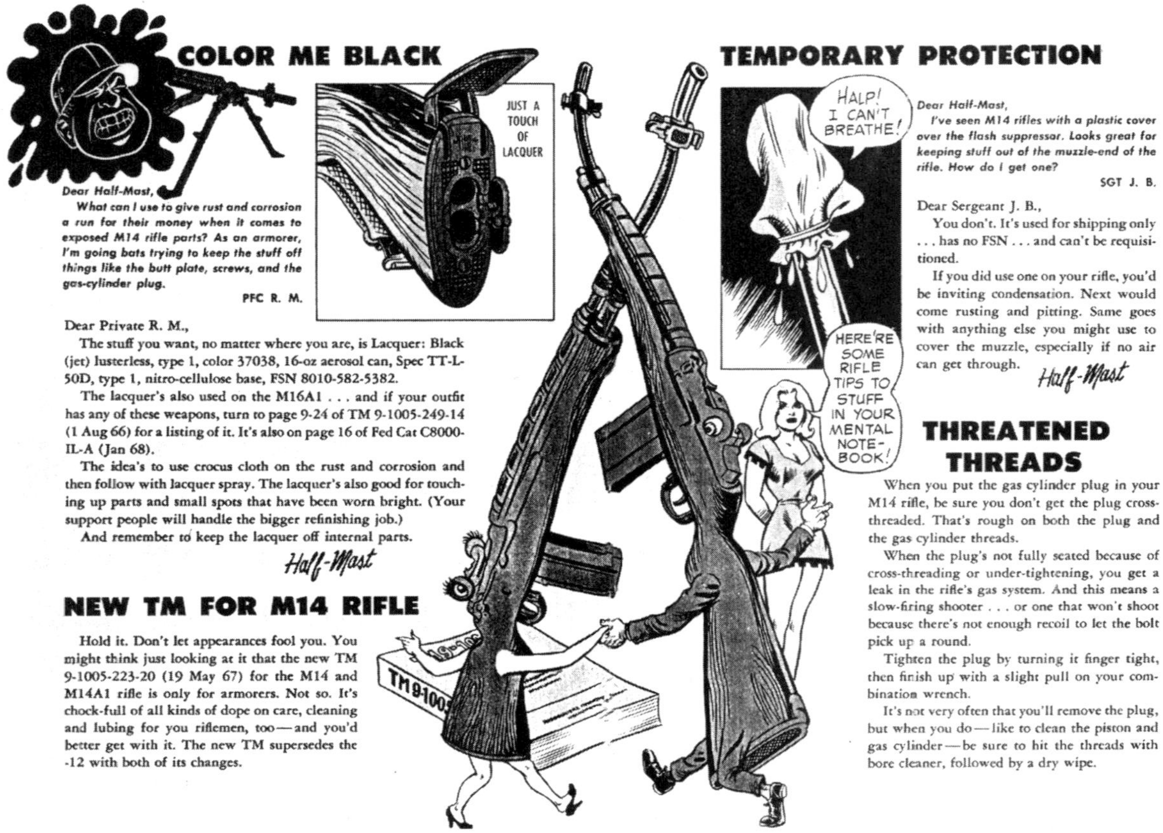

COLOR ME BLACK

Dear Half-Mast,
What can I use to give rust and corrosion a run for their money when it comes to exposed M14 rifle parts? As an armorer, I'm going bats trying to keep the stuff off things like the butt plate, screws, and the gas-cylinder plug.

PFC R. M.

Dear Private R. M.,
The stuff you want, no matter where you are, is Lacquer: Black (jet) lusterless, type 1, color 37038, 16-oz aerosol can, Spec TT-L-50D, type 1, nitro-cellulose base, FSN 8010-582-5382.

The lacquer's also used on the M16A1 . . . and if your outfit has any of these weapons, turn to page 9-24 of TM 9-1005-249-14 (1 Aug 66) for a listing of it. It's also on page 16 of Fed Cat C8000-IL-A (Jan 68).

The idea's to use crocus cloth on the rust and corrosion and then follow with lacquer spray. The lacquer's also good for touching up parts and small spots that have been worn bright. (Your support people will handle the bigger refinishing job.)

And remember to keep the lacquer off internal parts.

Half-Mast

NEW TM FOR M14 RIFLE

Hold it. Don't let appearances fool you. You might think just looking at it that the new TM 9-1005-223-20 (19 May 67) for the M14 and M14A1 rifle is only for armorers. Not so. It's chock-full of all kinds of dope on care, cleaning and lubing for you riflemen, too—and you'd better get with it. The new TM supersedes the -12 with both of its changes.

TEMPORARY PROTECTION

Dear Half-Mast,
I've seen M14 rifles with a plastic cover over the flash suppressor. Looks great for keeping stuff out of the muzzle-end of the rifle. How do I get one?

SGT J. B.

Dear Sergeant J. B.,
You don't. It's used for shipping only . . . has no FSN . . . and can't be requisitioned.

If you did use one on your rifle, you'd be inviting condensation. Next would come rusting and pitting. Same goes with anything else you might use to cover the muzzle, especially if no air can get through.

Half-Mast

THREATENED THREADS

When you put the gas cylinder plug in your M14 rifle, be sure you don't get the plug cross-threaded. That's rough on both the plug and the gas cylinder threads.

When the plug's not fully seated because of cross-threading or under-tightening, you get a leak in the rifle's gas system. And this means a slow-firing shooter . . . or one that won't shoot because there's not enough recoil to let the bolt pick up a round.

Tighten the plug by turning it finger tight, then finish up with a slight pull on your combination wrench.

It's not very often that you'll remove the plug, but when you do—like to clean the piston and gas cylinder—be sure to hit the threads with bore cleaner, followed by a dry wipe.

A typical article for the M14 taken from a PS manual. The cartoon format of the PS articles was popular with troops and was successful in getting soldiers to read them. (Courtesy of Bill Ricca)

tainment. To hold the soldier's attention the PS magazines had a comical slant and featured illustrations that included a pretty girl by the name of Connie Rodd and an old sergeant named Sgt. Half-Mast. Together the two characters answered questions, offered tips on maintenance and kept troops informed about new developments. The content of PS magazines was not limited to small arms, but covered a wide spectrum of equipment used by the Army.

The earliest known manual for the M14 rifle was published in 1958. In the manual the M14 was depicted in its early configuration, the most obvious were the M1 Garand buttplate and wooden handguard. One interesting aspect of the early manuals was that it listed the weight

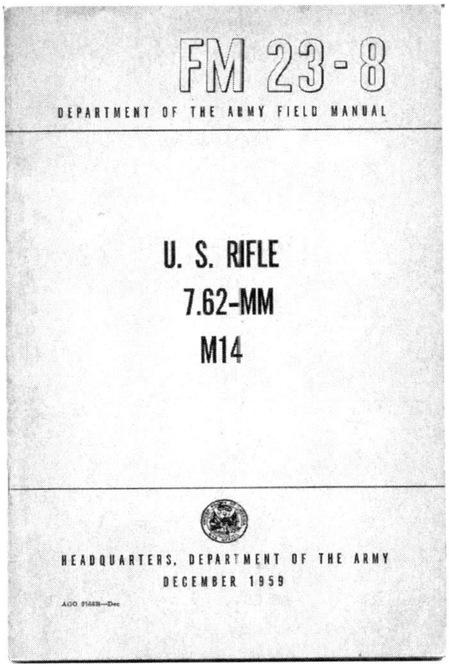
FM 23-8

DEPARTMENT OF THE ARMY FIELD MANUAL

U. S. RIFLE
7.62-MM
M14

HEADQUARTERS, DEPARTMENT OF THE ARMY
DECEMBER 1959

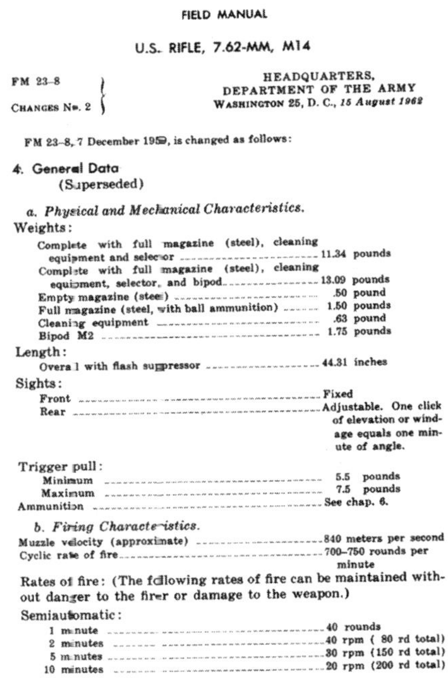
FIELD MANUAL

U.S. RIFLE, 7.62-MM, M14

FM 23-8
CHANGES No. 2

HEADQUARTERS,
DEPARTMENT OF THE ARMY
WASHINGTON 25, D. C., *15 August 1962*

FM 23-8, 7 December 1959, is changed as follows:

4. General Data
(Superseded)

a. Physical and Mechanical Characteristics.

Weights:

Complete with full magazine (steel), cleaning equipment and selector	11.34 pounds
Complete with full magazine (steel), cleaning equipment, selector, and bipod	13.09 pounds
Empty magazine (steel)	.50 pound
Full magazine (steel, with ball ammunition)	1.50 pounds
Cleaning equipment	.63 pound
Bipod M2	1.75 pounds

Length:

Overall with flash suppressor	44.31 inches

Sights:

Front	Fixed
Rear	Adjustable. One click of elevation or windage equals one minute of angle.

Trigger pull:

Minimum	5.5 pounds
Maximum	7.5 pounds
Ammunition	See chap. 6.

b. Firing Characteristics.

Muzzle velocity (approximate)	840 meters per second
Cyclic rate of fire	700–750 rounds per minute

Rates of fire: (The following rates of fire can be maintained without danger to the firer or damage to the weapon.)

Semiautomatic:

1 minute	40 rounds
2 minutes	40 rpm (80 rd total)
5 minutes	30 rpm (150 rd total)
10 minutes	20 rpm (200 rd total)

A 1959 dated manual published for the M14 rifle. An early supplement for the manual is shown on the right. (Courtesy of Don Thomas and Bill Ricca)

of the M14 rifle with an aluminum magazine rather than the steel one.

There were manuals published for all levels of maintenance and repair. Although there were a large number of updated manuals that superceded earlier publications they all basically had the same information in them. Changes often involved the introduction of a new part and the older part deemed obsolete. Improvements and changes were ongoing for U.S. military equipment, usually based on problems reported from the field.

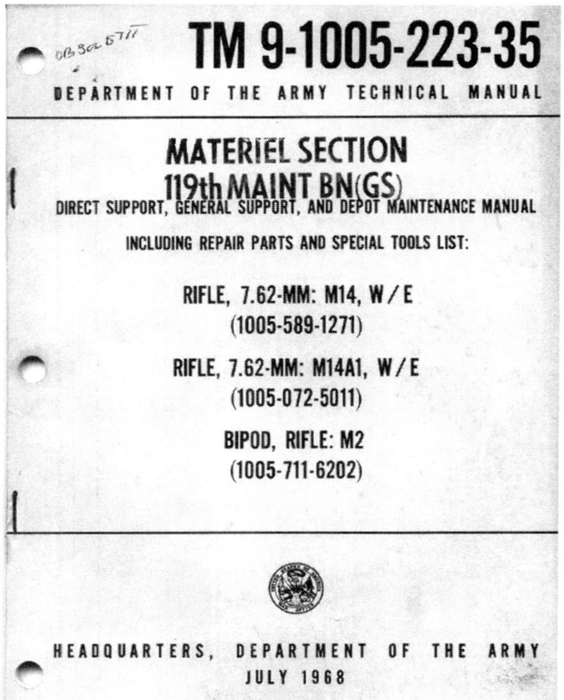

TM 9-1005-223-35

DEPARTMENT OF THE ARMY TECHNICAL MANUAL

MATERIEL SECTION
119th MAINT BN(GS)

DIRECT SUPPORT, GENERAL SUPPORT, AND DEPOT MAINTENANCE MANUAL

INCLUDING REPAIR PARTS AND SPECIAL TOOLS LIST:

RIFLE, 7.62-MM: M14, W/E
(1005-589-1271)

RIFLE, 7.62-MM: M14A1, W/E
(1005-072-5011)

BIPOD, RIFLE: M2
(1005-711-6202)

HEADQUARTERS, DEPARTMENT OF THE ARMY
JULY 1968

A Technical Manual with a title ending in a number 35 indicates that the publication was intended for levels 3, 4 and 5. This was a high level publication. (John M. Miller)

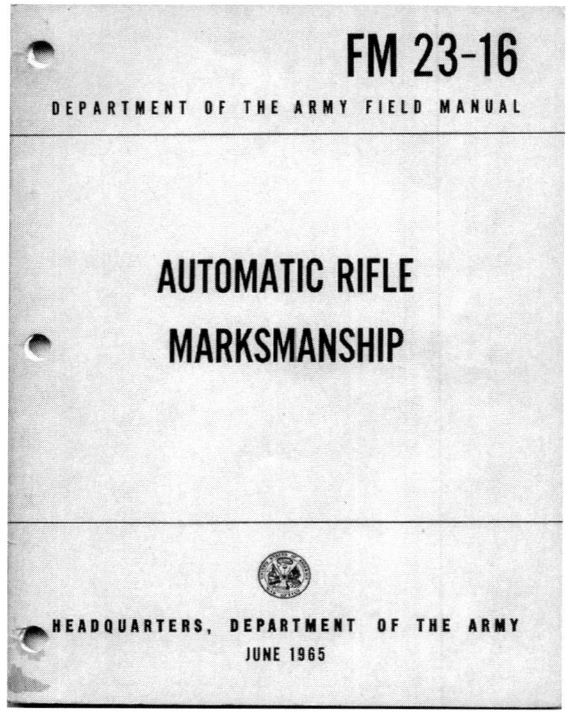

FM 23-16

DEPARTMENT OF THE ARMY FIELD MANUAL

AUTOMATIC RIFLE
MARKSMANSHIP

HEADQUARTERS, DEPARTMENT OF THE ARMY
JUNE 1965

Field manuals generally contained information for the soldier actually using the equipment or weapons covered. Small Arms manuals were assigned number 23. The manuals were periodically updated but would retain the same designation. For example there were a number of FM-8 manuals printed for the M14 rifle with different titles, like the one here for Automatic Marksmanship. (Donald G. Thomas).

Chapter 16
The M1A™ and Commercial M14-Type Rifles
"Necessity is the Mother of Invention"

The M1A™ and the commercial M14 type semiautomatic-only rifles have proven to be extremely popular with civilian collectors and shooters both in the United States and abroad. There have been an estimated total of 350,000 M14 type rifles made for the civilian market by as many as twelve different manufacturers, since the first M1A™ receiver was made in 1971. The rifles have a dedicated fan base that will adamantly defend the 7.62x51 mm M14 as the best military cartridge and rifle combination ever fielded, particularly when compared to the 5.56 mm M16 series of military rifles.

Springfield Armory 7.62 mm, M1A™ rifle. (Courtesy of Gary Osif)

The recent rediscovery of the 7.62 mm M14 rifle by the U.S. military has caused a resurgence of interest in the civilian M14 and M1A™ rifles. New civilian models are being offered based on the upgraded U.S. military issue versions of the M14 rifle being fielded in Iraq and Afghanistan.

The M1A™ rifle was the very first civilian-legal copy of the venerable military M14 rifle. The Springfield Armory Inc. M1A™ rifle has been in virtually continuous production since 1971 with an estimated total of over 200,000 rifles made to date. The M1A™ has become such a recognizable icon that when shown an actual military M14 rifle many enthusiasts will recognize it as an M1A™ rather than an M14!

The M1A™ rifle was the inspiration of Mr. Elmer Balance, owner of the L.H. Gun Company then located in Devine, Texas. Mr. Ballance obviously saw the potential civilian market for a semiautomatic version of the U.S. M14 rifle that was unavailable to civilians. Particularly interested in the proposed new

SPECIFICATIONS:

WEIGHT:	8.5 lbs. (Empty.)
OVERALL LENGTH:	42.5 inches
BARREL LENGTH:	24 inches
SIGHTS:	Front - fixed Rear - Adjustable for elevation and windage
MUZZLE VELOCITY:	2850 fps. (140 gr. bullet)
MAXIMUM RANGE:	3200 yds. (140 gr. bullet)

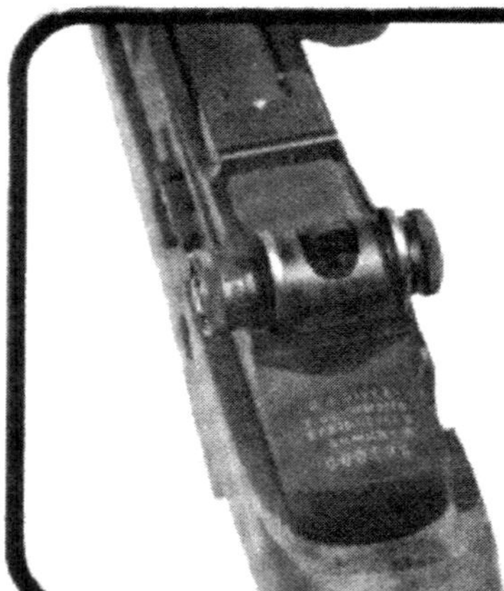

L.H. Gun Company guarantees the M1A for one full year against any defect in workmanship or materials. Three years of planning have gone into the making of this rifle. Each is manufactured using Government Surplus machinery we have purchased. All tools and tolerance guages are original government material. Our Match Grade rifles are being used by some of the country's top riflemen.

Place your order now! We're in full production. Allow 30 days for delivery. Cashier checks and Money Orders have top priority over personal checks as they take considerable time to clear.

7.62 ammunition is available, as well as ammunition in other calibers. **SPRINGFIELD ARMORY** F.F.L. #74-5563

~~L.H. GUN COMPANY~~
Rt. 1, Box 210
Devine, Texas 78106

A/C 512 663-5676

SPRINGFIELD ARMORY
RT. 1 BOX 210
DEVINE, TEXAS 78016

Flyer from the L&H Gun Company of Texas, the first commercial Springfield Armory. (Courtesy of Donald G. Thomas)

An early Devine, Texas M1A™ receiver, note that the letter "M"s are in large case letters like the original GI receivers. (Courtesy of Robert Billotti)

M14 type rifle would be High-Power match shooters, which were generally using the M1 Garand rifle to compete in their matches. On 1 January 1974 the M1A™ and commercial M14 type rifles were accepted for use in NRA sanctioned High Power rifle matches.

Since the original government operated Springfield Armory was closed in 1968, Mr. Ballance was able to obtain the name Springfield Armory as a trademark. The BATF at the time would not allow the receivers to be marked with the original M14 or M-14 designation, thus they were marked M1A™. The first M1A™ rifles left the Devine, Texas facility with GI parts and newly manufactured investment cast receivers. The Texas M1A™ rifles were available in a number of configurations including: a standard model, a National Match version and in the M14A1 (E2) configuration

Early M1A™ receiver, note that the two letter "m"s designating millimeter have been changed to a lower case. (Tom Okamura)

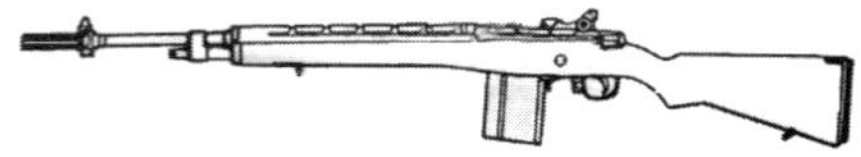

with a GI M14A1 pistol grip stock and GI M2 bipod. The prices started at $200.00 for the standard model fitted with a fiberglass GI stock. Original GI walnut and birch stocks were also offered. Less than 3,000 M1A™ rifles were produced in Texas before the ownership of the company changed in 1974.

M1A™ Receivers

M1A™ receivers are investment cast from AISI 8620 steel alloy, the same material that the original M14 receivers were made from. The only difference is that GI M14 receivers were made by the drop forging process. The investment cast receivers have proven over the years to be extremely capable of withstanding the firing of thousands upon thousands of rounds of 7.62 mm ammunition with no problems.

Much like the M14 rifle, surplus M1 Garand rifles were at one time simply not available to civilians. Also like the M14, there was quite a great civilian demand for the M1 rifle. Melvin Smith of Valley Ordnance, Wilkes-Barre, Pennsylvania was one of several individuals that had begun welding demilled M1 Garand receivers back together to fulfill the growing demand for the rifles. After the M14 was authorized in 1963 for use along side the M1 Garand in High-Power matches, interest in the M14 as a match rifle began to increase. Mr. Smith realized the demand for the M14 was now just as great, or even greater, than the demand for the M1 rifle had been. Early in the 1960s era there were a few companies that were in the business of converting M1 rifles into expedient clones of the M14. The problem with the M1 to M14 conversions was that the receiver needed to be reworked and then reheat-treated, a tricky process for the novice gunsmith.

Since there were no demilled M14 receivers available at the time to weld back together, Mr. Smith realized that he would have to manufacture completely new receivers. During Mr. Smith's research into the manufacture of M14 receivers, he reportedly called Winchester's engineering department to inquire about the procedures and machinery that was used to make the government M14 rifles. During this time he discovered some of the machinery that

The late Mel Smith of Valley Ordnance.

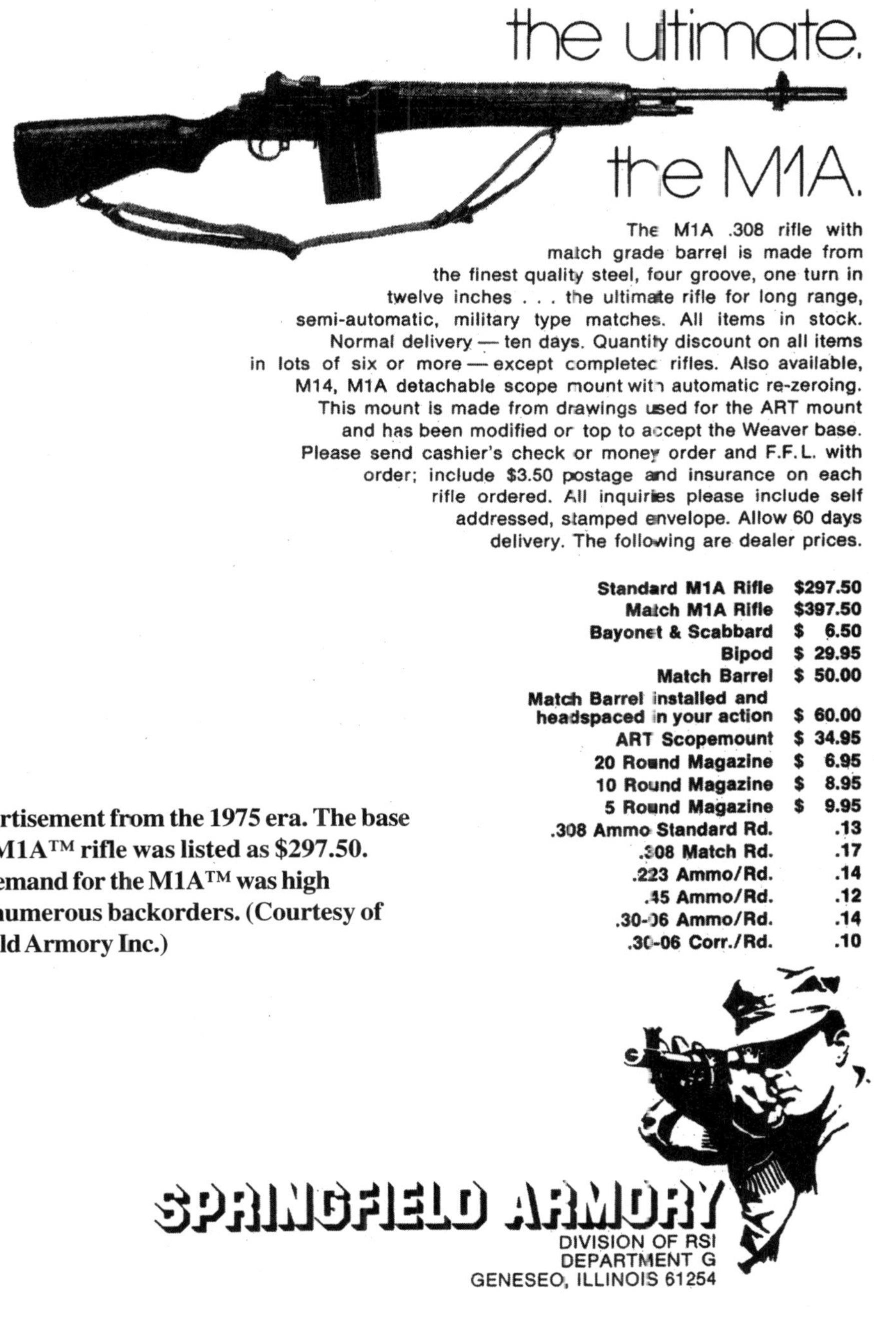

M1A™ advertisement from the 1975 era. The base price of the M1A™ rifle was listed as $297.50. Consumer demand for the M1A™ was high resulting in numerous backorders. (Courtesy of the Springfield Armory Inc.)

U.S. Rifle 7.62 mm M14

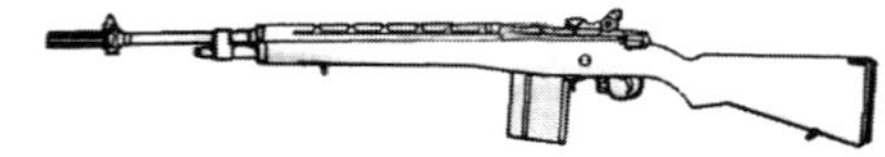

M-14 Machine Gun

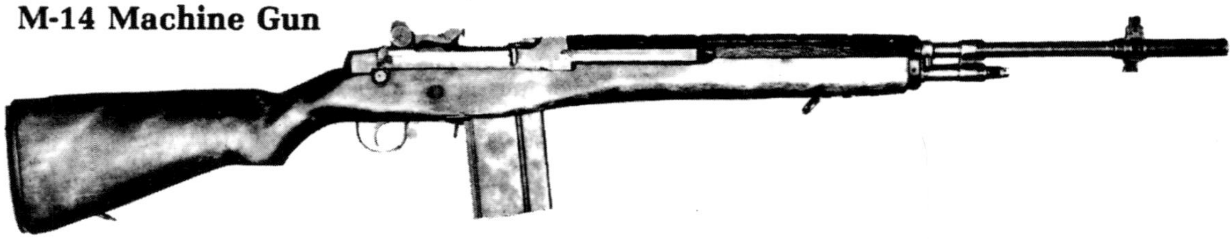

CLASSIFICATION: MACHINE GUN

DISTINCTIVE CHARACTERISTICS: This is the M-14 Rifle. It generally can be distinguished from the M-1 by shoulder plate, magazine and flash hider. The forestock is also different from that found on the M-1. The Selector Switch on the M-14 Machine Gun is located on the right rear side of the receiver, just above the trigger. All M-14's which are so marked are Machine Guns whether or not equipped with a Selector Switch.

SPECIAL NOTE: The Model M-1A rifle manufactured by the Springfield Armory, San Antonio, Texas or Geneseo, Illinois, is identical in appearance to the M-14, but is NOT a machine gun. The M-1A is a Title I weapon.

RATE OF TRANSFER TAX: $200.00

From an early BATF booklet. In the text it states that *"all M-14's which are so marked are machine guns, whether or not they are equipped with a selector switch".*

Winchester used for their M14 contract was about to be auctioned off. Smith was able to purchase much of Winchester's machinery and fixtures and set out to manufacture semiautomatic M14 rifle receivers. Since the original government process of manufacturing M14 receivers by forging was cost prohibitive, Melvin Smith decided to use the investment casting method to produce the new receivers. Smith incorporated a few changes in the original M14 receiver design. He increased the thickness of the web area between the legs, because he had noted that many M1 receivers had cracked in this area. Other changes implemented into the original design included a widened operating rod rail to stiffen the receiver and a few areas were machined for better clearance. Other changes were incorporated so that the select-fire components could not be installed, allowing the new receivers to be approved as a Title 1 firearm in accordance with the 1968 Gun Control Act. After an agreement was reached, Smith supplied his new receivers to Elmer Ballance in Texas for assembly into M1A™ rifles. Smith's Valley Ordnance would be responsible for manufacture of receivers and maintaining an adequate supply of parts, they would also be responsible for all quality control on the receivers (and barrels if they were required to make them). Elmer Ballance's L&H Gun Company would be responsible for all final assembly, testing and quality control of the completed rifles. L&H would also retain all sales rights to the new rifle designated as the M1A™ . Before long there were large num-

SECTION 5848.—DEFINITIONS [MACHINE GUNS AND CERTAIN OTHER FIREARMS]

26 CFR 179.28: Machine gun. Rev. Rul. 58–417

The Internal Revenue Service has had occasion to consider the classification status of the new U. S. Army M14 "rifle" chambered for the 7.62 mm NATO cartridge, in the light of the provisions of the National Firearms Act (Chapter 53 of the Internal Revenue Code of 1954). *Held*, (1) that since the M14 "rifle" operates either fully automatically or semi-automatically on a selective basis, it is in basic design and function a "machine gun" as defined in Section 5848(2) of the Internal Revenue Code of 1954 and, accordingly, subject to the provisions of the National Firearms Act, and (2) that the substitution of a selector shaft lock, in lieu of a selector lever assembly, on the M14 "rifle" is of temporary nature and, therefore, does not materially alter the classification thereof for purposes of the National Firearms Act.

1958 BATF ruling as applied to the M14 rifle. (Courtesy of John M. Miller)

bers of orders for the new M14 clone, causing backorders and many delays. Fortunately for the shooting community both Melvin Smith and Elmer Ballance were adamant about offering the public a well-made quality product.

Melvin Smith also supplied a few lesser-known manufacturers with his investment-cast receivers. In addition to receivers Valley Ordnance supplied a few newly made M14 parts when originals were unavailable. When the Springfield Armory Inc. changed hands, Melvin Smith continued to supply the company with his M1A™ receivers.

While an outside contractor supplied the receiver castings, all of the machine and finish work was performed by Valley Ordnance in Pennsylvania until July of 1996, when the company was sold to the Springfield Armory Inc. After 1996 the machines and equipment were moved to the Springfield Armory Inc. facility in Geneseo, Illinois. During the many years of M1A™ production several different vendors have been engaged to supply the raw castings for the receivers.

When production of the M1A™ first began in Texas the only part needed was a receiver. All of the other components were readily available as surplus parts at bargain prices. Most of the parts were available from the original government contractors, who became stuck with stockpiles of spare M14 parts after production was terminated with little warning in 1963. Much of the production machinery used

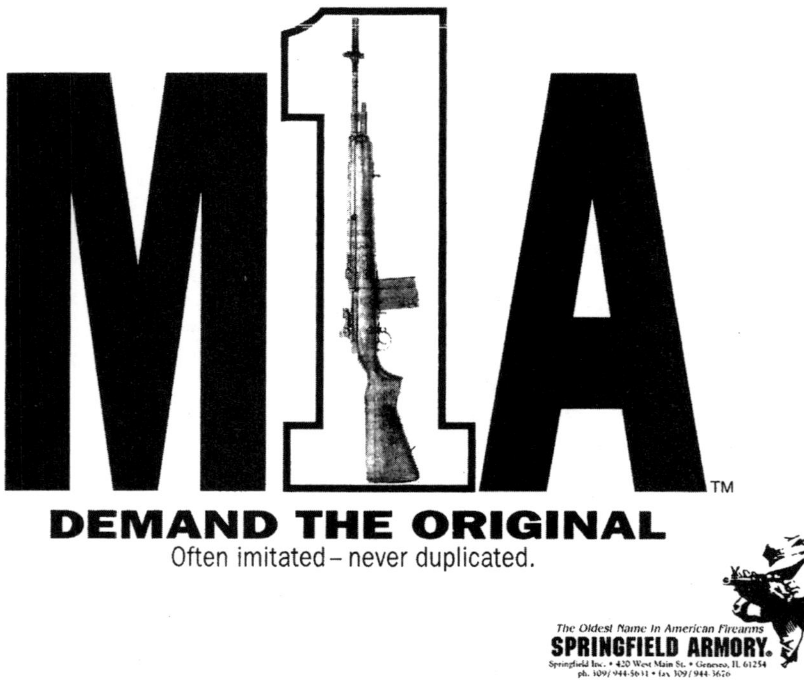

in the manufacture of original M14 rifles was eventually obtained by modern commercial companies, and used for manufacturing their M14 clones for the civilian market.

Parts and Components

Back in 1971 original (new) GI parts were very cheap and seemed to be available in such a number that the supply seemed inexhaustible at the time. As time went on more and more companies began to get into the M14 rifle business, and the supply began to slowly diminish. When this occurred companies were forced to manufacture the needed parts to maintain production. While investment cast receivers had proved rugged enough for the job, the same didn't hold true for other components like bolts, hammers and operating rods. During the early to mid 1980 era Israel began to sell off their large inventory of U.S. M14 rifles that had been obtained from the U.S. Government back in the 1970s. The rifles were stripped and the original select-fire receivers removed allowing the remaining parts to be imported back into the United States. This was quite a development for companies that were assembling M14 rifles, and the civilian M14 began a new era of popularity with the proliferation of new manufacturers entering the M14 rifle business.

Springfield Armory Inc.
Geneseo, Illinois

In 1974 the new owners of Springfield Armory Inc. relocated the entire operation to Geneseo, Illinois. After a few initial problems were overcome production of the popular M1A™ continued. Due to the rapidly growing demand for the M1A™ rifles backorders became common.

After April of 1991 the 7.62 mm caliber markings were no longer placed on the receiver heel of M1A™ rifles. (Courtesy of Gary Osif)

Springfield Armory Inc. offered a host of spare parts, accessories and services for the M1A™ and M14 rifles. Over the years a number of variations have appeared in their catalogs. Some of the more recent offerings have been;

The Standard M1A™ Rifle Model

SOCOM II

SOCOM 16 Rifle

Scout Squad M1A™ Rifle

Loaded Standard M1A™ Rifle

National Match M1A™ Rifle

Super Match M1A™ Rifle

M21 Tactical Rifle

M25 White Feather Model

Receiver Markings

Eventually some M1A™ rifles were offered in calibers other than the standard .308 (7.62 mm), including .243 and 7mm-08. Originally all M1A™ receivers were marked 7.62-mm. After April of 1991 the caliber markings were no longer placed on the receiver heel. Other markings on M1A™ receivers that were different than GI receivers; when first produced the letter M's used to designate millimeter's were marked in large case letters as were original receivers. Sometime during early M1A™ production the letters were changed to lowercase m's. There was a dash - between the digits 7.62 and the letters mm (7.62-mm.) All original receivers, except for the original gov-

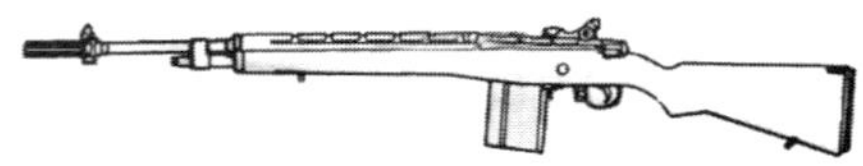

ernment Springfield Armory, lacked this dash mark.

The modern Springfield Armory Inc. receivers were marked M1A™ not M14. An encircled letter T (trademark) followed the Springfield Armory name. All M1A™ serial numbers are six digits (normal production) and those below 100,000 begin with a zero (0). No GI M14 receiver's serial numbers begin with a zero digit, except for a few presentation grade M14 rifles made at the government Springfield Armory in the early 1960 era. The Springfield Armory Inc. made a limited amount of M1A™ rifles with special serial numbers by customer request.

Other accessories and accouterments offered by the Springfield Armory Inc. include: original GI birch, walnut and fiberglass buttstocks, wood and fiberglass handguards, GI M14 barrels and custom barrels in a variety of different calibers. Scopes and mounts, military M2 and commercial bipods, magazines of varying capacity, pouches, bayonets, tools, winter triggers, cleaning kits and maintenance manuals.

During the 1980s among the accessories offered by the Springfield Armory Inc. were folding stocks for their M1A™ rifles. The stocks were produced by incorporating parts from the Italian BM59 rifle with modified M14 stocks. The company also offered newly made

The Springfield Armory occasionally made M1A™ rifles with special serial numbers (Courtesy of Scott Duff)

walnut stocks without the notch cut for the full-automatic fire selector lever.

The Springfield Armory Inc. currently offers a substantial variety of models based on their M1A™ rifle.

The Standard Model M1A™ Rifle

The basic M1A™ is much like the original "as issued" military 7.62 mm M14 rifle, but in a semiautomatic configuration. Currently the

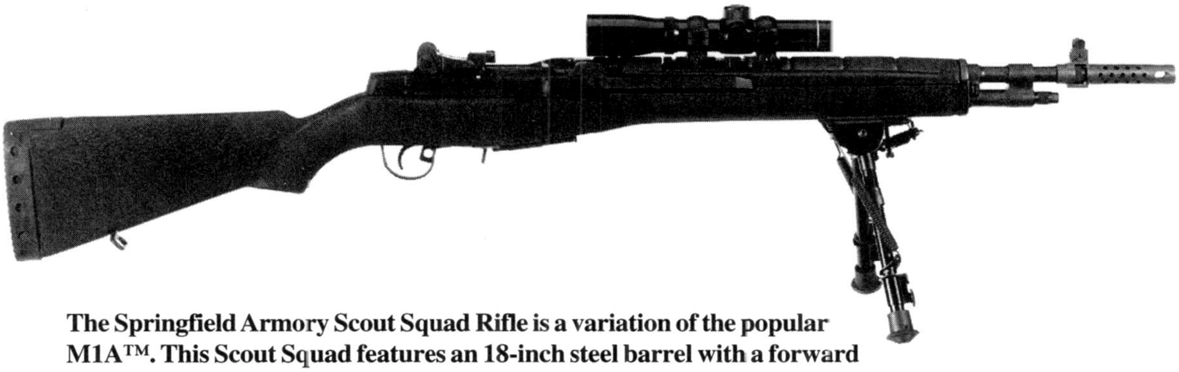

The Springfield Armory Scout Squad Rifle is a variation of the popular M1A™. This Scout Squad features an 18-inch steel barrel with a forward optical sight base, which provides an excellent platform for mounting of the Aimpoint military sighting system. The barrel is fitted with a proprietary muzzle stabilizer. (Courtesy of the Springfield Armory Inc.)

stock is available in American walnut with a military-type of hinged buttplate or black fiberglass fitted with a rubber buttplate. The standard M1A™ like its military counterpart has a 22-inch steel barrel, weighs 9.2 pounds, and has an overall length of 44.3 inches.

SOCOM II

One of the most recent M14 based 7.62mm rifles being offered by Springfield Armory Inc. is the SOCOM II semiautomatic-only rifle. The SOCOM is a compact close quarter combat (CQB) rifle featuring a shorter than standard 16.25-inch barrel fitted with a high-efficiency muzzle brake. The gas system has been retuned to operate with the shortened barrel. The SOCOM II rifle also features a new Cluster Rail System than enables the mounting of optics, lights or other accessory that was designed to fit a standard Picatinny rail.

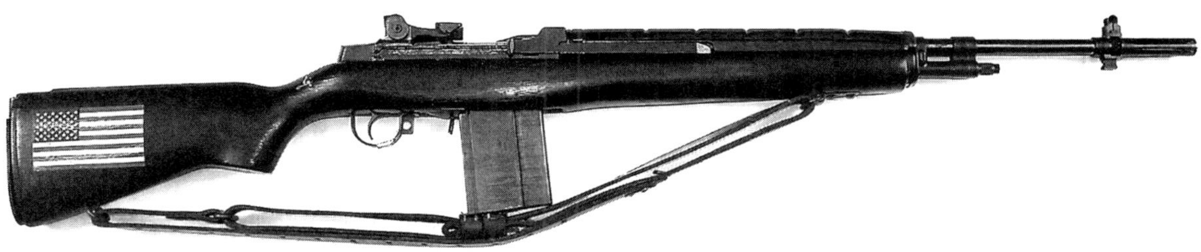

Springfield Armory M1A™ National Match rifle. (Courtesy of James Pecori)

SOCOM 16

Another short-barreled M14 variation offered by Springfield Armory Inc. is the 7.62 mm SOCOM 16 Rifle. The SOCOM 16 features a 16.25-inch barrel and proprietary muzzle brake. The front sight is tritium for sighting targets during low-light conditions. A scope base is forward-mounted for the mounting of electronic sighting devices or accessories. A two-stage military style trigger with a 5 to 6 pound pull is standard. The fiberglass stock is available in black or urban camo. The overall-length of the SOCOM 16 is 37.25-inches, the weight (without scope) is 8.9-pounds.

M1A™ rifle fitted with a National Match rear sight. (Courtesy of James Pecori)

During the late 1970 and early 1980 period Springfield Armory Inc. offered their customers a select-fire version of their popular M1A™ rifle. One of the major obstacles to purchasing a select-fire M1A™ rifle was the then seemingly hefty, $200.00 transfer tax along with the accompanying BATF paper work involved.

Scout Squad Rifle

An interesting variation of the M1A™ rifle is the Scout Squad Model. This rifle features an 18-inch steel barrel with a forward optical sight

1985 Shotgun News ad for select-fire M1A™ rifles. During the era there was not a big demand for select-fire M1A™ rifles. Only 125 were manufactured by the Springfield Armory. The manufacture of select-fire arms was banned in 1986, resulting in an ever-increasing prices for such firearms. (Courtesy of the Springfield Armory)

base, which provides an excellent platform for mounting of the Aimpoint military sighting system. The benefits of the design are improved target acquisition and unobstructed access to the receiver mechanism for reloading or manipulation of the operating rod handle. The barrel is fitted with a proprietary muzzle stabilizer. The Scout Squad rifle is available with either a walnut or black fiberglass buttstock.

Loaded Standard M1A™ Rifle

This M1A™ rifle offers upgrades usually added by enthusiasts to their "as issued rifles". The "Loaded" M1A™ features an air-gaged medium weight National Match grade barrel in stainless or chrome-moly steel. Other National Match components include front and rear sights and flash suppressor. The stock is available in walnut or black fiberglass.

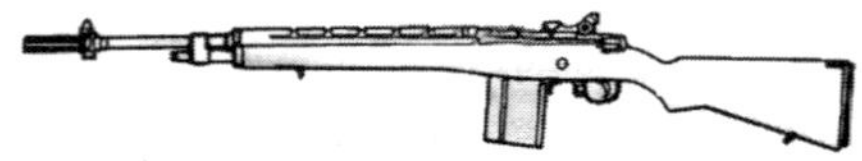

National Match M1A™ Rifle

The Springfield Armory Inc. National Match M1A™ is fitted with a many National Match features including a medium weight National Match: barrel, gas system, recoil spring guide, front and rear sights, trigger and flash suppressor. The action is custom glass bedded in a match grade American walnut stock. A stainless steel barrel is optional.

Super Match M1A™ Rifle

The 7.62 mm Super Match M1A™ Rifle by Springfield Armory Inc. features a rear lugged M1A™ receiver; an ultra heavy air-gaged Douglas National Match stainless steel barrel and a custom oversize operating rod guide. The trigger is match tuned with a two-stage 4.5-pound pull. Heavy match grade stocks available in either American walnut or a fiberglass model made by McMillan available in black or camo. A Hart or Krieger manufacture stainless steel match barrel is available by special order.

The M21 Tactical Rifle

The Springfield Armory Inc. M21 Tactical Rifle is advertised as the ultimate tool for the law enforcement sharpshooter. The 7.62 mm M21 is equipped with a walnut stock that features an adjustable cheek-rest for perfect alignment of the rifleman's eye with the scope. The iron sights are National Match configuration with a 26.75 inch sight radius. A Douglas Premium air-gaged heavy match barrel is standard. Available by special order are a Hart or Krieger manufacture stainless steel match barrel. The Tactical Rifle has a specially tuned 4.5 pound pull two-stage military match trigger.

The M25 White Feather™ Tactical Rifle

The M25 White Feather™ Tactical Rifle is a special M1A™ model designed to honor the late Carlos Hathcock, famed Marine sniper of the Vietnam War. The M25 is built on a Springfield Armory Inc. rear lugged receiver fitted with a Krieger Carbon heavy-match barrel and low-profile custom designed muzzle brake-stabilizer. Each M25 receiver is engraved with a likeness of Carlos Hathcock's signature, and White Feather logo. The trigger is a match-adjustable 1.5 to 4.75 pounds. The M25 has no provisions the mounting of standard iron sights.

Vietnam War Commemorative M14 (M1A™) Rifle

During 2003 Springfield Armory Inc. built a limited run of 500 Special Vietnam Commemorative M1A™ rifles. The semiautomatic rifles were specially built by Springfield Armory Inc. for the American Historical Foundation, who often offers commemorative fire-

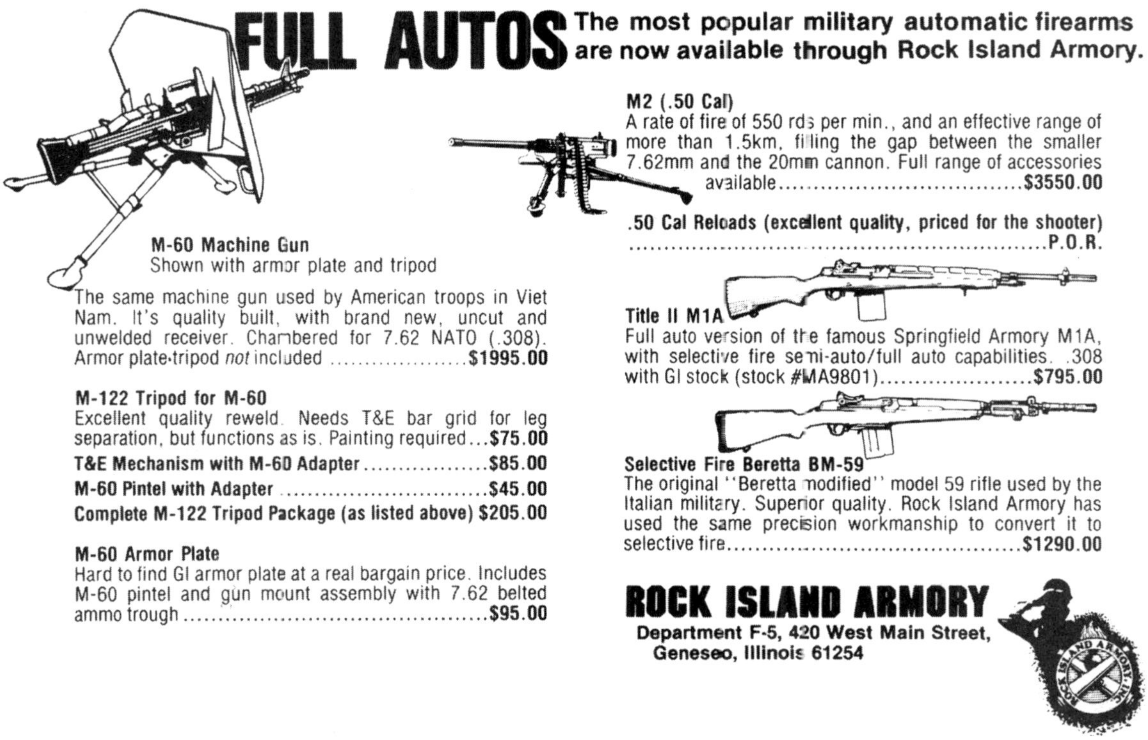

The Rock Island Armory was an affiliate of the Springfield Armory Inc. The company concentrated on the Class III portion of the business, offering M60 machine guns, M2 carbines and semiautomatic to select-fire M1A™ conversions. The company also offered many associated parts and accessories. (Courtesy of the Springfield Armory Inc.)

arms to the public. The 500 rifles were serial numbered from 001 to 500 with a VME prefix that stood for Vietnam-Museum Edition. Serial number VME 001 was presented to General William Westmoreland, who in turn presented the rifle to the Military Museum at West Point. Sixteen components of the commemorative rifle were polished and plated with 24-Karat Gold. The gold plated components included; the front sight; safety; magazine release; front and rear sling swivel assemblies (5 pieces); rear sight (6 pieces); the trigger; and the flash suppressor. The rest of the rifle was polished and blued to a gloss black color by the heat immersion process. The operating rod and receiver were engraved and filled with gold-gilt. The rifle was fitted with a high-gloss Reinhart Fajen American walnut stock. Included with the rifle were a 10-round magazine, field manual, a black leather sling and a certificate of authenticity. Available at extra cost was an American walnut display case. The case was lined with black felt and had an acrylic glass lid. The retail cost of the Vietnam War Commemorative M14 Rifle was $3,995.00, the optional display case listed for $249.00. For another $25.00 the rifle could be further personalized with additional engraving.

During the late 1970s and early 1980s period Springfield Armory Inc. offered their customers a select-fire version of their popular M1A™ rifle. One of the major obstacles to purchasing a select-fire M1A™ rifle was the, then seemingly hefty $200.00 transfer tax, along with the accompanying BATF paper work involved. The paperwork included obtaining the signature of the local authorities, fingerprinting and worst of all, the wait for the transfer to be approved, which often took over 90 days to process. Only 125 M1A™ rifles were originally manufactured in a select-fire configuration by the company.

Markings on a Smith M14 National Match forged receiver. (Courtesy of Smith Enterprises Inc.)

Rock Island Armory

The Rock Island Armory (not to be confused with the U.S. Government Rock Island Arsenal) was an affiliate of the Springfield Armory Inc. The Rock Island Armory, also located in Geneseo, Illinois, concentrated on the Class III portion of the business, offering M60 machine guns, M2 carbines and semiautomatic to select-fire M1A™ conversions. The company also offered many associated parts and accessories.

One of the services offered by the Rock Island Armory was the (then legal) conversion of Springfield Armory's standard M1A™ rifle to a select-fire configuration. At the time not many enthusiasts were interested in paying the $200.00 tax plus conversion costs, as only 35

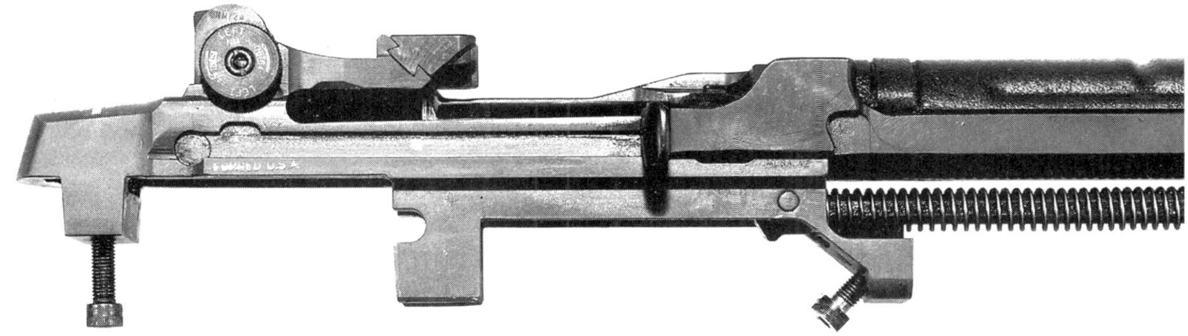

Smith Enterprises double-lug match M14 receiver. (Courtesy of Smith Enterprises Inc. of Arizona)

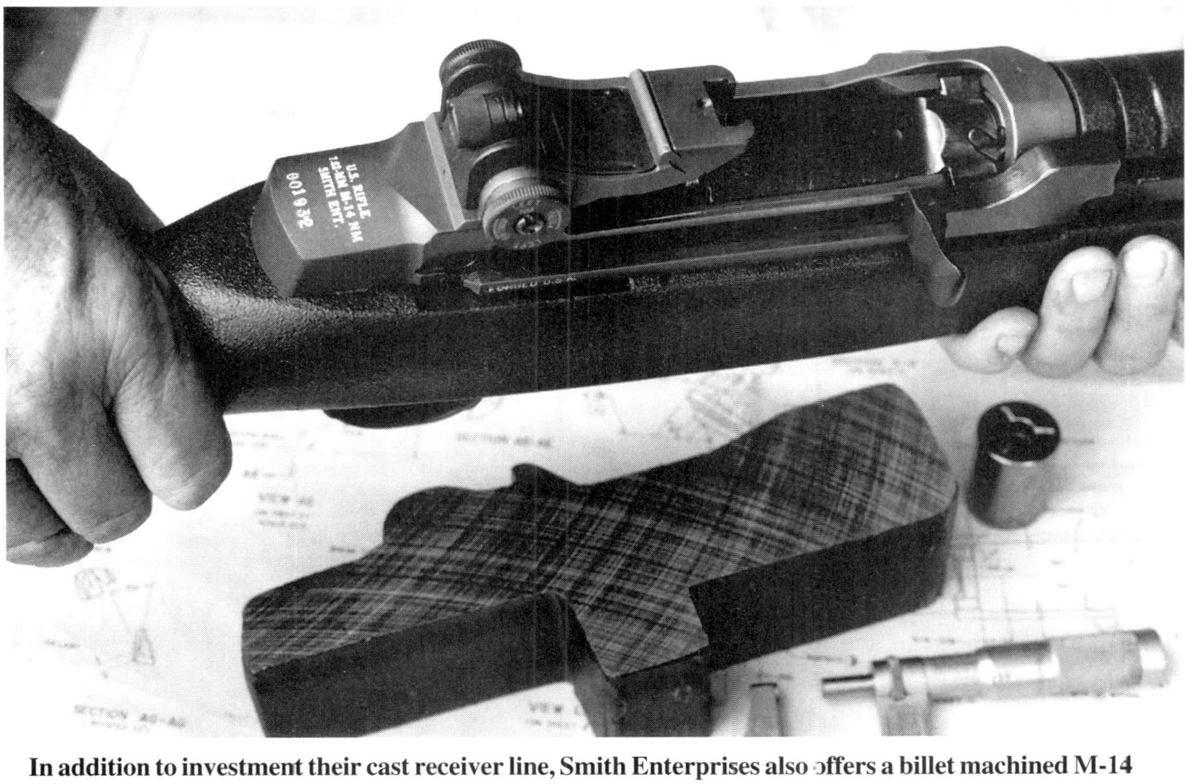

In addition to investment their cast receiver line, Smith Enterprises also offers a billet machined M-14 receiver. These receivers are easily identified by the marking FORGED USA on the operating rod rail of the receiver. Smith has been in the M14 rifle business since the 1980 era. (Courtesy of Smith Enterprises Inc.)

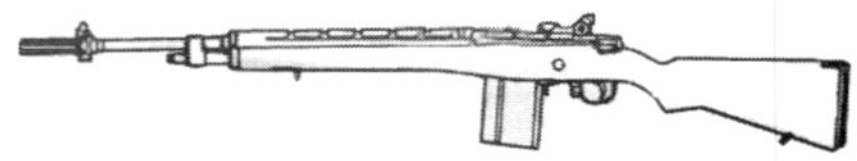

M1A™ rifles were converted by the Rock Island facility prior to the May 19, 1986 ban on such activities. At one point the $200.00 transfer tax was nearly one-half the cost of a new M1A™ rifle. M1A™ rifles that were converted to select-fire at the facility were marked RIA on the receiver.

Additional M1A™ and civilian M14 type rifles were legally converted to select-fire by a number of Class II manufacturers prior the May 1986 date at which time laws were enacted banning future conversions of semiautomatic firearms or the manufacture of new machine guns. Original GI M14 rifles are quite rare in the hands of civilians, there are estimated to be less that forty original M14 rifles in the NFA registry. In addition to the conversions and factory guns, there were a number of M14 rifles constructed using demilled M14 receivers, and registered prior to 1986.

Smith Enterprises Inc.

The Arizona based Smith Enterprises began life as the Western Ordnance International Corporation. The innovative company has done work for the United States military, many foreign governments as well as the civilian market. The company today is best known to the shooting community for their M-14 receivers. The company produced the receivers after they incorporated a number of improvements to the original design. The first receivers the company produced were manufactured by investment casting of AISI 8620 steel, the same material used in the manufacture of GI M14 receivers during the 1960 era. Smith Receivers are marked with their company name, a serial number and either M-14 or M-14 NM. In addition to investment cast receivers Smith also offers a billet machined M-14 receiver. These receivers are easily identified by the marking FORGED USA on the operating rod rail of the receiver.

During the 1980s Smith Enterprises offered their M-14 rifles in a select-fire version. The receivers were marked with the Smith name and M-14 designation. The serial numbers on factory select-fire rifles began with the prefix FA. Smith produced a total of 176 factory select-fire models. (Courtesy of Drew Paskovitch)

During the late 1980s Smith Enterprises in-

A select-fire Smith-LaFrance M14K rifle. This short-barrel M14 uses an M60-type of gas system. (Courtesy of Smith Enterprises Inc.)

troduced a carbine version of the M14 rifle, designated as the M14K. The rifle was a short barreled version of the M14 rifle. Experiments with the short barrel 7.62 mm weapons conducted by the Ordnance Corps during the late 1950s era had proven that a short-barrel 7.62 weapon had excessive report, muzzle flash and a violent recoil. Intrigued by the challenge of perfecting such a design, Richard and son Ron Smith working with Tim LaFrance set out to find a solution to the problems associated with a short M14 rifle. After considerable research and development efforts it was discovered using a variation of the gas system designed for the M60 machine gun made the concept workable. A new muzzle device was designed to control the flash and recoil generated by the shortened thirteen-inch barrel of the M14K carbine. To enhance the accuracy potential of the weapon a National Match barrel and sights were fitted. The M14K carbine was available in a semiautomatic-only or a select-fire configuration. The semiautomatic version featured a permanently affixed muzzle device to maintain a legal sixteen-inch barrel length. A conventional wood M14 stock or a side folding stock with a rear pistol grip was available. The company offered complete M14K rifles as well as a conversion service on a customer's M1A™ or M14 rifle.

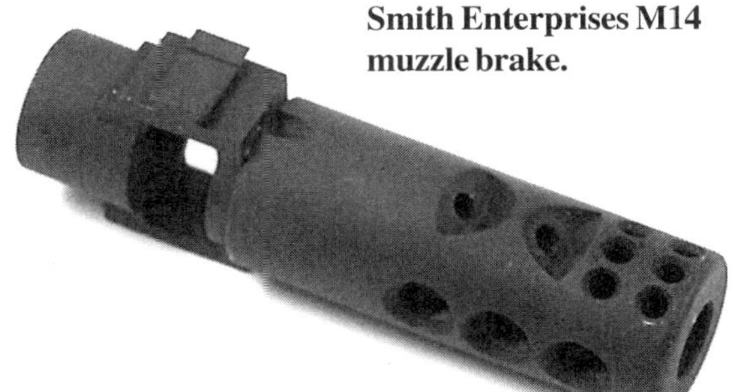

Smith Enterprises M14 muzzle brake.

The Smith Enterprises M14SE (the SE suffix designates Smith Enterprises) was originally developed by the Smith organization to provide the U.S. Army with an updated 7.62 mm M14 rifle for U.S. Army Squad Designated Marksman or SDM. Smith Enterprises has been supplying the U.S. military, as well as many other government entities with their modified M14 rifles. Smith offers a similar rifle to the civilian market designated as the Semiautomatic Sniper System (SASS) Crazy Horse Rifle. (Photograph courtesy of Robert Silvers)

Smith Enterprises "Crazy Horse" M14 rifle with optional Sage International stock assembly. (Courtesy of Robert Silvers)

Not associated with Smith Enterprises of Arizona was Smith Manufacturing LTD of Holland, Ohio. Smith LTD made a small number of semiautomatic M14 rifles back in the 1970-1980 era for the civilian market. The Smith LTD rifles were assembled using original GI parts on an investment cast receiver.

During the 1980s Smith Enterprises offered their M-14 rifles in a select-fire version. The receivers were marked with the Smith name and M-14 designation. The serial numbers on factory select-fire rifles began with the prefix FA. Smith produced a total of 176 factory select-fire models. There were also a limited number of Smith Enterprises M-14 rifles converted to a select-fire configuration by other Class II manufacturers not affiliated with Smith Enterprises.

Smith Enterprises Inc. manufactures and sells a number of accessories for the M14. One quite effective item is their muzzle brake that is machined from 8620 steel. The Smith brake replaces the original flash suppressor and attaches in the same manner, being secured with a castle nut and locked into place by a set-screw. Other items sold by Smith are flash suppressors, gas cylinder locks, extended bolt latches, and target sights. The United States Government has purchased a large number of Smith Enterprises parts for incorporation into the latest Enhanced Battle Rifle (EBR) version of the M14 resurrected for duty.

Chinese/Smith M14 Rifles

Ron Smith, the current company president has done a great deal of research and testing of

Markings on a Smith Manufacturing LTD M-14 receiver.

Armscorp Inc. 1985 era receiver. (David Fassinger)

the Chinese M14 rifles that were imported into the U.S. The company offers many services to upgrade and improve the safety and performance of both Norinco and Polytech M-14S rifles.

Smith Enterprises Inc. Crazy Horse™ M14 Rifle

The Smith Enterprises M14SE (the SE suffix designates Smith Enterprises) was originally developed by the Smith organization to provide the U.S. Army with an updated 7.62 mm M14 rifle for U.S. Army Squad Designated Marksman or SDM. Smith Enterprises has been supplying the U.S. military, as well as many other government entities with their modified M14 rifles. Smith offers a similar rifle to the civilian market designated as the Semi-automatic Sniper System (SASS) Crazy Horse™ Rifle.

The Smith M14SE is not designed as a target rifle, but an accurate, rugged military weapon. The stocks are not glass bedded or modified. If an alternate stock is requested it would need

Armscorp of America Inc. manufacturer's markings located on the side of the receiver. (Courtesy of James Pecori)

Armscorp of America Inc. offers a complete line of semiautomatic M14 rifles from basic field to competition match grade. Armscorp M14 rifles are built on receivers made from 8620 certified steel investment castings. The receivers are heat-treated and all tolerances correct to the most recent U.S. GI specifications. Armscorp semiautomatic receivers are available in standard, rear lugged and double lugged configurations. (Courtesy of Drew Paskovitch)

to be fitted to the rifle. There are several configurations of the semiautomatic M14SE rifles. The basic weapon features a medium heavy chrome-moly steel barrel with a 1:10 twist 4-lands and grooves chambered for the U.S. M118 Long Range 7.62 mm ammunition; a Smith Enterprises gas system with special design gas plug, lock, front band and gas piston; Smith melonite heat-treated scope mount and rings; heavy duty M1913 Picatinny rail; Smith

The Federal Ordnance Company of California began the manufacture and assembly of their M14 SA type rifles in 1988. Federal Ordnance receivers were produced by the investment casting process. The receivers were marked with the Federal Ordnance name on the receiver heel, although some were marked on the receiver leg. Originally the rifles were assembled using original GI parts. (Courtesy of Glenn Whitehouse)

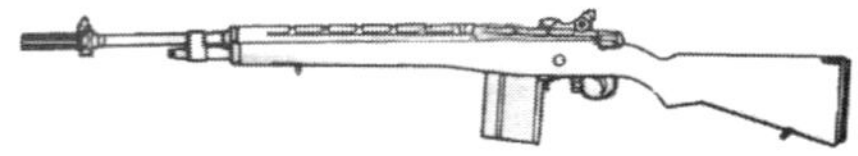

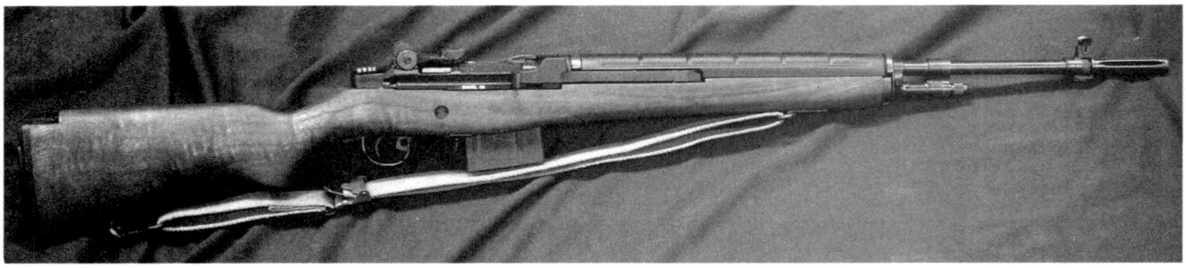

Fulton Armory M14. The company offers a complete line of M14 rifles in a number of configurations from a basic field grade to a competition-grade. (Photo courtesy of Fulton Armory)

Enterprises Direct Connect Vortex flash eliminator, extended bolt stop; 4.5 pound tuned trigger assembly, with special S-7 tool steel trigger and hammer pins.

Crazy Horse™ options include; an 18-inch medium heavy barrel in chrome-moly or stainless steel; Tritium front sight; a fully adjustable from 2.5 to 5.0 pound pull trigger assembly; full range of Leupold Mark 4 optics; Sage International modular collapsing stock (the USN Mark 14 Mod 0 configuration); a sound sup-

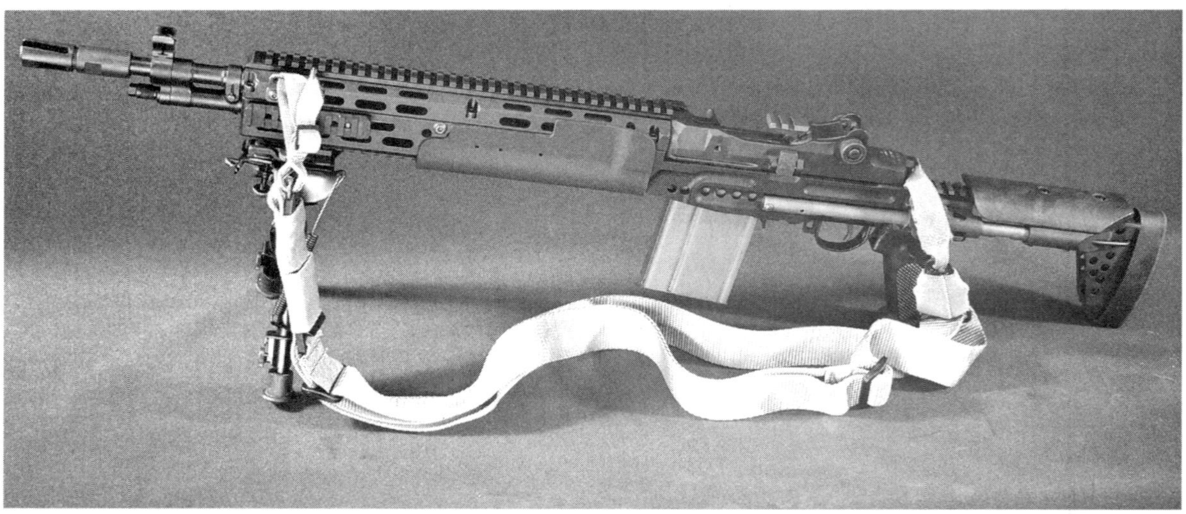

One of Fulton Armory's latest offerings is the MARK 14 Ultimate Battle Rifle a semiautomatic, civilian legal version of the SOCOM Crane Mk 14 Mod 0 rifle, EBR, being fielded by U.S. military forces in Afghanistan and Iraq. The Futon Mark 14 rifle includes a Mk14 stock by Sage, extended bolt catch, 18.5-inch Fulton Armory barrel, a Vortex Direct Connect flash suppressor and a Tritium front sight. (Photo courtesy of Walt Kuleck, Fulton Armory)

The Fulton receivers are marked with the Fulton Armory logo and several model designations to include M14, XM25 or M21, depending on the options chosen by the customer. On most models the company guarantees their rifle's accuracy capabilities. (Photo courtesy of Walt Kuleck, Fulton Armory)

pressor. Slings and other accessories are also available. The Smith Enterprises rifle features a laser alignment process for barrel installation. The Smith M14 sniper barrels are designed for the 7.62 mm M118 LR rounds not, M80 ball ammunition

Smith Manufacturing LTD of Ohio

Not associated with Smith Enterprises of Arizona was Smith Manufacturing LTD of Holland, Ohio. Smith LTD made a small number of semiautomatic M14 rifles back in the 1970 era for the civilian market. The Smith Manufacturing LTD rifles were assembled using original GI parts on an investment cast receiver.

Armscorp of America Inc.

Armscorp of America Inc. is a company located in Silver Springs, Maryland that began assembly of M14 rifles using their semiautomatic receivers and used surplus GI part sets back in the 1980s. The Armscorp receivers were made from investment castings, although there was an early attempt by the firm to produce forged receivers around 1985. Smith Enterprises at one time supplied Armscorp with receivers. The firm was the first to market match-grade lugged receivers for competition shooters. In addition to M14 type rifles the company once offered Israeli FN Rifles with newly manufactured semiautomatic receivers.

Armscorp currently offers a complete line of semiautomatic M14 rifles from basic field to competition match grade. The company also offers custom-built rifles to the customer's requirements. Armscorp M14 rifles are built on their receivers made from 8620 certified steel investment castings. The receivers are heat-treated and all tolerances correct to the

A.R. Sales of California was one of many companies that attempted to break into the civilian market for the M14 rifle over the years. (Photo courtesy of Walt Kuleck)

Photo courtesy of the Fulton Armory

most recent U.S. GI specifications. Armscorp semiautomatic receivers are available in standard, rear lugged and double lugged configurations.

Federal Ordnance

The commercial Federal Ordnance Company began the manufacture and assembly of their M14 SA type rifles in 1988. Federal Ordnance receivers were produced by the investment casting process. The receivers were marked with the Federal Ordnance name on the receiver heel, although some were marked on the receiver leg. Originally the rifles were assembled using original GI parts. After the supply of GI parts became to diminish reproduction parts were used to maintain production of the rifles. A number of M14 type semiautomatic rifles were assembled with Federal Ordnance cast receivers and parts imported from China. The imported parts were the same as those used in the manufacture of the Chinese M14 rifles. The Federal Ordnance/Chinese rifles were marketed by a large U.S. surplus dealer.

The Fulton Armory

The Fulton Armory was established in 1987. The company offers parts, accessories and complete shop services for a number of U.S. military firearms to include the AR-15, M1 Garand rifle, the M1 Carbine and the M14 rifle. They also offer a service for upgrading the Chinese M14S rifles. In 2003 Fulton Armory began to offer their own semiautomatic M14 rifle. The Fulton Armory receivers are investment cast and machined to their own unique specifications by a subcontractor. The receivers are marked with the Fulton Armory

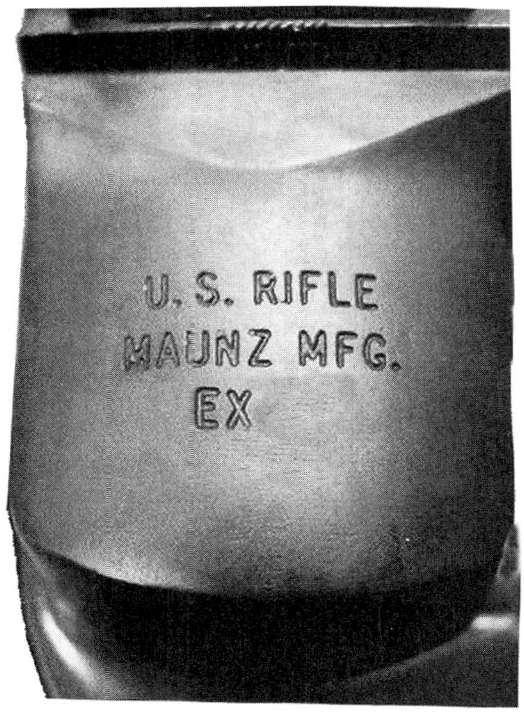

Maunz Manufacturing of Ohio assembled a small number of semiautomatic-only M14 rifles using GI parts and commercially produced cast receivers. The receivers were marked with one of several names to include: Maunz, H&R Gun Co. and Smith Ltd.

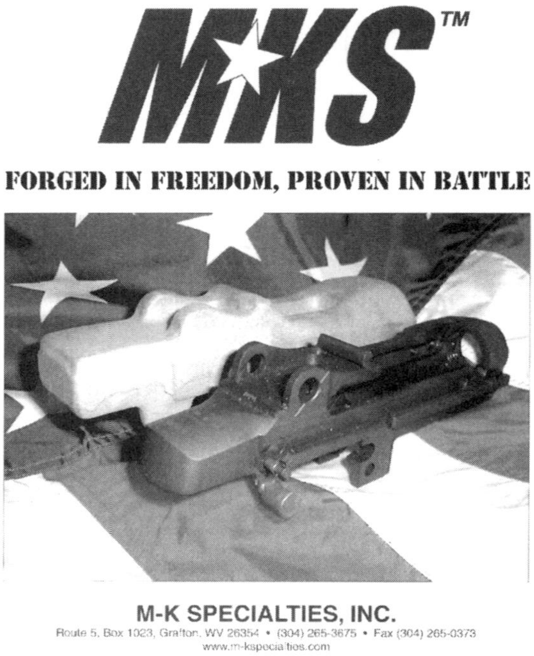

LRB Arms began in the M14 receiver business by marketing and distributing receivers from a company known as MKS. The MKS receivers were forged overseas and machined into a finished state in the U.S. by MKS. (Donald G. Thomas)

LRB of New York began producing commercial M14SA type receivers in 2003. Rather than use the investment casting method, LRB uses AISI 8620 steel that is heated and drop forged into shape, the same way original GI contract receivers were made. The remaining machine work is performed on modern CNC machines. (Courtesy of LRB)

logo and several model designations to include M14, XM25 or M21, depending on the options chosen by the customer. On most models the company guarantees their rifle's accuracy capabilities.

The company offers a complete line of M14 rifles in a number of configurations from a basic field grade to a competition M14 rifle. An interesting option offered by the Fulton Armory is a non-functioning GI selector switch for those wanting the as-issued look of the military M14 rifle. For those who prefer to build their own rifle, Fulton Armory offers bare receivers, including a match grade with a rear lug, barreled receivers with a headspaced bolt and rifle kits.

An unmarked Norinco receiver heel. Many M14 rifles imported from China were marked on the side of the receiver.

The Chinese clones of the U.S. M14 rifles offered a low cost alternative to M1A™ and M14 rifles then being sold by U.S. companies. (Courtesy of Donald G. Thomas)

One of Fulton Armory's latest offerings is the MARK 14 Ultimate Battle Rifle, which is a semiautomatic, civilian legal version of the SOCOM Crane Mk 14 Mod 0 rifle, EBR, being fielded by U.S. military forces in Afghanistan and Iraq. The Fulton Armory Mark 14 rifle includes a Mk14 stock by Sage, extended bolt catch, 18.5-inch Fulton Armory barrel, a Vortex Direct Connect flash suppressor and a Tritium front sight that is mounted

1990 era ads from the *Shotgun News* featured Norinco select-fire dealer samples and Federal Ordnance M14 rifles. (Courtesy of Donald G. Thomas)

Type-57 M14 rifle that were made in Taiwan (The Republic of China). The Type-57 rifles produced in Taiwan were manufactured on U.S. machinery for the arming of the Taiwan military by the state arsenal.

on the gas cylinder lock and includes an extended bolt catch. Options include a 22-inch barrel, scope mounts, chromium bore, Harris bipod and special optional receiver markings. The Fulton Armory is located in Savage, Maryland.

Entréprise Arms

In 1996 Entréprise Arms of Irwindale, California began to offer their customers a different option for a semiautomatic M14 receiver. In place of the traditional investment cast method used by most manufacturers, Entréprise began to produce an M14 receiver that is CNC machined to GI specifications from a 12-pound billet of AISI 8620 steel. The company also offers limited production of single or double lug M14 receivers by special order. After machining the receivers are heat-treated. Entréprise offers complete shop services to assemble the customer's parts on their receivers. Entréprise Arms also offers their own line of M14 barrels in a variety of configurations. The company offers an unconditional lifetime guarantee on their receivers.

Heese Arms

Heese made M14-style receivers between the years of 2000 and 2003. The receivers were manufactured by a subcontractor in Italy by the investment cast method.

Maunz Manufacturing

Maunz Manufacturing of Ohio assembled a small number of semiautomatic-only M14 rifles from 1970 to 1980 using GI parts and commercially produced cast receivers. The receivers were marked with one of several names to include: Maunz, H&R Gun Co. and Smith Ltd. Total production was very limited, making the Maunz rifles very uncommon today.

LRB Arms

LRB Arms began in the M14 receiver business by marketing and distributing receivers from a company known as MKS. The MKS receivers were forged overseas and machined into a finished state in the U.S. by MKS. After a problem with the BATF receiver blanks for MKS were no longer imported. LRB of New York began producing commercial M14SA type receivers in 2003. Rather than use the investment casting method, LRB uses AISI 8620 steel that is heated and drop forged into shape. The remaining machine work is performed on modern CNC machines. The company currently offers complete rifles, stripped receivers, barreled receivers and numerous parts and accessories.

The Chinese M14 Rifle Clones

During the 1980 era Mainland China was granted most-favored-nation trading status. The Chinese quickly discovered the lucrative U.S. civilian arms market, and collectors who were anxious to obtain previously unavailable Communist small arms, like the AK-design rifles and Tokarev T33 pistols. The BATF implemented several restrictions and modifications, which were quickly implemented, and exporting to the U.S. began. Before long the Chinese began to expand their line of small arms to include weapons of U.S. design like the M1911A1 pistol and M14 rifle.

Two Mainland Chinese semiautomatic versions of the U.S. M14 rifle have been imported into the United States. One was manufactured by Poly Technologies the other by China North Industries better known in the U.S. as Norinco. Both of these Chinese entities were

The value of the Chinese M14 rifles gradually began to increase as the supply began to diminish. (Ad from the *Shotgun News*)

Markings and serial number on a Poly Tech Chinese M14 receiver (Courtesy of Walt Kuleck, Fulton Armory)

best known for their semiautomatic AK-47 clones that have been imported into the country for civilian shooters.

Poly-technologies is a subsidiary of the People's Liberation Army and manufactures a large assortment of weapons for that organization. Norinco or China North Industries is part of the Chinese Ministry of Ordnance. Norinco also manufactures weapons for the Mainland Chinese military. Both of the Chinese organizations manufactured and imported weapons and ammunition to the armies of other countries.

The mainland Communist Chinese (The People's Republic of China) M14S rifles are not to be confused with the 7.62 mm Type-57 M14 rifle that were made in Taiwan (The Republic of China). The Type-57 rifles produced in Taiwan were manufactured on U.S. machinery for the arming of the Taiwan military by the state arsenal. Most of the machinery was the former U.S. Government owned equipment used by Harrington and Richardson during their M14 production in the U.S. The Taiwanese rifles are the equivalent of the U.S. M14 rifles, except for the Chinese characters marked on the receiver heel. Taiwan adopted the Type 57 rifle in 1968 and began production in 1969. The Type 57 rifles had select-fire capability. The TYPE 57 designation originates from Taiwanese dating system that begins with the Nationalist Revolution of 1911.

The primary advantage of the imported M14S Chinese made clones were their original price, which at the time of their original importation was running at about one-half the cost of a U.S. equivalent commercial civilian M14 or M1A™ rifle.

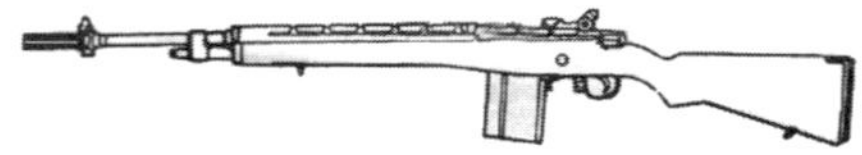

The wooden stocks on the Chinese M14 rifles were usually oversized and crudely made. Fortunately for the owners of the rifles, the Chinese wood could easily be replaced by original GI M14 stocks.

Ever since the first Chinese M14S rifles were exported to the United States the rifles made by both manufacturers, Polytechnologies and Norinco, have come under a lot of scrutiny and discussion on both quality and safety issues. While the Chinese clones had a few desirable features like a drop-forged receiver, other issues such as proper heat-treating and headspacing have come under suspicion.

The barrels on the Chinese imports proved to be nearly the equivalent of a genuine U.S. GI production M14 barrel. For some reason the Chinese choose to incorporate two metric threads on the barrel assembly. One on the castle nut that secures the flash suppressor, the other on the gas cylinder lock. Like U.S. barrels, the bores and chambers were chromium plated. The barrel threads are the same as U.S. made rifles, but the Chinese barrels are secured by a staked set-screw extending through the side of the receiver.

The rear sights fitted to the Chinese rifles are basically poorly made, and typically will not stay adjusted. The rear sight aperture is usually larger than those found on GI equivalents. Unfortunately GI issue sight assemblies or components will not interchange with those on the Chinese M14 rifles.

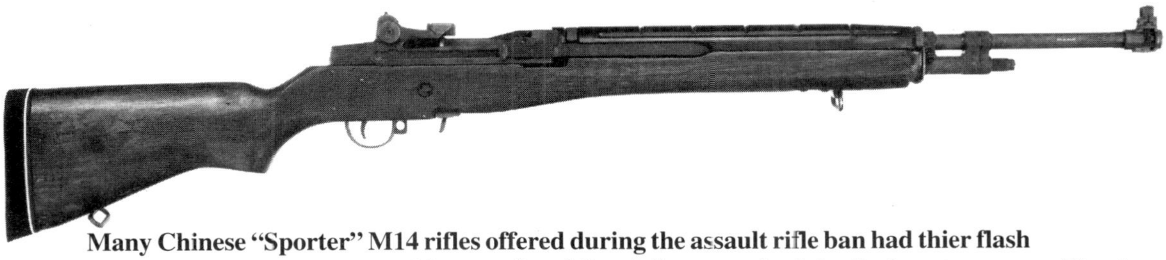

Many Chinese "Sporter" M14 rifles offered during the assault rifle ban had thier flash suppressors removed and a rubber recoil pad fitted. Some retained the flash suppressors without the slots milled into it making it a mere barrel extension and thus not affected by the ban.

The bolts of the Chinese rifles appear to be the major problem area with the Chinese M14, resulting in excessive headspacing, a dangerous situation. Smith Enterprises of Arizona highly recommends that the Chinese bolt be replaced. Smith offers a "bolt conversion" service.

The Chinese gas cylinder is well made of chromoly-steel but the threads for the gas cylinder plug are metric. The inside diameter of the gas cylinder and accordingly the steel-chrome plated gas piston is larger than that used in U.S. GI M14 rifles and its civilian counterparts.

The Operating rods are forged and in most cases fully serviceable, however Smith Enterprises recommends that the operating-rod spring be replaced with one of GI manufacture. Chinese parts are generally not marked with any type of identifying numbers or logo except for possibly the trigger housing with a 5 or 6 digit number. If marks were applied they were usually crudely hand engraved by a vibrating electro-pencil.

Smith Enterprises of Arizona performed a lot of testing and evaluation of the Chinese M14 rifles. The company began to offer owners of the imported M14 rifles many services to upgrade their guns. The Fulton Armory also offers a number of services to bring Chinese M14 clones up to spec.

As the first Chinese M14s were offered all of the gun periodicals began testing them for articles. Surprisingly most of them turned in quite respectable scores on the range.

Most of the wooden stocks on the Chinese M14 rifles were generally fitted quite poorly being oversized and bulky in many areas. Fortunately for the owners of the Chinese rifles U.S. GI stocks available in walnut, birch and even fiberglass are interchangeable (with some fitting required) and are abundant on the surplus market.

U.S. Rifle 7.62 mm M14

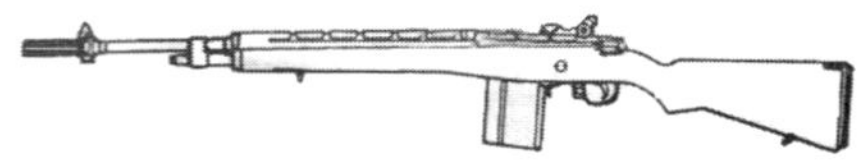

The ever-growing list of laws regulating the importation of rifles into the United States resulted in a number of different configurations of Chinese M14 rifles in various stages of politically correctness. Some versions featured a flash suppressor without the slots machined into them, and later the flash suppressor was cut off flush with the muzzle.

The Chinese M14 rifles have been imported and offered by a number of U.S. companies. Century Arms was a large importer of Norinco M14 rifles and as the new U.S. laws on importing foreign-made militaria were implemented, Century began assembling the Chinese parts kits onto U.S. receivers provided by Federal Ordnance. Before long more laws were enacted prohibiting the assembly of "assault rifles" from foreign parts. Century Arms shifted gears and began importing "Sporter" version's of the M14 that were made "politi-

M-14-THREE AVAILABLE. Orig. Win., H&R, TRW. Overlapping halves used by one of the best manufacturers in the business. Assembled with orig. new G.I. parts including bbl. Stocks used but hand picked/exc. plus. Three day inspection with full refund if not satisfied, $1525. each; 9-9 PM EST,
2-2

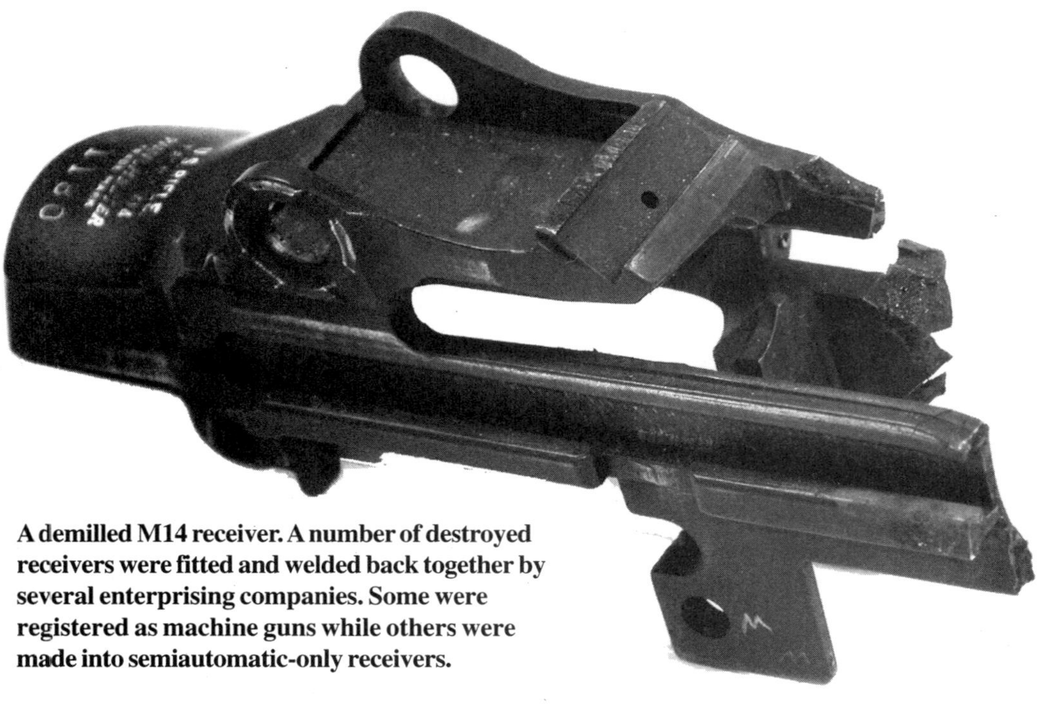

A demilled M14 receiver. A number of destroyed receivers were fitted and welded back together by several enterprising companies. Some were registered as machine guns while others were made into semiautomatic-only receivers.

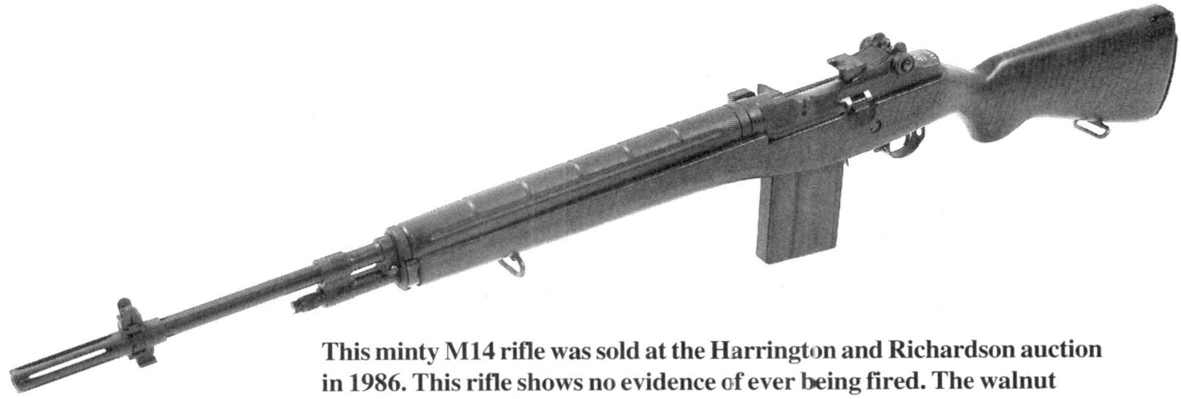

This minty M14 rifle was sold at the Harrington and Richardson auction in 1986. This rifle shows no evidence of ever being fired. The walnut stock has a beautiful tiger stripe pattern. (Steve Brousseau collection)

Approximately 15 transferable original GI M14 rifles came from the U.S. Department of Energy. Most of the DOE rifles had an inventory number engraved on the side of the receiver. The DOE rifles, the H&R auction guns and a few amnesty-registered M14 represent the total of the original M14 rifles in the NFA registry, believed to be less than 50. (Photo courtesy of Prokopios Ziros and Robert Armitano)

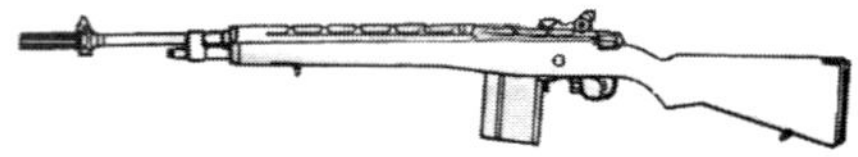

U.S. MARTIAL MACHINE GUNS

- M16-A1 - Assault Rifle - Mfg.. H&R Arms Co./U.S.A. Marked: "U.S. GOVERNMENT PROPERTY". Not a Dealer Sample. New/Unfired **$2750.00**
- M14 Selective Fire Assault Rifle - Mfg.. H&R Arms Co./U.S.A. - Not a Dealer Sample - From the H&R Factory Collection - Limited at **$4000.00**
- Many Other Non-restricted Machine Guns Currently Available - Write For Free Listing.

R.J. PERRY & ASSOCIATES, INC.
P.O. Box 151

During the early part of 1986 the H&R factory had a bankruptcy auction. A handful of unissued, transferable M16A1 and M14 rifles were sold at the event. The rifles were later offered by R.J. Perry and associates. Note the $4000.00 asking price for the M14 rifles! To place that price into prospective, just prior to the 1986 ban brand new Colt M16A1 rifles were available for approximately $500.00. The H&R auction M14s, a few amnesty guns and a small number released from the Department of Energy are the only original GI-un-remanufactured M14 rifles in the system, making them quite rare. (Courtesy of Donald G. Thomas)

cally correct" by eliminating the evil features of the M14. The resulting Century Arms "Sporter" model was functionally the same but lacked a flash suppressor and bayonet lug, and was fitted with a rubber recoil pad in place of the former hinged M14 buttplate. The Sporter rifles were originally shipped without a magazine, but five-round capacity magazines were eventually made available. Ironically the Sporter was allowed to retain its Chinese-made receiver because of the implementation of the other modifications.

Government Contract M14 Rifles

Original GI contractor M14 rifles are unavailable to the general public and the chances of them ever becoming available are virtually non-existent. The reason is that the original GI M14 rifle was manufactured with a full automatic capability. The BATF's stance is "once a machine gun always a machine gun", regardless of any efforts to render the rifle to a semiautomatic-only configuration, it would be illegal. Civilians in most states can own full automatic weapons after going through a transfer process with the BATF. Any transferable machine gun would have to have been in the system prior to 1986. If you are in possession of an unregistered firearm with select-fire capability it cannot be registered, and would be considered contraband by the BATF. The BATF aggressively enforces this statute. Few of the original M14 rifles remain in the stores of the U.S. Government. Many of them have been given or sold to foreign countries while many more have been destroyed by the government after being declared obsolete. More recently the M14 has been pressed back into service

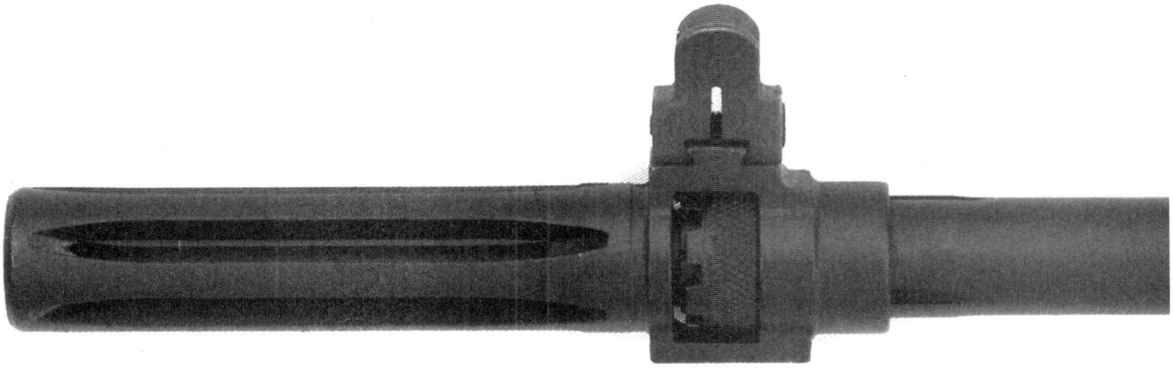

During the infamous assault rifle ban period that lasted from 1994 until it expired in 2004, M1As were sold with their bayonet lugs removed in order to comply with the law. (Gary Osif)

A new recruit is issued (among other equipment) an M14 rifle. (NARA)

for U.S. troops fighting in Iraq and Afghanistan.

New Manufacture (welded) GI Receivers

Many Class II manufacturers helped to satisfy some of the market's desire for a select-fire M14 rifle by manufacturing a "new" M14 receiver from the demilled pieces of several original receiver remnants. Once a receiver was demilled (torch cut into pieces) it was no longer considered a firearm, only pieces of scrap steel. A Class II manufacture would take several over-lapping pieces from destroyed receivers and manufacture a "new" receiver by carefully fitting them and then welding them together. The manufacturer would file the required paper work to legally register it with the BATF (prior to 1986). Many question the quality or integrity of a demilled

The availability of most original M14 rifles was restricted to law enforcement or Class III dealers. Many of these rifles had their receivers destroyed and ended up as parts sets. (Courtesy of Donald G. Thomas)

receiver that had been welded back together, although there have been very few failures reported after years of them being used. Of course the quality would depend upon what procedures were used in the welding process and the skill of the manufacturer involved. There were several well-known Class II manufacturers performing the process over the years. In addition to select-fire receivers, there were several firms welding demilled GI receivers' back together to construct a semiautomatic-only receiver. Some of these M14 receiver remnants that were resurrected as semiautomatic-only receivers came under the close scrutiny of the BATF.

When production of the M1A™ first began in Texas the only part needed was a receiver. All of the other components were readily available as surplus parts at bargain prices. Many of the parts were available from the original commercial government contractors who were stuck with them after M14 production was suddenly terminated in 1963. Much of the production machinery used in the manufacture of original M14 rifles was eventually obtained by modern commercial companies, and used for manufacturing their M14 clones for the civilian market.

The Assault Weapons Ban of 1994

On September 13, 1994 the infamous Assault Weapons Ban was passed in to law. The only logical portion of the law was it was scheduled to expire in ten years. The Ban affected the M1A™ and commercial M14 type rifles because of their detachable magazine with capacity of over ten-rounds, the flash suppressor and bayonet lug.

The Springfield Armory Inc. has reported that their M1A™ rifles with a serial number over 084,000 were most likely manufactured after September 13, 1994 making them post-ban rifles, stripped receivers that were assembled into rifles after that date would also have been considered post-ban. Fortunately the ban was allowed to expire In September 2004 and the Federal directives no longer apply. However local and state laws will still apply.

GI Match M14 Rifles

The original U.S. Government Springfield Armory in Massachusetts was famous for producing match grade standard military rifles for

Original 1960 era Springfield Armory National Match receiver. Match rifles had their selector locks permanently welded to fire semiautomatic-only. Despite the disabling of the M14's full-automatic function, laws were enacted to keep M14 rifles from being distributed to civilian match shooters. (Courtesy of the Springfield Armory Museum)

many years. The problem that the Armory encountered was that the M14 was the first standard infantry rifle that had select-fire (full-automatic) capability. Feasibility studies were conducted at the Armory in an effort to find a way to permanently reconfigure the issue M14 rifle to function semiautomatic-only. The proposals that were considered ranged from a complete redesign of the receiver to simply removing the parts that gave the weapon its

The ownership of an M1A or M14 rifles for many rekindles memories of days long past, when as young men they carried the weapon, often into harm's way. (U.S. Army)

full-automatic capability. While preparing the first M14 rifles for the 1963 National Matches at Camp Perry, Ohio it was decided to weld up the select-fire components to permanently render the rifle as a semiautomatic-only firearm.

The U.S. Army considered match rifle competition an important program and directed the Springfield Armory to design and assemble some M14 rifles permanently modified for semiautomatic-only fire. The modified rifles were then to be presented to the Field Services Division, Office of the Chief of Ordnance for approval.

By 1967 the M14 rifle (top) was being phased out as the "Standard A" infantry rifle, replaced by the 5.56 mm M16A1.

The 7.62 mm Rifle, U.S. M14, Modified

The Engineers at the U.S. Springfield Armory designed a modified M14 that had several design changes implemented. The standard M14 receiver was redesigned to eliminate the slot for the full-automatic connector lever, a relocated dismount notch for the operating rod and lastly the lug that was normally attached to the right rear side of the receiver was removed. The buttstock lacked the notch located on the right side, normally used to clear the selector lever (or lock). The hinged M14 buttplate was replaced by the flat M1 rifle type buttplate. The cutoff valve located in the gas cylinder, used for launching rifle grenades was eliminated. The notch located on the operating rod normally used for tripping the connector lever was eliminated. The M14 sear was replaced by a sear from the M1 rifle. The barrels for the M14 match rifles would be manufactured to match-grade specifications. The rear sight assembly was replaced by the 1/2 minute of angle sights used on the Nation Match M1 Garand. Three M14 receivers were manufactured at the Springfield Armory to the new semiautomatic design specifications. The receivers were assigned special serial numbers X500, X501 and X502. Two of the receivers were assembled in to complete M14 rifles, minus the full-automatic components, and delivered to the Office of the Chief of Ordnance in Washington, D.C.

The details of redesigned M14 rifles were explained to Major General Nelson M. Lynde,

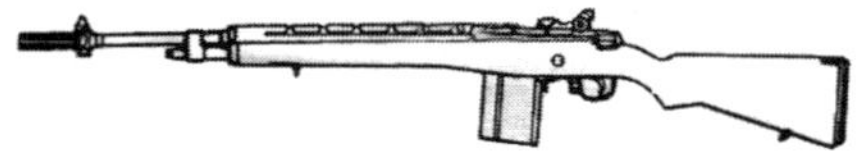

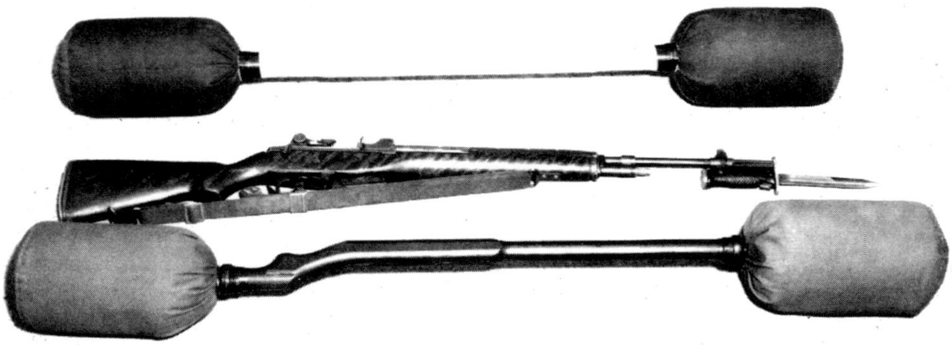

A unique Marine pugil stick designed like an M14 rifle. The older style with a straight handle is displayed at the top of the photograph. (NARA)

Chief of the U.S. Army Field Forces. After examining the weapons and the manufacturing deviations required to produce them, the General consulted with the Director of Civilian Marksmanship (DCM). At the meeting between the Director and the General a decision was made that would have a profound effect on future civilian ownership of military M14 rifles. General Lynde decided that manufacture of the match rifles deviated too far from the standard issue M14 and that they could not be easily converted back to select-fire military weapons should the need ever arise. Ironically, the semiautomatic-only M14 rifle design had been approved by the Bureau of Alcohol, Tobacco and Firearms, and the vast majority of M14 rifles were issued with selector locks in place that rendered them to a semiautomatic only function. The locks could be removed if needed, but removal of the locks seldom occurred. Years later the famous M1A™ version of the M14 would utilize many of the features incorporated into the original Springfield Armory's semiautomatic receiver and a new commercial industry was born.

The Gun Control Act, implemented in 1968, contained new laws that would keep M14 rifles from ever being sold to civilians, regardless of any and all attempts to render them to a semiautomatic-only configuration.

Chapter 17
Replacing the M14 rifle

The 5.56 mm M16A1 rifle with 20-round magazine. During the war in Vietnam M16A1 weapons were manufactured by the firms of Colt, Harrington & Richardson and the General Motors Hydramatic Division.

The M14 was basically an improved M1 Garand rifle. The .30 caliber M1 Garand was first adopted by the U.S. Ordnance Department back in 1936. While the M1 had proven to be the premier battle rifle during World War II, by the Korean War era its design was dated.

The early AR-15 Colt rifles were designated as the XM 16 E1 indicating experimental status. The improved and more reliable M16A1 was officially adopted in 1967. (U.S. Marine Corps Museum)

Shortly after the M14 rifle was adopted the many individuals in the U.S. armed services became interested in a new small caliber lightweight rifle called the AR-15. The AR-15 would eventually be adopted as the

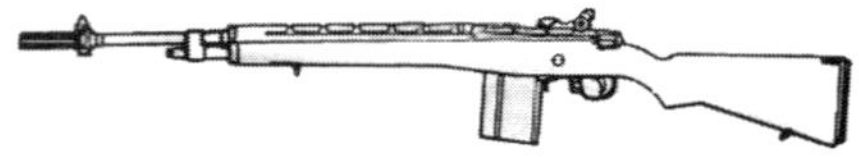

M16 rifle and upgraded to the M16A1 configuration just a short time later.

One of the primary differences between the M16A1 and the M14 weapons was weight, not only of the rifles themselves, but the ammunition. A standard M14 rifle and a combat load of 100 rounds of 7.62 mm ammunition weighed seventeen pounds. For the same weight a soldier could carry an M16A1 weapon and 280 rounds of 5.56 mm ammunition. The question was what would be more effective, an infantryman carrying 100 rounds of 7.62 mm ammunition or 280 rounds of 5.56 mm ammo?

One of the primary differences between the M16A1 and the M14 rifles was weight. Not only of the rifles themselves, but the ammunition. A standard M14 rifle and a combat load of 100 rounds of 7.62 mm ammunition weighted seventeen pounds. For the same weight a soldier could carry an M16A1 weapon and 280 rounds of 5.56 ammunition.

Military studies had shown that most infantry engagements occurred at ranges of less than 400 meters, at that range the 5.56 cartridge compared to the more powerful 7.62 round. Many argued that the powerful 7.62 was more effective, but you can only kill the enemy so dead. Another argument was; just why do most infantry engagements happen at less than 400 meters? How can you see a man sized, camouflaged target at over 400 meters that is trying to conceal himself from you? At ranges over 400 meters the situation is best handled by a general-purpose machine gun or mortar fire, if available.

During the Vietnam War long-range firefights were not very common. The M16A1 rifle (arguably) was probably better suited to the environment and elusive guerilla style of combat generally encountered during the Vietnam War.

The M14 rifle has many, many proponents, but it has often been stated (arguably) that anyone who is fond of the M14 probably never had to carry one in a hot humid environment. A combat infantryman spends 99.98 percent of his time carrying his weapon and the remainder of the time (.02 percent) firing it at the enemy.

The less violent recoil of the 5.56 mm round made rapid follow-up shots with the M16A1 more accurate. And of course the small 5.56 round made full-automatic fire much more effective, which often proved life-saving in an ambush situation. (NARA)

The less violent recoil of the 5.56 mm round made rapid follow-up shots with the M16A1 more accurate, and of course the small 5.56 round made full-automatic fire much more effective. During the Vietnam War if a unit ran into an ambush situation, the ability to return rapid automatic fire was usually quite advantageous to survival.

The initial AR-15 rifle program was privately developed by Armalite, the smallarms division of the Fairchild Engine and Aircraft Company. The company had tried to market the .223 caliber AR-15 rifle to the U.S. Army, but soon

U.S. 5.56 mm M16A2 Rifle. (Courtesy of the U.S. Marine Corps Museum)

discovered that bureaucratic redtape and resistance from both the Army and Ordnance Corps made the adoption of a rifle that was developed "outside" of the established system unlikely. Armalite grew tired of the resistance and finally in December of 1959 sold the entire AR-15 package to Colt's Patent Firearms Manufacturing Company. Colt had a longtime and successful relationship with the U.S. Army and Ordnance dating back to before World War I. Ironically at the same time period the first large shipments of the M14 rifles were being delivered to the Army.

With Colt heading the AR-15 rifle program, aggressive marketing was able to get the AR-15 fielded in limited numbers for testing in Vietnam during 1962. The only U.S. personnel operating in Vietnam at that time period were military advisors. The first reports on the AR-15 from the field were very enthusiastic (although later discovered to be exaggerated). The small caliber AR-15 was heralded because of its killing power, handling, lightweight

During the 1980 era the M16 was developed further into the M16A2 model. The M16A2 featured improved sights, a heavier barrel with a faster twist and a new pistol and foregrip. The M16A2 has largely been replaced by the M4 carbine another variation of the M16 series. (Courtesy of the Marine Corps Museum, Quantico)

and its suitability for the small-stature Vietnamese soldiers.

Based on the early reports from the field, computer-aided statistical documents were generated and sent to the Secretary of Defense. In the report it was stated that the .223 (5.56 mm) caliber AR-15 rifle was (in theory) superior to the current Standard M14 rifle in areas of reliability, ruggedness, performance and ease of maintenance. Additionally the AR-15 weapon and its .223 caliber ammunition could be produced at less cost than the current M14 and its 7.62 mm ammunition. The Secretary of Defense Robert S. McNamara instructed Cyrus Vance, Secretary of the Army to personally oversee a new unbiased study to evaluate both the M14 and AR-15 rifles. The benchmark weapon used in the study would be the enemy's AK47 rifle being effectively used against U.S. and friendly Vietnamese forces in Vietnam.

Aberdeen Proving Center was one of many agencies involved in the evaluation of the AR-15 and M14 rifles. The trials at Aberdeen included their normal routine of examining, measuring, weighing and photographing every component of each weapon. The testing at Aberdeen included many accuracy evaluations in which the AR-15 was often deemed superior to the M14, most notably during full-automatic firing. It was determined that the combined low-recoil generated from the .223 cartridge and the in-line design of the AR-15's buttstock contributed to a target group size less then one-half of that of the M14 rifle.

After being issued in large numbers problems soon began to occur with the M16 rifle during combat in Vietnam. The infantryman in the field was losing all confidence in the weapon. An investigation was launched and before long an improved model, the M16A1 rifle was introduced. Although the M16 initially encountered

A modern M16 SOPMOD variation with telescopic sight, muzzle device Picatinny rail system, and Harris bipod. (John M. Miller)

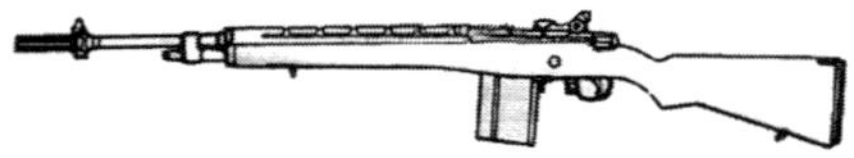

a number of problems, some were attributed to a switch in the propellant powder used in the 5.56 mm cartridges, and others were due to lack of proper maintenance. Despite its early problems the M16 series has endured. During 1982 the further improved M16A2 was adopted. The U.S. Army now fields the M4 series weapon of the basic M16 design, the M4 was adopted in 1994. The M16 series of weapons is rapidly approaching their 40th year as the Standard Infantry rifle, its longevity is unmatched by any other U.S. rifle.

Chapter 18
RifleMaintenance

The M14 rifle was designed, built and tested to function in the worst possible combat environments, making most of today's civilian use mild by comparison. Despite its ruggedness there are a few important points in maintaining the M14/ M1A™ rifle the civilian owner should be aware of. In addition to the text presented in this chapter, there are useful tips that appeared in the original U.S. Army PS magazines reprinted throughout this book. However well a rifle is made, it is still a mechanical device and thus requires periodic inspection, cleaning, maintenance and lubrication to keep it functioning as designed.

The 7.62 mm U.S. M14 rifle was designed, built and tested to function in the worst possible combat environments, making most of today's civilian use mild by comparison. (NARA)

Always wear safety glasses when working on or around firearms and magazines. Be careful using cleaning chemicals and solvents. Always follow the manufacturers' instructions and dispose of used materials in an environmentally safe manner.

Spare Parts and Components

Original M14 parts were very well made to strict specifications from premium steel and heat-treated for the correct hardness. It is always best to use original GI replacement parts when they are available. Many enthusiasts maintain their own spare parts supply, prudently purchasing various parts when they are available. Anyone who has followed the surplus parts industry over the years will have observed that availability of most surplus parts come and go. Just when the supply seems to have dried up, more appear, but don't always assume this will happen. If you need spare parts it is recommended that you buy them when they are available. Many military rifles, like the M1 Garand, were produced in enormous numbers; this was not the case with the M14. Though once plentiful, many original M14 rifle parts have been used for the assembly of M1A™ rifles and many other M14 clones over a period of many years.

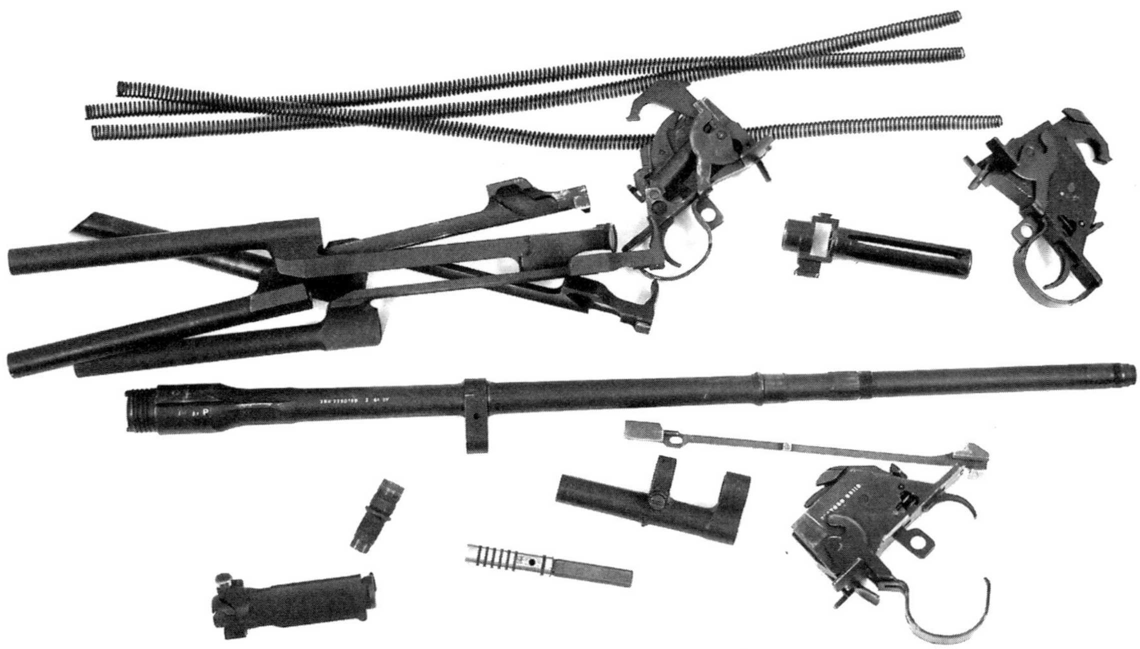

There are a few parts from the U.S. M1 rifle that will interchange with those of the M14. These parts include; the extractor; extractor spring; extractor spring plunger; hammer; hammer pin; hammer spring; hammer spring plunger; hammer spring housing; stock sling swivel; lower buttplate screw; safety lever; rear sight aperture; rear sight base; sight cover; trigger; trigger pin and handguard clip. The elevation knob and pinion are also physically interchangeable, but the knob and pinion of the M1 sight is calibrated in yards, while the M14 knob is calibrated in meters. The M14 knob is marked with a capital letter M for identification.

A few years ago there were a number of M14 operating rods on the market that had been demilled (cut into pieces) and welded back together. The welds can usually be detected upon close inspection. Look for variations in the color of the finish, pin-holes or misalignment.

In 1994 the assault weapons ban was implemented (now expired). One of the provisions of that law banned bayonet lugs on rifles manufactured during the ten-year ban period. As a result many M14 flash suppressors had their bayonet lugs milled off and new ones were made without the lug. Since the ban expired, a great demand for GI flash suppressors with a bayonet lug intact has been generated. This has caused the price of original flash suppressors to skyrocket. A number of aftermarket cast flash suppressors have been manufactured to help meet the demand. A cast flash suppressor can usually be recognized by joints or seams running along the top and bottom of the part.

All original government contract parts will have a letter or number code on them, and most were machined from steel forgings. Markings on GI parts are usually distinct and sharp. Faint or shallow marks are generally observed on castings or possibly a well-used GI part that has been sandblasted and refinished. Many aftermarket parts have been manufactured overseas from investment castings. An aftermarket cast part can usually be identified by a close examination. Cast parts will usually have a seam or joint present from the mold that it was cast in. Look for rounded corners and edges, GI parts will have sharp, square corners and milling or machine marks on them. Avoid Chinese made parts like those used in their M14 clone rifles imported during the 1990 era.

Excessive wear or breakage is obvious on many parts, but others require the use of military or commercial gages to determine their serviceability. While some of the gages are quite easy to use, they can be expensive to purchase for occasional use by an individual. In many cases where a functioning or safety

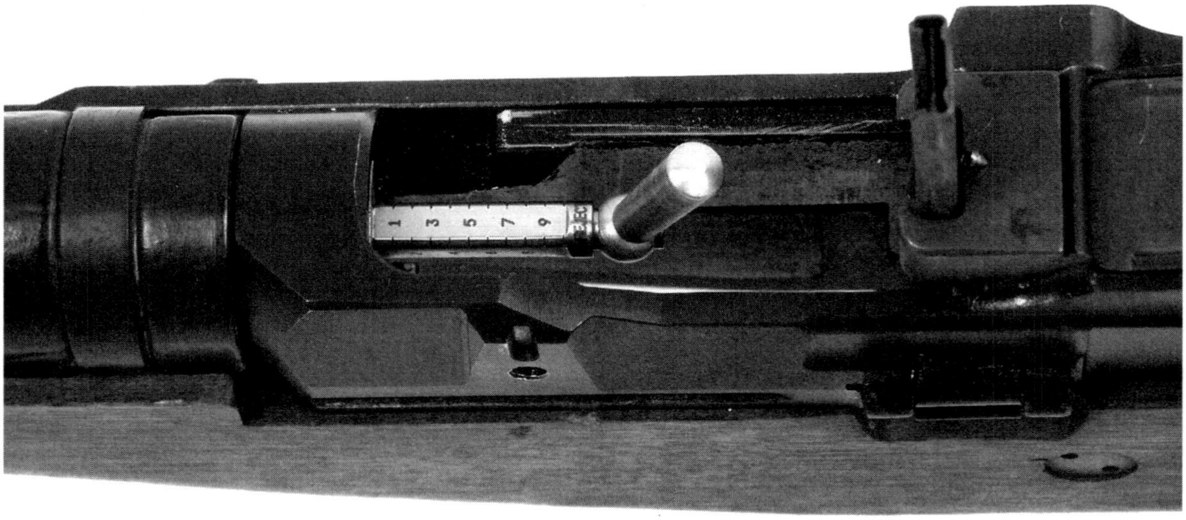

A barrel inspection begins with the chamber end of the barrel. This critical area is measured with a throat erosion gage, this measurement is often referred to as the barrel TE. (Courtesy of Scott Duff)

problem exists it is best left to a professional gunsmith familiar with the mechanics of the M1A™ and M14 rifle.

An adequate supply of spare parts largely depends on how often your M1A™ or M14 rifle is used or its anticipated use in the future. A minimum parts supply would consist of a complete set of springs; an extractor; ejector and firing pin. More aggressive shooters should have all of the minimum parts plus; a complete bolt assembly (headspaced to your rifle); a complete trigger group; gas cylinder and piston. For serious long-term use a complete parts set would be ideal, but short of that, in addition to all the aforementioned items include; an operating rod; barrel and a flash suppressor with lock screw and castle nut.

The Barrel

One of the primary components that is inspected for wear and serviceability is the barrel, this is particularly true when the purchase of a used rifle is being considered. A barrel inspection begins with the chamber end of the barrel. This critical area is measured with a throat erosion gage, this measurement is often referred to as the barrel TE. There are GI and commercial equivalent throat erosion gages available. The gage is graduated in readings from 0 to 10. The lower the reading the less wear the barrel has. A reading of 5 is considered the maximum wear limit for a barrel to be considered serviceable. To use the gage, the bolt is retracted and locked rearward. The gage is then inserted into the throat of the bore until it stops. The top edge of the receiver is used as a reference point to take the reading. Be extremely careful that the bolt is not released while the gage is inside the barrel. The TE gage will determine wear, but will not guarantee the accuracy of the barrel.

NO LUBE

Of course, you know you don't ever put any lube at all in the M14's gas system parts. You assemble 'em absolutely dry—and clean. These parts are made of corrosion-resisting steel that don't need lubing. Matter of fact, oil'll only invite trouble—dirt, corrosion, sluggishness, no-go!

But sometimes it's pretty hard to keep oil out of there when you're cleaning and lubing the rest of the weapon. So, here's a little trick to protect the cylinder, piston and plug before you start cleaning and coating the inside of the bore with prescribed lubricant:

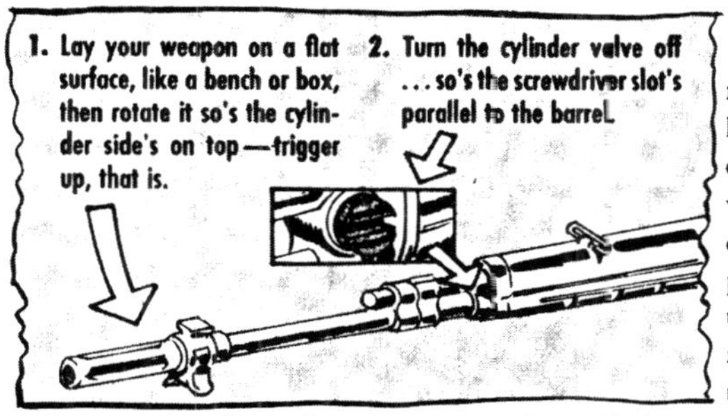

This way, if any oil from the patch should happen to sneak into the opening in the cylinder valve, Gravity Gerty'll draw it back down so's the patch can wipe it off next time through. No oil will reach the cylinder plug this way, that's for sure.

But, again, don't forget to turn the cylinder valve upright when you're through.

Your TM 9-1005-223-12 (Feb 65) is chockful of dope on correct methods for cleaning the M14's gas system parts — pay hard attention to it. Especially, don't use steel wool, a wire brush or scrapers of any kind on these parts. You might change some of the critical dimensions and ruin your fire.

Use CR bore cleaner (FSN 6850-682-6835—2-oz can) for normal cleaning and carbon removing compound P-C-111a (FSN 6850-620-0610—5-gal can) for real tough carbon deposits. In a pinch, you can try dry-cleaning solvent or mineral spirits paint cleaner, but never use gasoline, benzene or high-pressure water, steam or air.

When visually inspecting a barrel's bore and chamber a bore light is needed. The bore should be shiny and have sharp distinct lands and grooves. A barrel with a chromium-plated chamber and bore should be checked for signs of the chromium-plating flaking or peeling off. The condition of the barrel's muzzle is paramount to good accuracy; many barrels are ruined by improper or careless cleaning. On the M14 the flash suppressor needs to be removed for proper inspection of the muzzle. The lands and grooves of the muzzle should appear sharp and unworn.

Headspace gages are used to indicate the proper fit of a rifle's bolt in relationship to the barreled receiver. There are both GI and commercial equivalent gages to measure headspace. GI gages have a relieved area to allow the gage to be used with an assembled bolt. The use of commercial headspace gage in an M14 or M1A™ rifle requires the removal of the ejector and extractor.

Dear Private J. P. B.,

Prevention's the best cure in a case like this, of course, but for that, hold one.

First off, if you have a frozen or over-tightened plug that won't yield to your combination tool, don't mess around with it. Turn it in quick to your company armorer. He has a vise and brass clamps to hold the M14 while he works on it.

And now, prevention:

When you're installing the plug, tighten it to snug only—just tight enough to hold. No muscle required.

Half-Mast

Flash Suppressor

A flash suppressor alignment gage is used to detect a bent or misaligned flash suppressor. The inside of a GI flash suppressor is only slightly larger then the barrel's bore diameter. A loose or misaligned suppressor can affect the exiting projectile's accuracy. Periodically inspect the inside of the flash suppressor for signs of copper, indicating that the projectiles are striking the device.

Also carefully inspect the flutes and end of the flash suppressor for cracks or signs of erosion. A simple test is to suspend the flash suppressor from a string and then strike it with a steel tool like a pin-punch, if an extended ringing sound is heard there are probably no cracks present. If a dull thud is heard it is most likely cracked or broken and needs replaced.

Gas Cylinder and Piston

A critical area to insure reliable functioning of the M14 and M1A™ rifle is the gas cylinder and piston assembly. The serviceability of these components is determined with a no-go type of gage. The minimum diameter of the gas piston is 0.4968-inches. The gas piston will generally become unserviceable before the gas cylinder.

Gas Cylinder Plug

Another problem that frequently occurred with the M14 was the over-tightening of the gas cylinder plug. Over-tightening of the plug, especially on a hot barrel, would make it extremely difficult to loosen, often requiring it to be sent to an armorer for removal.

Bolt Assembly

Firing pin protrusion is the distance that the tip of the firing pin extends from the face of the bolt. This dimension is measured with a firing pin protrusion gage. The minimum protrusion is 0.044-inch, the maximum 0.060-inch. Too little protrusion will result in misfires, excessive protrusion will result in pierced cartridge primers. The locking lugs of the breech bolt should be carefully inspected for cracks or damage. The face of the bolt should be inspected for cracks and excessive erosion.

Trigger Pull

The maximum trigger pull specified by the military for the M14 rifle is 7.5-pounds, the minimum is 5.5-pounds. Trigger pull of less than 4.5-pounds is unsafe and requires immediate attention by a qualified gunsmith.

Safety

One of the most repeated articles published in the U.S. Army PS magazines with regard to the M14 rifle was the improper use of the M14's safety feature. Troops would consistently try to place the safety on when the weapon was not cocked. The nose of the safety lever rests on the bottom slot of the hammer when it is in a forward or uncocked position. Any pressure applied to the safety tab inside the trigger housing would usually damage or break the nose of the safety lever off. In some cases the broken safety was not immediately apparent until the dismembered piece jammed up the trigger mechanism.

Rear Sight

The rear sight assembly should be periodically inspected, cleaned and lubricated. The edges of the aperture and the top of the aperture that contacts the rear sight cover should be lightly coated with oil or grease. On the National Match rear sight a small amount of oil should be placed on the rotating part of the sight hood. The rear sight aperture should be periodically elevated. Also check the screw in the elevation pinion to insure it is tight

Cleaning Tips

The M14 and M1A™ rifles should always be cleaned in a horizontal, upside-down position.

U.S. Rifle 7.62 mm M14

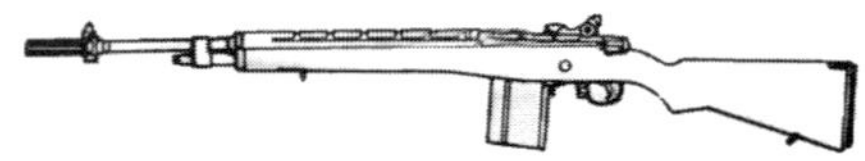

This will prevent bore cleaner from entering into the gas system, which should always be kept free from cleaning solutions and lubricants. For routine cleaning, the M14 should be cleaned while assembled to minimize component wear and damage. To retain the bolt in a rearward position insert an empty stripper-clip into the receiver guide then release the bolt, the clip will hold the bolt back without fear of it being accidentally released.

Most barrel damage and wear occurs through the improper use of a cleaning rod. Due to the M14 rifle's design it is necessary to clean the barrel from the muzzle end of the barrel. To minimize barrel damage, a cleaning rod guide such as the one produced by the J. Dewey Company is highly recommended. The guide is used at the muzzle end to keep the cleaning rod from contacting the bore. Also recommended is the one-piece Dewey cleaning rod that is coated with nylon to help prevent barrel wear. Avoid using the sectioned GI steel cleaning rod. Improperly used the rod can cause damage to your rifles bore.

Apply bore cleaner to a correct size bronze bore brush and make several passes through the entire length of the barrel. Attach the patch holder to the cleaning rod and make several passes changing the soiled patch each time, and repeat until the patch emerges clean. Copper Fouling; GI rifle bore cleaner (RBC) while

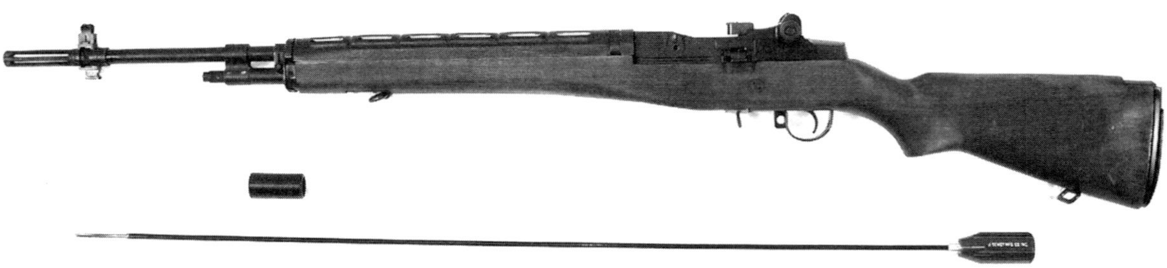

A commercial cleaning rod guide such as the one produced by the J. Dewey Company is highly recommended. The guide is used at the muzzle end to keep the cleaning rod from contacting the bore. Also recommended is the one-piece Dewey cleaning rod that is coated with nylon to help prevent barrel wear.

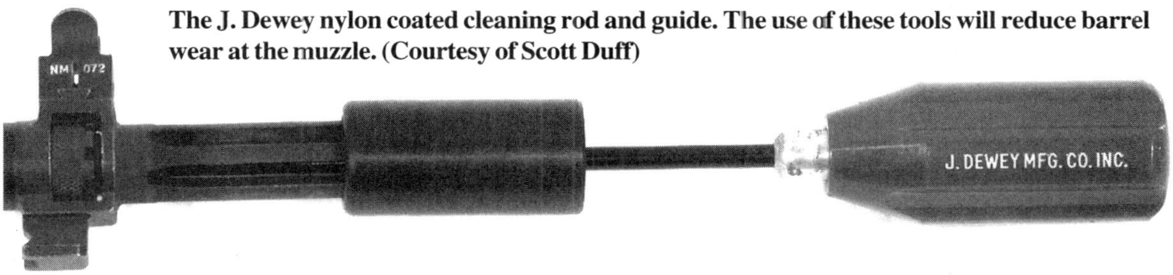

The J. Dewey nylon coated cleaning rod and guide. The use of these tools will reduce barrel wear at the muzzle. (Courtesy of Scott Duff)

effective in removing powder fouling, will not remove deposits from copper-jacketed projectiles. Excessive copper fouling will affect the accuracy of the rifle. Special chemicals are available to the shooter to remove these deposits. A patch soaked with a suitable cleaner specifically formulated to remove copper fouling should be used every fourth cleaning. Follow the manufacturer's instructions for proper use.

The best tool for cleaning the chamber of the M14 rifle is the GI ratcheting chamber brush.

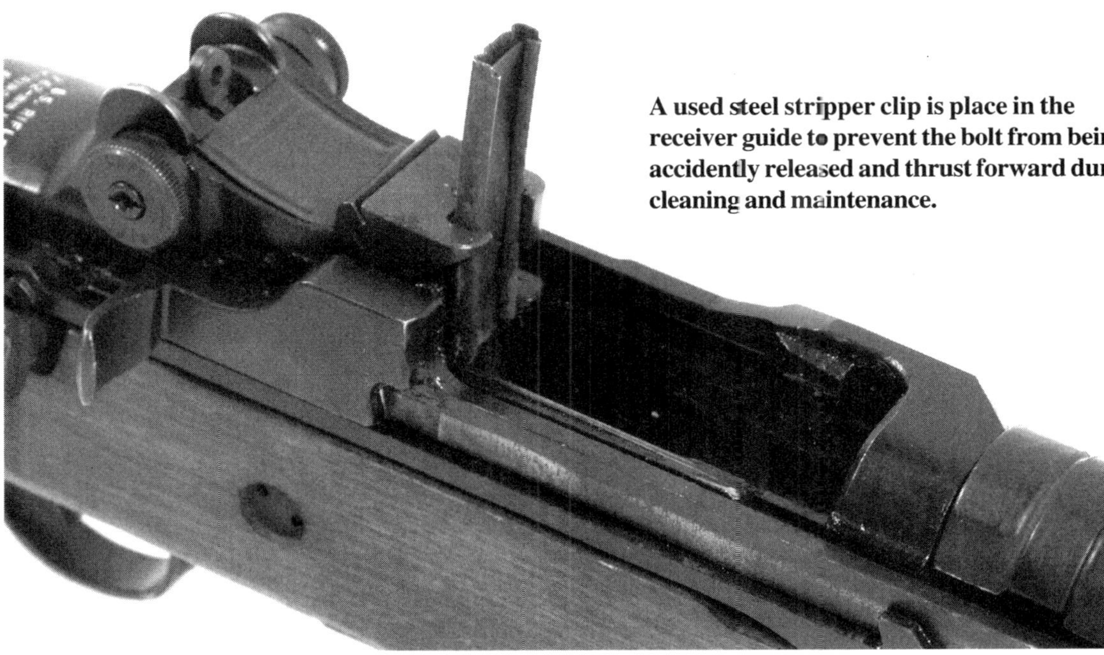

A used steel stripper clip is place in the receiver guide to prevent the bolt from being accidently released and thrust forward during cleaning and maintenance.

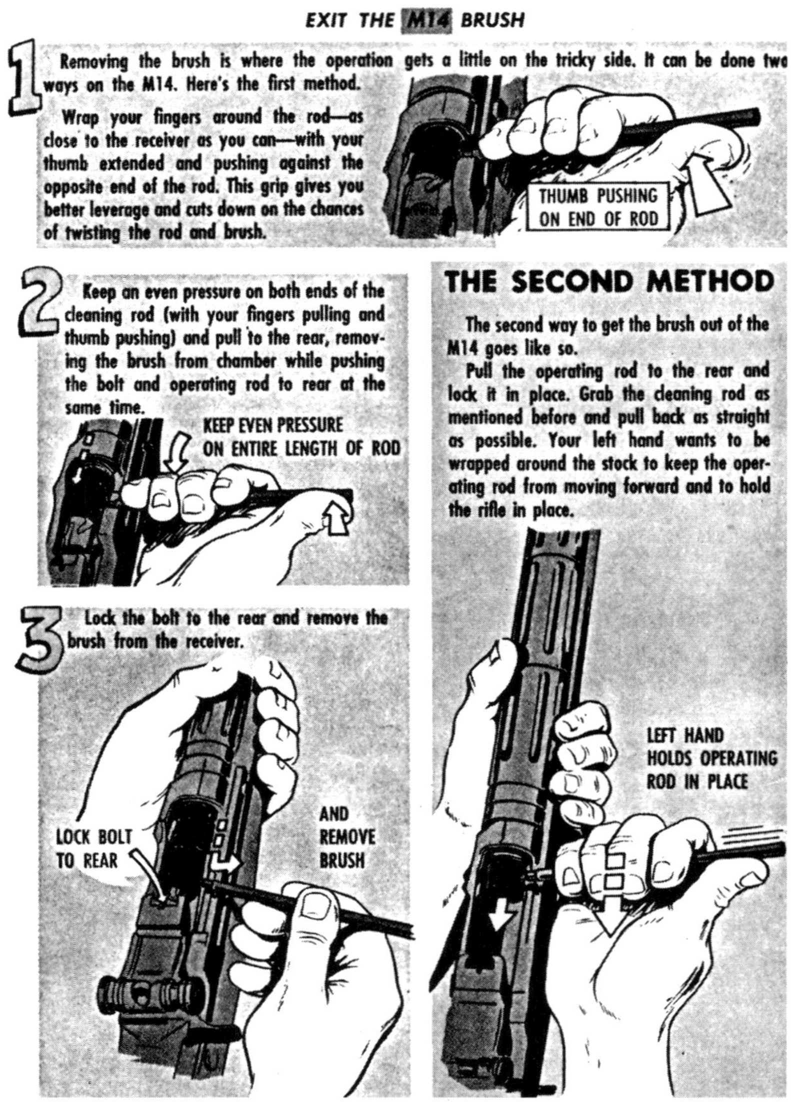

Insert the ratcheting brush inside the chamber; slowly close the bolt onto the brush. Screw a section of the GI cleaning rod into the plastic base of the brush, and use the brush's ratcheting feature to rotate the brush several times inside the chamber Remove the chamber brush and follow-up with dry patches until the chamber is clean and dry.

The gas system of the M14 is critical for reliable functioning; it should be periodically maintained by disassembly and cleaning after firing approximately 300 rounds. There are several precautions that need to be taken when cleaning the gas system. When removing or installing the gas cylinder plug, a gas cylinder wrench should always be used. The gas cylinder and plug are both made of 416 stainless steel and as such have a tendency to stick and gall making removal difficult at times. The gas cylinder wrench is used in conjunction with the combination tool. The wrench of the combination tool is used to remove (or tighten) the plug. The gas cylinder wrench is used to hold the gas cylinder in place preventing its twisting or movement from the transfer of torque being applied to the nut with the combination tool.

One of the best tools for cleaning the chamber of the M14 rifle is the issue GI ratcheting chamber brush. Insert the ratcheting brush inside the chamber; *slowly* release the bolt onto the brush. A section of the steel GI cleaning rod is used as a handle to rotate the brush.

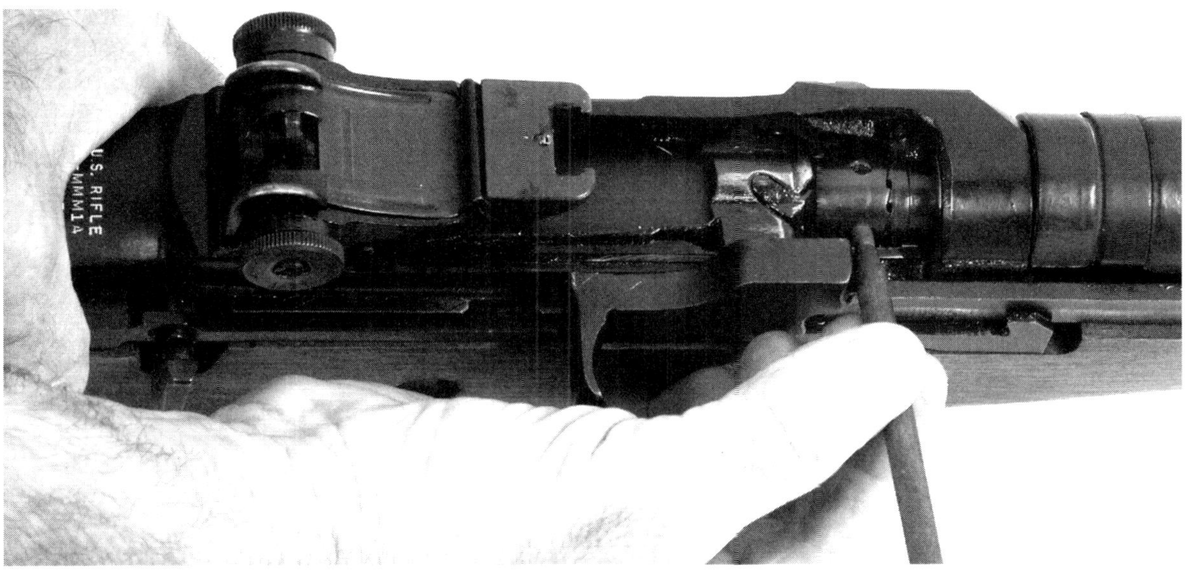

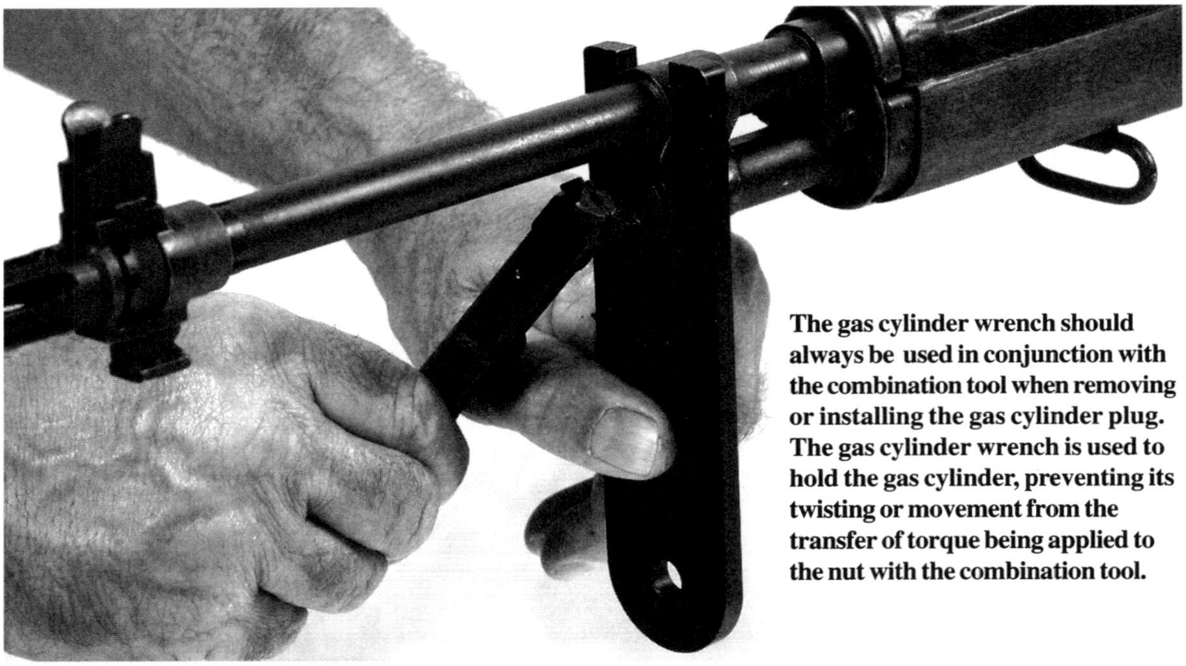

The gas cylinder wrench should always be used in conjunction with the combination tool when removing or installing the gas cylinder plug. The gas cylinder wrench is used to hold the gas cylinder, preventing its twisting or movement from the transfer of torque being applied to the nut with the combination tool.

Once the gas plug is removed the gas piston can be removed from the gas cylinder by tilting the muzzle in a downward position. There will generally be a substantial amount of carbon build-up inside of the gas plug and gas piston, these areas can be cleaned with a bore brush and powder solvent. The outside of the gas piston should only be cleaned with a nylon brush or rag and bore cleaner. Abrasive materials such as steel wool or Scotchbrite pads should never be used to clean carbon from the gas piston.

The inside of the gas cylinder can be effectively cleaned with bore cleaner and a .45 caliber bronze bore-brush, followed by running patches through it until it is clean. Insure that the inside of the gas cylinder and the entire gas piston are both completely dry before reassembly. Rotate the spindle valve to loosen carbon deposits. A small amount of anti-seize compound can be placed on the threads of the gas cylinder plug to help prevent galling. Caution must be used when installing the plug. The threads are very fine and the plug can easily be cross-threaded, this will usually result in irreparable damage to both the gas plug and cylinder. Be careful not to over-tighten the gas cylinder plug.

The gas piston can be removed from the gas cylinder by tilting the muzzle slightly downward. The outside of the gas piston should only be cleaned with a nylon brush or rag and bore cleaner. Abrasive materials such as steel wool or Scotchbrite pads should never be used to clean carbon from the gas piston. The inside of the gas piston and plug can be cleaned with a bore brush and powder solvent.

Lubrication

The M14 was designed to be primarily lubricated with grease. Before applying new lubricant remove all traces of the old grease. A thin film of grease should be applied to all contact surfaces in particular the locking lugs and roller of the bolt; the cam area of the operating rod and at the rear of the bolt that contacts the hammer. Grease should also be applied to the channel of the receiver where the lug of the operating rod travels. On select-fire rifles lube the end of the connector rod that contacts the operating rod. The inside of the operating rod tube should be lightly oiled, using an oil soaked patch on a cleaning rod. Over lubrication is common and attracts dirt, a properly lubricated M14 rifle will be good for several thousand rounds. GI rifle grease Lubriplate 130A or a commercial equivalent is recommended. The barrel and chamber should be dry, except in cases where the rifle will be stored for a long period of time.

A properly installed gas piston.

Magazines

Magazines require little maintenance under normal conditions. A periodic disassembly, cleaning and minimum lubrication with gun oil are normally all that is required. CAUTION:

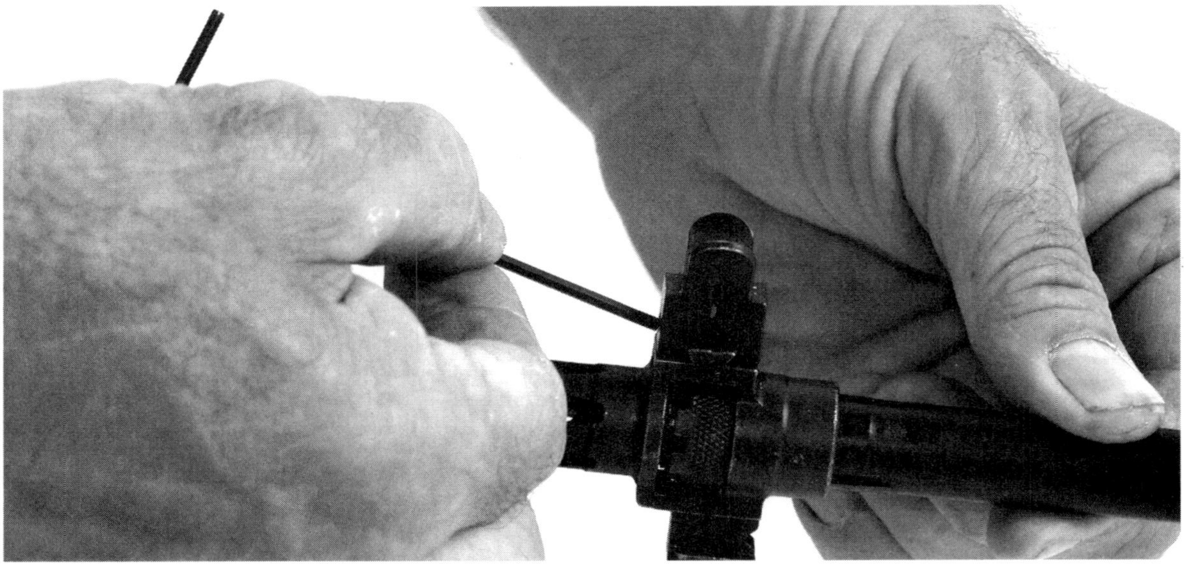

To remove the M14 flash suppressor loosen the hex-head locking screw. After removing the lock-screw use the flash suppressor pliers to unscrew the castle-nut from the barrel. Slide the flash suppressor off of the barrel. Upon reinstallation the castle-nut may need to be turned slightly to accommodate the locking screw.

always wear safety glasses when disassembling and reassembling magazines. The follower is under spring tension and can cause injury when the floor plate is removed, or while the magazine is being reassembled.

Storage

First visually inspect the rifle to insure that it is not loaded! To prepare an M1A™ or M14 rifle for storage perform normal cleaning procedures. Coat all metal surfaces with a light coat of oil. Release the trigger guard to relieve the clamping pressure on the stock. Never store a firearm in a vinyl gun case, cases made of such materials will cause the firearm inside to sweat and will cause corrosion. The best place to store any firearm is in a locked cabinet or safe equipped with a dehumidifier. After removal from storage the rifle should be cleaned and all traces of oil removed from the barrel and gas system.

Ammunition

The use of commercial or reloaded 7.62 mm ammunition is not recommended for use in a M14 or M1A™ rifle. Military 7.62 mm ammunition was originally designed to be used in the select-fire M14 rifle and M60 machine gun, the feeding, firing and extraction of ammunition in these weapons is rigorous. In order to withstand rough military use, the brass cartridge cases were designed with thicker walls and reinforced heads. This is not the case for commercial .308 caliber ammunition designed for sporting use. Additionally due to their thicker cases military 7.62 mm brass has a smaller capacity for powder than commercial .308 cases.

It is always best to use NATO specification

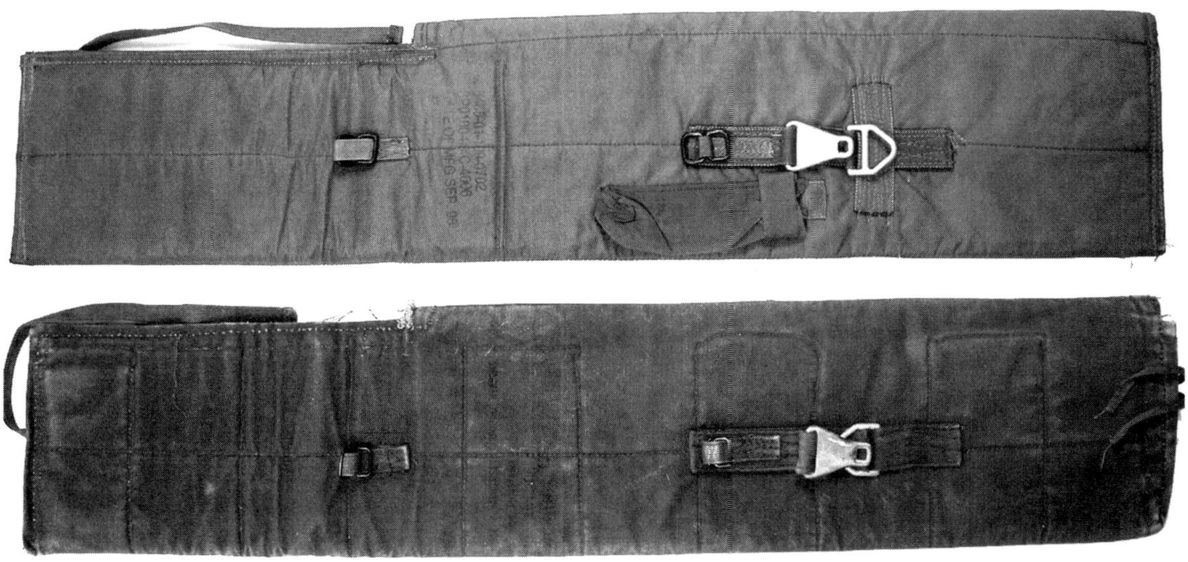

For transporting military rifles to and from the range military M1950 rifle cases are ideal. These were made of both nylon (top) and cotton duck material. (Courtesy of Jim Pecori)

surplus ammunition or a commercial equivalent manufactured to military specifications. Currently 7.62 mm ammunition is available at a lower cost than it can be economically reloaded. Resist the temptation to use the cheapest ammunition that is available, it is usually inexpensive for a reason, which could include age, poor storage or frequent misfires. Damage to you or your expensive rifle may result. For obtaining the best accuracy from the M1A™ or M14 rifle, U.S. GI M852 Match or Federal Cartridge Gold Medal Match ammunition is recommended.

Slam Fires

The M14 rifle was designed to feed from its magazine. NEVER insert a round into the chamber by hand and release the bolt; a dangerous slam-fire could occur injuring the operator and damaging the rifle.

Another condition that can cause a dangerous slam-fire is when the hammer doesn't remain in a cocked position and follows the bolt into battery position, firing the rifle without pulling the trigger. Some of the conditions that cause slam-fires are; defective or broken hammer hooks, sear-trigger assembly or a broken trigger pin. To perform a function check; first remove the magazine and visually check the chamber to insure the rifle is not loaded. Point the muzzle in a safe direction. Pull the operating rod rearward and release it. The M14 was designed with a two-stage trigger pull. While squeezing the trigger you should feel the slack of the first stage and then a stop, continuing to squeeze the trigger you should be able to detect the metallic sound of the hammer falling. While holding the trigger rearward, recock the operating rod and release it. Release finger pressure on the trigger and squeeze it again. You should once again hear the hammer fall. If not, DO NOT fire the rifle until it has been inspected and repaired. The hammer is following the bolt into battery instead of being held rearward until the trigger is pulled, a VERY DANGEROUS condition.

Basic Field Stripping Procedures

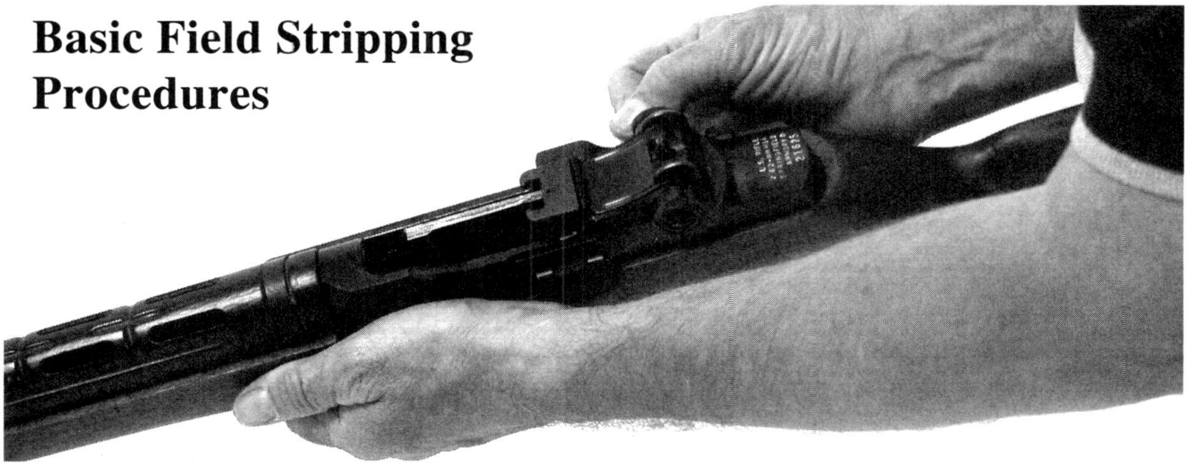

1. Clear the rifle. Remove the magazine and retract the operating rod, visually inspect the chamber to insure that the weapon is not loaded.

2. Remove the trigger group. To remove the trigger group, grasp the rear of the trigger guard, pull it up and rearward to unhook it. Lift the assembly upward to remove it. Becareful not to open the trigger-guard more than 90-degrees.

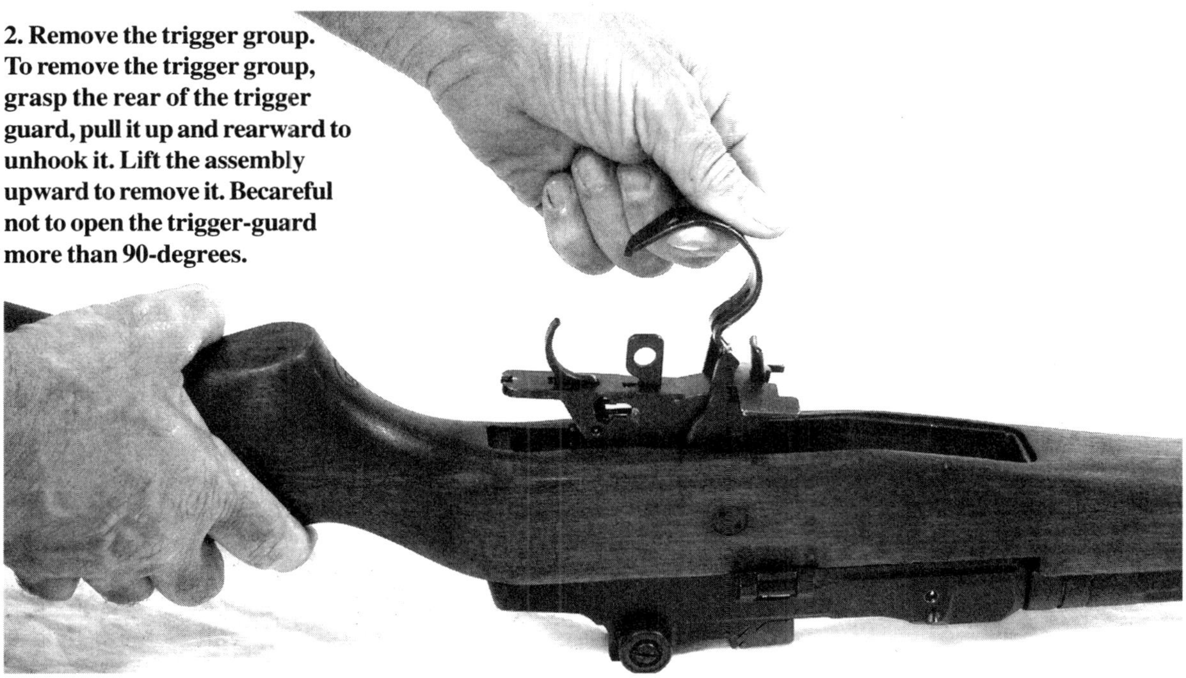

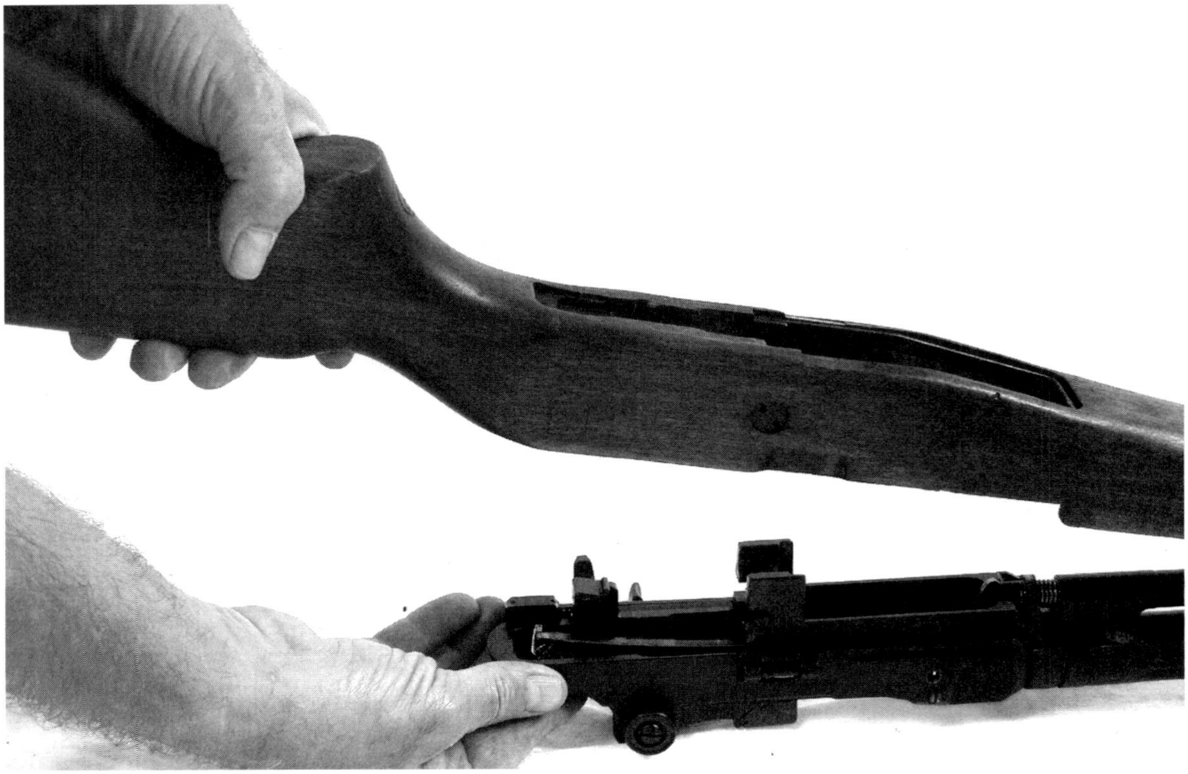

3. Separate the receiver from the stock. Grasp the receiver and stock and separate them, it may be necessary to lightly strike the stock with your hand to get them apart.

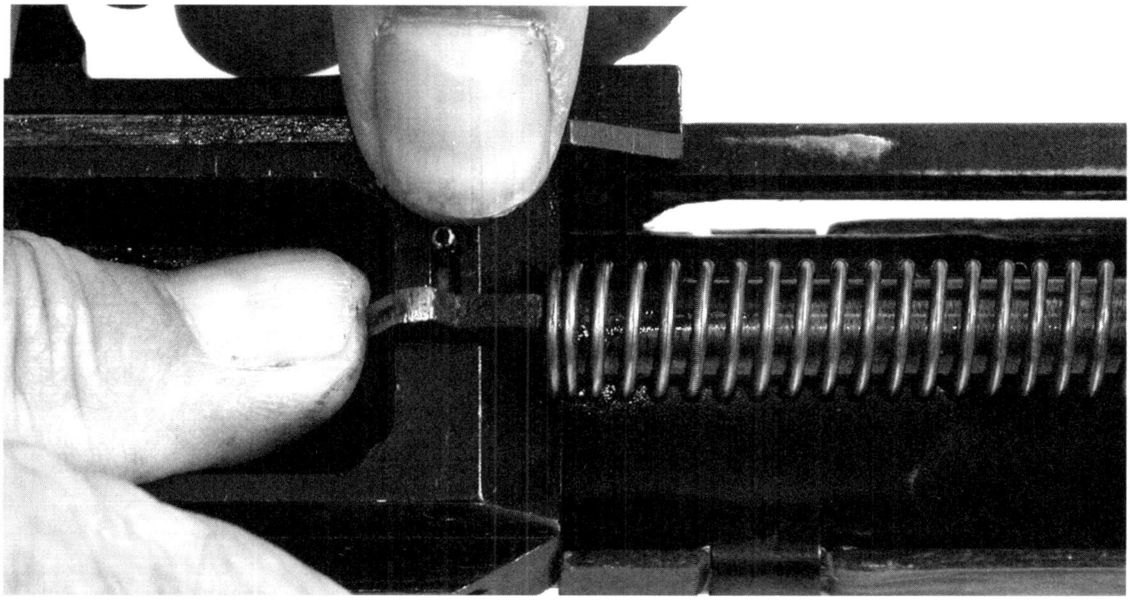

4. Removing the recoil spring; CAUTION wear safety glasses! Using thumb pressure push the rear of the guide rod forward. At the same time slide the connector lock outward towards the operating rod to disengage the guide rod. Slowly remove the recoil spring and guide.

5. On M1A™ or M14 rifles with select-fire capability the next step is to remove the connector rod assembly. If a selector lock is present it is already in position for removal, if equipped with a selector device; turn the lever until the letter "A" is facing toward the rear sight's windage knob. With the bolt closed, place your thumb at the rear of the connector assembly and push forward until the forward end of the connector shaft can be lifted off of the connector lock shaft. Turn the lever clockwise until the elongated hole is aligned with the stud on the sear release. Remove the connector.

Rotate the selector until the letter A on the face is towards the rear sight's windage knob.

Removing the connector lever.

6. Removing the operating rod. On M14 rifles; pull the operating rod handle rearward until its guide lug is aligned with the disassembly notch in the receiver. Rotate the op-rod down and outward until it is disengaged from its track in the receiver.

On M1A™ rifles the disassembly notch is located at the top of the operating rod track making the removal procedure slightly different. On the M1A™ rifle the operating rod is aligned with the disassembly notch and lifted upward to remove it.

7. After the operating rod has been removed, the bolt can be lifted out of the receiver by sliding it forward and rotating it slightly.

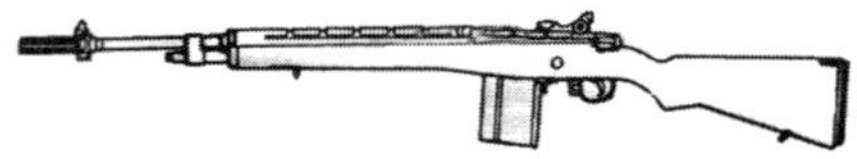

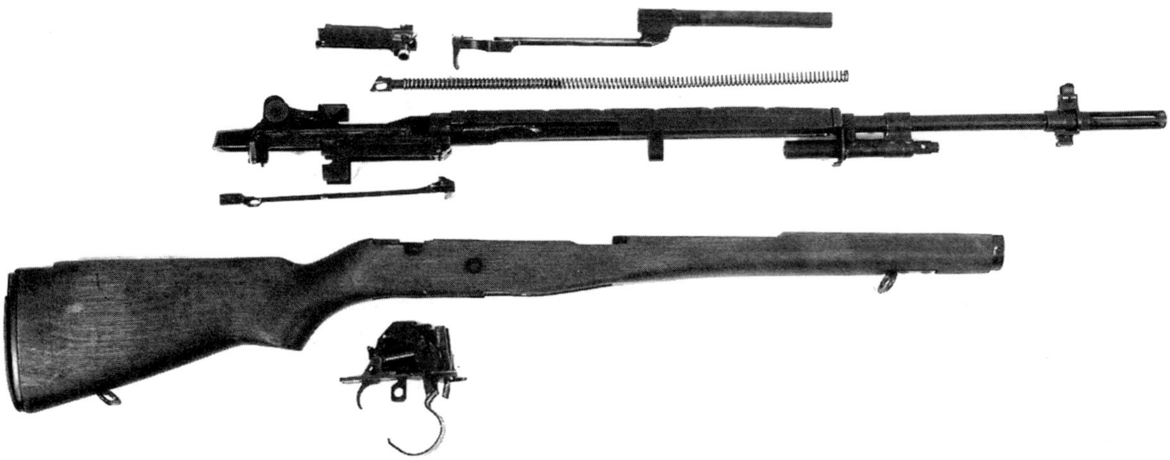

The M14 / M1A™ rifle field stripped for routine maintenance.

Index

B

C

D

Index

U

V

Index

W

X

Z

U.S. Rifle 7.62 mm M14

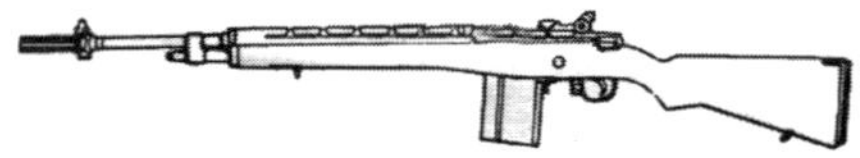